PREBIOLOGICAL SELF ORGANIZATION OF MATTER

The Eighth College Park Colloquium on Chemical Evolution

Prebiological Self Organization of Matter, 8–10 April 1987

Sponsored by

The Laboratory of Chemical Evolution, Department of Chemistry, University of Maryland, College Park, Maryland 20742

Supported by

The National Aeronautics and Space Administration

PREBIOLOGICAL SELF ORGANIZATION OF MATTER

Edited by

Cyril Ponnamperuma
University of Maryland
College Park, Maryland

and

Frederick R. Eirich
Polytechnic University
Brooklyn, New York

Proceedings of

The Eighth College Park Colloquium on Chemical Evolution

A. DEEPAK Publishing 1990
A Division of Science and Technology Corporation
Hampton, Virginia USA

A. DEEPAK Publishing
A Division of Science and Technology Corporation
101 Research Drive
Hampton, Virginia 23666–1340 USA

Library of Congress Cataloging-in-Publication Data

College Park Colloquium on Chemical Evolution (8th: 1987)
Prebiological self organization of matter/edited by Cyril Ponnamperuma and Frederick R. Eirich.
p. cm.
"Proceedings of the Eighth College Park Colloquium on Chemical Evolution."
Held 8–10 April 1987 at University of Maryland, College Park, Maryland.
Sponsored by the Laboratory of Chemical Evolution, Department of Chemistry, University of Maryland, College Park.
Includes bibliographical references and index.
ISBN 0–937194–18–2
1. Chemical evolution—Congresses. 2. Self-organizing systems—Congresses. I. Ponnamperuma, Cyril, 1923– . II. Eirich, F.R. (Frederick R.) 1905– . III. University of Maryland, College Park. Laboratory of Chemical Evolution.
QH325.C725 1987
577—dc20 90–3640
CIP

Printed in the United States of America

CONTENTS

PAPERS PRESENTED

PREFACE

Since its inception, the Laboratory of Chemical Evolution has been involved in research varying from model prebiotic syntheses to the analyses of meteorites and Missions to Mars. As a natural extension, it has also been engaged in providing platforms of discussion for the many aspects of Exobiology and the corresponding Prebiotic Chemistry. The present volume presents the papers and abstracts of the Eighth Maryland Colloquium on "Prebiological Self Organization of Matter," covering a field that focuses on our main area of ignorance: the transition from non-living to living matter, a field that, because of its extra dimension of difficulty, has been notoriously long on speculation and short on good science. The purpose of this Colloquium has, thus, been to provide a forum for discussions to make inroads into this twilight area, and trying to define specific questions. If good science is the effort to find keys where the light is, then this Colloquium may be seen as an attempt to widen the area of light shining onto the age-shrouded origins of life, and thereby on the nature of life itself.

Cyril Ponnamperuma
F.R. Eirich

INTRODUCTION

Frederick R. Eirich

Polytechnic University Brooklyn, New York
and
Laboratory of Chemical Evolution
University of Maryland
College Park, Maryland

The motions of the cosmic bodies, the laws of chemistry and physics, the orchestration of the parts of a living cell, the existence of animal societies, and the notion of order conceived by our minds impresses on us that our Universe has a tendency to form interacting and cooperative organizations. Yet, of all levels of ordering, we know next to nothing about the chemical processes that became self-maintaining and self-replicating, i.e., became alive.

This Colloquium, arranged so that a number of workers in Chemical Evolution may present and discuss their views and their research, will consist of four sessions: 1) on the systems and conditions that will allow self-organization; 2) and 3) concerning the same kind of molecules that are at the center of the life processes today, there will be talks on the self-organization of peptides and nucleotides. In session 4) we will hear of the probability that life needed to become organized within the boundaries of a cell. Together with the recent COSPAR sessions on the "Minimal Requirements for the Emergence of Life" and on "The Limits of Life," the Transactions of this Colloquium should represent a substantial part of the State of the Art.

Trying to summarize the proceedings, we have some more evidence that the type of experiments in Molecular Evolution and the type of chemistry on which life is based, are the same. This knowledge though, is not sufficient for us to link non-living to living processes. Nor is it enough to know that life operates within the confines of irreversible thermodynamics to enable us to define "what constitutes life." In our search for the origin of life and an understanding of its nature, we have not discovered the fire yet.

At present, any definition has to be based on our understanding of life on this planet, i.e., we have to explain "life" by <u>our</u>

life. Such self-reference must render any description incomplete, a lack of guidance that complicates our experimental approaches. Moreover, science and logic are based on causality, that is, on observing the reproducible one-to-one sequences which we derive from our macroscopic perception of events which themselves are the resultants of billions of molecular processes. This logic of macroscopic causality may not be appropriate in the range of molecular (or atomic) dimensions where according to our way of thinking indeterminancy rules, i.e., where sequences in time may appear accidental. The chemistry of life, and its beginning, may well be governed by the "logic" of molecular and not by that of macroscopic events. The uniqueness of the shuffling of genes during replication is one example. The "creative" (associative) activities of our minds, too, are based on molecular activities. In the realm of life, consequences may not have roots that we can trace readily, or can always be repeated reproducibly.

Mathematical biologists directed their efforts at a formalism that could describe how an up to this point continuous function could suddenly become qualitatively different. Other formalisms attempt to cover the organization of groups of biological reactions whose interaction makes the total function larger than the sum of its parts. Whatever the merits of these theoretical approaches, they do not enable us to reconstruct the hardware of life. The only way to bridge the gap between macroscopic and microcopic events seem the experimental attempts to develop multimolecular interacting, cyclic, self-maintaining and environmentally responsive, units which eventually develop an instruction system capable of replication. The emphasis is on individual units not only because single cells are the units of life, and not only because of the protection that encapsulation affords, but because of the order impressed by boundary conditions on physical or chemical processes. Thus we may visualize cellular entities containing a chemistry fuelled by gradients of energetic and molecular fluxes across the membranes. Cellular entities will have passed the threshold from non-living to living when they reach a "self-maintaining" and "self-regulating" state in the face of environmental changes. The papers we are going to hear relate to our knowledge, and lack thereof, of the chemistry leading to this threshold.*

----

*It was not possible to reproduce all presentations as full papers. These will appear in the order of the Program, in full, or as abstracts. Dr. Przybylski was unable to attend, though the abstract of his paper is included. The allotted time was filled by Dr. Kanavarioti's talk; see her abstract.

MOLECULAR COEVOLUTION OF PHOSPHATE COMPOUNDS AND POLYPEPTIDES

Herrick Baltscheffsky
Arrhenius Laboratory
University of Stockholm
S-106 91 Stockholm Sweden

ABSTRACT

In bacterial photophosphorylation light energy may be utilized for the formation of two alternative phosphorylated products, adenosine triphosphate (ATP) and, as we have shown, also inorganic pyrophosphate (PPi). They can then be used to drive various energy requiring reactions. The final step in the light-induced formation of PPi in bacterial chromatophores is catalyzed by a membrane bound H^+ -PPi-synthase and that of ATP by a membrane bound H^+-ATP-synthase. PPi may well have preceded ATP as (pre)-biological energy carrier, and early membrane bound energy conversion may have involved Pi, PPi, and a PPi-synthase, later complemented with a more complex system involving Pi, ADP, ATP, and an ATP-synthase. An obvious question is whether any evolutionary relationship can be found between enzymes involved in PPi and ATP metabolism.

Earlier observed similarities between enzymes binding and metabolizing nucleotides include amino acid sequences and 3-D patterns. Recently we have found that such similarities extend to the soluble inorganic pyrophosphatase from yeast (H. Baltscheffsky and M. Baltscheffsky, Origins of Life, 1986, <u>16</u>, 377; H. Baltscheffsky <u>et al.</u>, <u>Acta. Chem. Scand.</u>, 1987, B41, 106). These observations have led us to introduce the new concept of molecular coevolution of phosphate compounds and polypeptides. The phosphate compounds embrace simple inorganic phosphates such as Pi, PPi, and also low and high molecular weight polyphosphates, the more complex nucleoside phosphates such as mono-, di-, and polynucleotides, as well as carbamoylphosphate, various sugar phosphates, cyclic nucleo-side monophosphates, guanosine 5'-diphosphate 3'-diphosphate, 2'-phospho-AMP, and P^1, P^4-bis-(5'-adenosyl)-tetraphosphate.

Arguments concerning the validity and usefulness of this new concept will be presented and attention will also be given to phosphate compounds such as hydroxylapatite, inorganic trimetaphosphate (Acevedo and Orgel, <u>Nature 321</u>, 1986, 790), and zinc

PREBIOLOGICAL SELF ORGANIZATION OF MATTER
C. Ponnamperuma and F.R. Eirich (Eds.)

ISBN 0-937194-18-2

tripolyphosphate, which may constitute important links between chemical and biological evolution.

INTRODUCTION

Among the several phosphate compounds which may serve as energy donors in living organisms, ATP (adenosine triphosphate) has long been recognized as "universal energy currency." It is common knowledge that the energy required for the formation of ATP can be obtained from a variety of energy liberating cellular reactions, which can be coupled to the reactions involved in the actual formation of ATP. Conversely, when ATP is broken down, the energy of this "energy-rich" compound can be used to drive various energy requiring reactions in all known types of living matter.

In photosynthesis and respiration, membrane-bound oxidation-reduction systems provide the energy for the formation of ATP. The actual reactions where ATP is formed from ADP and Pi (inorganic phosphate) are catalyzed by a membrane-bound ATP synthase (or F_1F_0ATPase). The integrally membrane-bound "F_0" part of this complex enzyme serves as a proton pump for the coupling of ATP formation to oxidation-reduction, in agreement with the chemiosmotic theory of Mitchell.

In the photosynthetic bacterium *Rhodospirillum rubrum* a simple, inorganic alternative to ATP, namely PPi (inorganic pyrophosphate), can also be formed at the expense of light energy (H. Baltscheffsky *et al.*, 1966; H. Baltscheffsky and von Stendingk, 1966). In addition, PPi can drive several energy requiring reactions of this organism (M. Baltscheffsky, 1967; Keister and Yike, 1967). In fact, as has been discussed recently (M. Baltscheffsky and P. Nyren, 1984; M. Baltscheffsky *et al.*, 1985; H. Baltscheffsky *et al.*, 1986), the photosynthetic formation of PPi which occurs with an integrally membrane-bound PPi synthase (or PPase) (M. Baltscheffsky *et al.*, 1982), shows many fundamental similarities to that of ATP, and also some differences.

Very soon after our discovery of the bacterial photosynthetic PPi formation we endeavored to speculate about the possible evolutionary significance of this first alternative electron transport phosphorylation system (H. Baltscheffsky, 1967). The simplicity of the PPi molecule, which may well have been formed prebiotically from Pi in a suitable hot environment, in connection with its involvement in bacterial photophosphorylation and in other biological energy transfer reactions (Siu and Wood, 1962), would appear to make this inorganic compound an extremely plausible candidate for being a first, or at least very early participant in the molecular evolution of prebiological and early biological conversion and conservation of energy.

Clearly, the more complex and more versatile ATP molecule must have taken over any leading role of PPi in energy metabolism at a very early stage of the origin and early evolution of life on earth. Whereas PPi can only provide energy and phosphorylate, ATP is able to provide energy, to phosphorylate, pyrophosphorylate, and adenylate. Considering the "principle of continuity" in evolution, one is led to the question: has there been a stepwise transition from a very early energy metabolism with PPi to that with ATP in the molecular evolution of life, and can such a transition be traced in current energy transfer systems? We have recently found what we interpret as possible indications of such a transition in the structures of certain polypeptides involved in phosphate compound metabolism (H. Baltscheffsky et al., 1987).

Here we will report, more specifically, on these results and, more generally, on a new concept based on them. This concept concerns the possibility of a more general molecular coevolution of phosphate compounds and polypeptides. It thus would not be restricted to the phosphate compounds PPi and ATP and the enzymes and other phosphate compounds (Pi, AMP, and ADP) involved in their primary energy transfer reactions, but rather embrace a great number of metabolites containing the phosphate residue. The development of this concept is further stimulated by the estimation (la Cour et al., 1985) that more than half of the reactions in living cells involve phosphate compounds.

SIMILARITIES BETWEEN POLYPEPTIDES INVOLVED IN PPi AND ATP METABOLISM

Many (but not all) nucleotide binding proteins show striking similarities in the three-dimensional pattern of α -helices and β -strands in the nucleotide-binding region. Particularly characteristic is the binding of FAD or NAD at the C-terminal ends of the parallel β -strands of Rossmann folds. (Rossman et al., 1974; Rossmann et al., 1975; Rossmann and Argos, 1978). Regions of amino acid sequence similarities have also been found in the region of the nucleotide-binding fold (Walker et al., 1982; Gay and Walker, 1983; Möller and Amons, 1985). In studies with adenylate kinase an additional, third segment with such similarities was recently described (Fry et al., 1986). This third segment was found in the suggested ATP-binding site of adenylate kinase, which was compared with corresponding regions in other nucleotide-binding enzymes, including the α -and β -subunits of F_1ATPase and the ADP/ATP transferase. The segment in rabbit muscle adenylate kinase contains a β -strand running from amino acid 114 to 119, which is an aspartate residue. This hydrophobic β -strand is part of a parallel β -pleated sheet and appears to flank the triphosphate chain of the bound MgATP, including thus the reaction center of the active site. It may well serve to exclude water and minimize hydrolysis and, also according to Fry et al., (1986), the aspartate 119 may accept a hydrogen bond from a water ligand of Mg^{2+} on MgATP or, alternatively, directly coordinate this Mg^{2+}.

In order to test the possibility that similarities might exist between ATP- and PPi-binding polypeptides we studied the only specifically PPi metabolizing enzyme which has been described in both 3-D and amino acid sequence, the soluble PPase from yeast (H. Baltscheffsky et al., 1987). Kuranova et al., (1983) had described metal-binding by the aspartate-119 of this enzyme. The bound metal is analogously situated at the PPi substrate as the Mg^{2+} at the ATP bound to the adenylate kinase. Furthermore, the amino acid sequence preceding and ending with aspartate-119 of the yeast PPase (Cohen et al., 1978) was remarkably similar to corresponding sequences in α- and β -subunits of F_1ATPases, in particular those from Rsp. rubrum and Rps. blastica (Walker et al., 1985). Thus some evidence had been obtained by us, both at the 3-D and the amino acid sequence levels, for similarities between, on the one hand, a PPi-binding polypeptide, the soluble PPase from yeast, and on the other, different ATP-binding polypeptides.

In Table I we compare the amino acid segment under discussion from rabbit muscle adenylate kinase, the α- and β-subunits of the F_1ATPase from Escherichia coli and Rsp. rubrum, a soluble PPase from yeast, and a possibly corresponding segment from E. coli alkaline phosphatase. The selected sequences of seven amino acids show that the PPase in this region is somewhat more similar to the α-subunits than to the β-subunits of the F_1ATPases, and that they all are markedly different from the adenylate kinase and the alkaline phosphatase, which here show some similarity to each other. It is an open question whether or not the similarities between the PPase and even the Rsp. rubrum α-subunit of the F_1ATPase, with five identities in a sequence of seven amino acids, are due to homology, i.e, divergent evolution from a common ancestor molecule. More sequences have to become known, and among these the amino acid sequence of the membrane-bound PPase from Rsp. rubrum, which has been solubilized and purified (Nyren et al., 1984), will be of particular significance also in this evolutionary context.

Not only the 3-D structure and the amino acid sequence of the membrane-bound and energy conserving PPase of Rsp. rubrum should be of immediate interest. More detailed knowledge of the active site of alkaline phosphatase of E. coli (Sowadski et al., 1985) should be particularly valuable, as this enzyme can hydrolyze PPi and ATP. To these pieces of lacking information may be added a number of points connected with the similarities discussed above. For example:

1. The fact that aspartate-119 in both adenylate kinase and PPase appear to be linked to metal at the reactive phosphate residues of the bound substrates is not accompanied by strong similarities in the sequences compared. In addition, there is no direct evidence that the sequences of the α- and the β-subunits of the F_1ATPases compared here are involved in the nucleotide binding;

2. Aspartate-119 may be binding to the purine and not the phosphate part of GTP in GTP-binding enzymes (Sigal et al., 1986); and

3. The proline residue in PPase and the ATPase subunits is a β-breaker, and the parallel β-sheet of the adenylate kinase does not exist in the PPase.

MOLECULAR COEVOLUTION OF PHOSPHATE COMPOUNDS AND POLYPEPTIDES

In connection with our recent comparison of polypeptides involved in the metabolism of PPi and ATP (H. Baltscheffsky et al., 1987) and the possibility that the similarities observed indicated common ancestry between the two enzyme types, we introduced a new concept, molecular coevolution of phosphate compounds and polypeptides. This was done to provide a hopefully useful framework for the search of molecular traces of such a process and to focus attention on a new angle of possible evolutionary reality. In addition, we referred to the analogy between this concept and that of Williams (1985), who had considered metal ions and proteins and stated: "The properties and functions of metal ions in proteins evolved with the evolution of proteins."

Some main types of phosphate compounds are shown in Table II: inorganic phosphates, adenosine phosphates, ADP-riboses, and dinucleotides, to facilitate the illustration of our molecular coevolution concept in more detail.

Due to its structure, the inorganic orthophosphate lacks the restriction of the other monomers of Table II which requires them to react in a head-to-tail fashion to give high molecular weight polymers, the dinucleotides being examples of the consequences of this restriction. Not only PPi but also high molecular weight polyphosphates are involved as alternatives to ATP in a great number of metabolic reactions, often found in prokaryotes (Kulaev, 1979; Wood and Goss, 1985; Pepin and Wood, 1986). Thus it is necessary to consider the possibility that different kinds of inorganic phosphate compounds played an early evolutionary role as donors of energy (PPi), phosphate (PPi and poly-(P)n) and may be PPi (poly-(P)n).

In Table II three sets of monomer-to-polymer phosphate compounds have been included: inorganic phosphates, adenosine phosphates, and ADP-riboses. The complexity of the compounds increases in this order, and also from left to right. One may ask if a corresponding increase in complexity of the active sites of the respective enzymes can be discerned, and, if so, to what extent stepwise coevolutionary pathways may be less difficult to recognize with comparatively complex substrates and active sites. The indications of similarities already found with PPi and ATP metabolizing enzymes are an encouraging source of stimulation in this search for such pathways.

Among the many important kinds of phosphate compounds which have not been included in Table II are nucleotides with bases other than adenine, with the exception of poly-XMP, or RNA (X being any of the four nucleotide bases of RNA). The extreme importance of RNA's, not least as unique carriers of very specific information and functions as ribozymes, is undisputed, but they have to be included here, heretically enough, as another group of phosphate compounds. It should be observed, however, that among those phosphate compounds which have been left out from Table II for the sake of clarity and simplicity, are DNA and its mononucleotides, sugar phosphates, phospholipids, acetyl phosphate, carbamyl phosphate, and many other phosphate compounds of metabolic importance.

The central role of metal ions in enzyme reactions and their evolution (Williams, 1985) has a counterpart in the reactions of phosphates, where as a rule a divalent cation (Mg^{2+}, Mn^{2+}, Ca^{2+}) is involved in the reaction of the phosphate compound. In the enzyme reactions where phosphate compounds are involved one must thus often consider a presence of metal ions in enzymes and with substrates. The aspartate-119 discussed above appeared to have metal ion rather than a substrate as its site of contact with the free carboxyl group in the reactions involving PPi and ATP, but not when GTP was involved. Clearly, these quite striking examples must be taken into account in considerations of different coevolutionary possibilities and restrictions.

Much of what has been discussed above may be collected into a simple scheme, which is focussed on PPi and ATP as early phosphate compounds, of which at least PPi may have been formed comparatively easily on the primitive earth. (See Scheme I) Some possible roles of the phosphate mineral hydroxyapatite in this context of the early evolution of life (Orgel, 1986) are also indicated, as are the links between phosphate compounds, polypeptides, and metals. Finally, a question mark indicates that the fundamental, central question, which has so far emerged in connection with the molecular coevolution hypothesis and which concerns the possibility of traceable evolution from energy metabolism with PPi to energy metabolism with ATP, notwithstanding the similarities which we have described, still remains to be answered.

ACKNOWLEDGMENT

This work has been supported by the Swedish Natural Science Research Council.

REFERENCES

Acevedo, O. L. and Orgel, L. E., 1986: Nature, 321, 790.

Baltscheffsky, H., 1967: Acta Chem. Scand., 21, 1973.

Baltscheffsky, H., Alauddin, M., Falk, G. and Lundin, M., 1987: Acta Chem. Scand. B, 41, 106.

Baltscheffsky, H. and Baltscheffsky, M., 1986: Origins of Life, 16, 377.

Baltscheffsky, H., Lundin, M., Luxemburg, C., Nyren, P. and Baltscheffsky, M., 1986: Chem. Scr., 26B, 259.

Baltscheffsky, H. and von Stedingk, L.-V., 1966: Biochem. Biophys. Res. Commun., 22, 722.

Baltscheffsky, H., von Stedingk, L. -V., Heldt, H. W. and Klingenberg, M., 1966: Science, 153, 1120.

Baltscheffsky, M., 1967: Biochem. Biophys. Res. Commun., 28, 270.

Baltscheffsky, M., Baltscheffsky, H., and Boork, J., 1982: in Electron Transport and Photophosphorylation, J. Barber, ed., Elsevier, Amsterdam, 249.

Baltscheffsky, M., Boork, J., Nyren, P. and Baltscheffsky, H., 1985: Physiol. Veg., 23, 697.

Baltscheffsky, M. and Nyren, P., 1984: in Bioenergetics, L. Ernster, ed., Elsevier, Amsterdam, 187.

Cohen, S. A., Sterner, R., Keim, P. S. and Heinrikson, R. L., 1978: J. Biol. Chem., 253, 889.

Fry, D., Kuby, S. and Mildvan, A., 1986: Proc. Natl. Acad. Sci., 83, 907.

Gay, N. J. and Walker, J. E., 1983: Nature, 301, 262.

Keister, D. L. and Yike, N. J., 1967: Arch. Biochem. Biophys., 121, 415.

Kulaev, I. S., 1979: The Biochemistry of Inorganic Polyphosphates, Wiley, New York, 266.

Kuranova, J. P., Terzyan, S. S., Voronova, A. A, Smirnova, E. A. and Vainstein, B. K., 1983: Bioorg. Khim., 9, 1611.

la Cour, T. F. M., Nyborg, J., Thirup, S. and Clark, B. F. C., 1985: EMBO J., 4, 2385.

Möller, W. and Amons, R., 1985: FEBS Lett., 186, 7.

Nyren, P., Hajnal, K. and Baltscheffsky, M., 1984: Biochim. Biophys. Acta, 766, 630.

Orgel, L. E., 1986: Origins of Life, 17, 27.

Pepin, C. A. and Wood, H. G., 1986: J. Biol. Chem., 261, 4476.

Rossman, N. G., Moras, D. and Olsen, K. W., 1974: Nature, 250, 194.

Rossman, N. G., Liljas, A., Bränden, C. -J. and Banazak, L. J., 1975: The Enzymes, 11, 61.

Rossman, N. G. and Argos, P., 1978: Mol. Cell. Biochem., 21, 161.

Sigal, I. S. et al., 1986: Proc. Natl. Acad. Sci. USA, 83, 952.

Siu, P. M. L. and Wood, H. G., 1962: J. Biol. Chem., 237, 3044.

Sowadski, J., Handschumacher, H., Krishna Murthy, H. M., Foster, B. and Wyckoff, H., 1985: J. Mol. Biol., 186, 417.

Walker, J. E., Saraste, M., Runswick, M. J. and Gay, N. J., 1982: EMBO J. 1, 945.

Walker, J. E. et al., 1985: J. Mol. Biol., 184, 677.

Williams, R. J. P., 1985: Eur. J. Biochem., 150, 231.

Wood, H. G. and Goss, N. H., 1985: Proc. Natl. Acad. Sci. USA, 82, 312.

ADDENDUM

Baltscheffsky, H., 1988: in The Roots of Modern Biochemistry (eds. H. Kleinkauf et al.), p. 917, de Gruyter, Berlin.

de Duve, C., 1988: ibid, p. 881.

Rouse, R.C., Peacor, D.R., and Freed, R.L., 1988: Am. Mineral., 73, 1988.

Taylor, W.R. and Green, N.M., 1989: Eur. J. Biochem., 179, 241.

TABLE I

Selected Amino Acid Sequences in Some Polypeptides Metabolizing ATP and/or PPi

Sequence	Polypeptide
113 119	
- T - L - L - L - Y - V - D -	adenylate kinase
- T - L - G - A - P - I - D -	E. coli α-ATPase
- V - L - G - E - P - V - D -	E. coli β-ATPase
- V - I - G - E - P - V - D -	Rsp. rubrum β-ATPase
- G - L - G - N - P - I - D -	Rsp. rubrum α-ATPase
113 119	
- G - D - N - N - P - I - D -	yeast soluble PPase
117 123	
- N - G - A - L - G - V - D -	E. coli alkaline phosphatase

TABLE II

Some Phosphate Compounds in Cellular Metabolism

	Pi	PPi	poly-P	
AMP	ADP	ATP	poly-AMP	(poly-XMP)
	ADP-ribose		poly-(ADP-ribose)	
	dinucleotides (NAD, NADP, FAD)			

x = base in mononucleotide

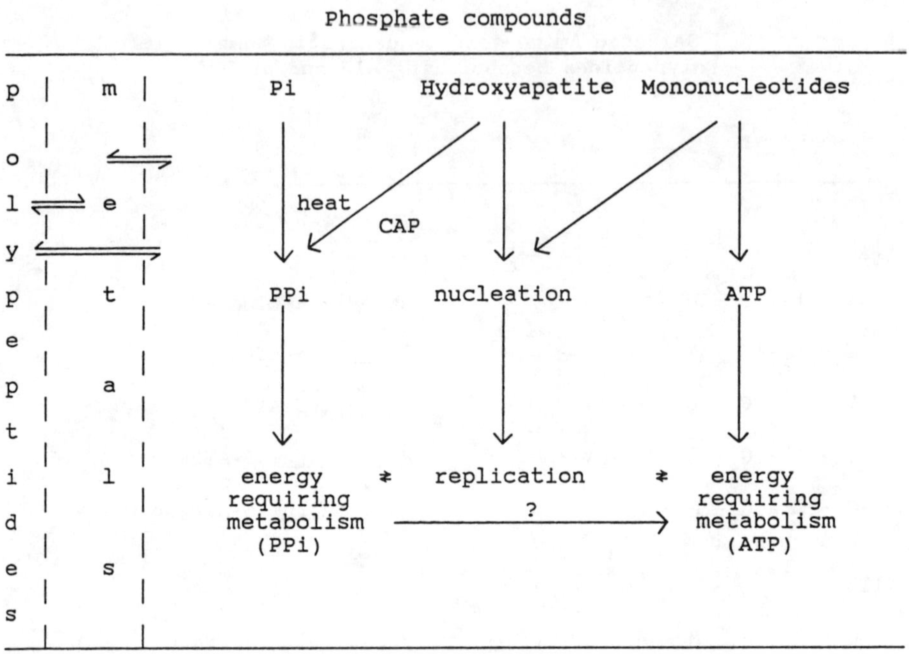

CAP = carbamyl phosphate

SCHEME 1

A HOLISTIC PRECELLULAR ORGANIZATION MODEL

J. Oro
Department of Biochemical & Biophysical Sciences
University of Houston
Houston, Texas 77004

and

A. Lazcano
Departamento de Microbiologia
Escuela Nacional de Ciencias Biologicas, IPN
Apdo. Postal 4-870, 06400
Mexico, D. F., Mexico

ABSTRACT

On the basis of contemporary cellular systems, it is argued that reproduction and a self-sustaining metabolic pathways could not have evolved until both activities were unified into a single integrated process. Therefore, it appears that the elucidation of the problem of the origin of life has to be sought in the molecular cooperation processes which led to the present-day synergistic interaction between nucleic acids and protein within a semipermeable membrane. It is argued that life in the absence of membranes is impossible. Membranes are essential to life not only because they allow the existence of an internal microenvironment separate from the external milieu, but also because any self-replicating and catalytic system lacking an enclosing membrane would be unable to undergo differential replication.

It is suggested that the interaction which led eventually to the establishment of a positive feedback system that included the ancestors of contemporary nucleic acids and proteins occurred within the boundaries of porous, lipidic liposomes. This possibility is supported by 1) the coexistence in carbonaceous chondrites of nitrogen bases, amino acids and other organic compounds (Miller _et al._, 1976) with boundary-forming molecules (Deamer, 1985, 1986); and 2) the encapsulation of polyribonucleotides, urea and Zn^{++} within liposomes made of dipalmitoyl phosphatidylcholine (Baeza _et al._, 1987 a, b). It will be argued

ISBN 0-937194-18-2

that further evolution of precellular systems led eventually to early cells which included 1) a RNA genome; 2) small proto-tRNA molecules which probably charged themselves with amino acids of abiotic origin; 3) a polyribonucleotide matrix ancestral to ribosomes; and 4) catalytic peptides of non-enzymatic origin. Further evolution of these systems led to the ancestors of contemporary prokaryotes.

INTRODUCTION

During the past three decades a large number of compounds of biochemical significance have been synthesized under plausible prebiotic Earth conditions (Oro and Lazcano, 1981; Lazcano et al., 1983). This list includes the monomeric components of proteins, nucleic acids and membranes. Furthermore, some of these compounds or their immediate chemical precursors have been detected in carbonaceous chondrites (Miller et al., 1976), cometary nuclei (Lazcano and Oro, 1981) and in the interstellar medium (Irvine and Hjalmarson, 1984). Although these observations and experiments constitute a significant development in the field of chemical evolution as a necessary prerequisite for the appearance of life (Oparin, 1924; Haldane, 1929), they do not in themselves answer a key question about the origin of life, which is, how was the first living system formed? In other words, whereas the past experimental work demonstrates how the organic molecules necessary for life could have been synthesized abiotically, it provides little insight on how these molecules could have assembled and interacted to generate the self-reproducing autopoietic (Varela et al., 1974; Fleischaker and Margulis, 1987), and evolving living system which is assumed to be the primordial cell ancestral to all forms of life on Earth (Oro and Lazcano, 1984).

The purpose of this paper is to discuss the possible ways in which the cooperative association of prebiologically synthesized oligomers and monomers leading to the first self-maintaining replicating entity took place. This precellular event (or events) is very little understood but certainly it is most significant, since it marks the major transition between non-living and living worlds, or between chemical and biological evolution (Figure 1). This transitional phase is sometimes referred to as precellular (Oparin, 1938 et seq.), or protobiological evolution (Oro, 1980).

Three hypothesis of precellular evolution are reviewed. The replication-first, the metabolism-first, and the membrane-first hypotheses. The latter is better defined as the molecular-cooperation hypothesis. In a way, it represents the combination of the first two hypotheses into one, since life could not start on Earth until a synergistic cooperation was established among the different molecular functions characteristic of life (catalysis, replication, translation and metabolism within a discriminating boundary), so that the generated primeval cell could survive, reproduce and evolve.

PRECELLULAR EVOLUTION HYPOTHESES

If we assume that monomeric and oligomeric molecules were present at an early stage on the Archean Earth, the next most important question has to do with the nature of the simplest form of life. In other words, how can a minimal living entity emerge from these randomly distributed molecules? This is the fundamental problem of precellular evolution (Oparin, 1938 et seq.). Several hypothesis have been made in attempts to understand this question using a number of different approaches.

Most of the hypotheses are characterized by a bias or preference for a certain biomolecular function over others, because it is perceived as the one that triggers the initiation of cellular organization (Shapiro, 1986). The two better studied and most frequently stated hypotheses are the ones that give preference either to replication, or to metabolism (Kamminga, 1987). Thus, if we assume that the boundary of a primitive cell can be provided by a coacervate (Oparin, 1938) or a liposome (Deamer and Oro, 1980) acting as a rudimentary cellular membrane, then the question can in principle be reduced to the elucidation of the relative importance of replication or catalysis (as molecular basis of reproduction and metabolism, respectively) for the initiation of biopoiesis. In what follows we summarize and discuss briefly these hypotheses.

REPLICATING-FIRST HYPOTHESIS

Attempts to explain the origin of life have relied on the assumption that life arose due to the spontaneous formation of a single substance capable of replicating itself. This hypothetical compound has been sometimes identified as an enzyme (Troland, 1914 et seq.), a gene (Muller, 1929) or a double-stranded DNA molecule (Muller, 1955). Although this reductionist approach is deeply rooted in the intellectual background of modern biology (Kamminga, 1980), there is clearly no such thing as a "living molecule." However, after the elegant experiments of Spiegelman and his coworkers (Mills *et al.*, 1967; Spiegelman, 1971), the replication-first hypothesis was advanced and re-stated by Eigen and coworkers (Eigen, 1971; Eigen and Schuster, 1977; Biebricher, 1983).

Spiegelman's (1971) experiments demonstrate that, if selection pressures from the external environment are exerted to a test tube mixture of RNA, mononucleoside triphosphates, the appropriate ions and Qβ-replicase, the RNA genome becomes shortened to about 220 nucleotides for maximum efficiency of replication. In the case of Eigen's (1984) experiments, a steady state population of de novo RNA "genomes" of about 120 nucleotides each, is eventually obtained.

The Qβ-replicase enzyme, which is claimed to function without an external template, shows a lag phase, indicative of the preformation of a small oligonucleotide (up to 5 subunits) which is then used as a de nova template initiator (Schuster, 1984). It may be speculated that the enzyme probably has several sites for binding proximal nucleotides, as in the case of RNA polymerase (Lazcano et al., 1987a) and RNase A (de Llorens et al., 1987). The de novo experiments with Qβ-replicase have been questioned by Hill and Blumenthal (1983) who have argued that a small molecule of RNA may have been present in one of the Qβ-replicase enzyme subunits. Furthermore, it is clear that the experiments of Spiegelman and Eigen have been presented as models only, since the conditions are not really prebiotic. In both cases, a bacteriophage enzyme, the Qβ-replicase, has been used.

In an attempt to further elucidate potential prebiotic models of replication, Orgel and coworkers (Inoue and Orgel, 1981) were able to demonstrate the template-directed synthesis of oligonucleotides by preparing highly activated monomers (e.g., 2-Me ImpG) and allowing them to interact with homopolymer templates (e.g., poly C). Under appropriate conditions, the chemical derivatization of the monomer, or activation, takes the place of the enzyme.

Orgel's experiments clearly demonstrate the molecular basis of the template-directed synthesis of oligonucleotides and of information transfer in the absence of enzymes (Joyce et al., 1984). However, the prebiotic significance of these experiments is limited by the fact that the activation of monomers has been accomplished in totally anhydrous non-prebiotic conditions (Inoue and Orgel, 1981, 1982). Another problem is brought about by the fact that activated pyrimidine nucleotide monomers do not readily self-assemble by stacking as their purine counterparts do, so that the complete replication cycle is never achieved, or only partially realized (Joyce et al., 1984). However, it is conceivable that appropriate modification of the activation and template-directed reactions may eventually lead to the demonstration of the replication process under plausible prebiotic conditions. Template-directed polymerization reactions including "nucleic acid-like" molecules in which ribose and deoxyribose have been substituted by glycerol or other molecules (Schwartz and Orgel, 1985) support the idea that replication of both RNA and DNA was preceded by that of other simple linear molecules in which genetic information was coded in nitrogen bases (Schwartz, 1987). Recently it has been suggested (Joyce et al., 1987) that prior to the appearance of RNA as a cellular genetic material, a polymer made of flexible, acyclic, probably prochiral nucleotide analogues was used to store genetic information.

Following a different approach, experiments carried out by Zaug and Cech (1986) demonstrate the possibility that RNA itself can act in both capacities, as a template as well as a catalytic molecule. Thus, it has been shown that the L-19 intron of Tetrahymena thermophyla pre-ribosomal RNA has the ability to cut and rejoin together pentacytidylic acid molecules to eventually generate a polycytidylic acid containing as many as 30 nucleotide units. The reaction occurs by attack of the guanosine 3'-OH of the L-19 intron 3' end on the pentacytidylic acid, which is partially based-paired to the other 5' end of the L-19 intron (Cech, 1986). The process is repeated several times, adding one nucleotide at a time, until the oligonucleotide has been elongated to about a tridecacytidylic acid. Since the L-19 intron is generated in each cycle intact in the fashion of an enzyme, it is considered by many as an enzyme (Zaug and Cech, 1986). The only difference with ordinary protein enzymes is that the reaction catalyzed by the RNA enzymes such as the L-19 intron is very slow, the turnover number being only 1.7 (Zaug and Cech, 1986); i.e., the L-19 intron can cut and splice the chain of pentacytidylic acid about 100 times an hour, or 1.7 times per minute.

In summary, the two types of reactions studied by Orgel and Cech and their coworkers produce relatively large oligonucleotides by template-directed synthesis and by elongation, respectively, but both stop short of replication. On the other hand, the two reactions studied by Spiegelman (1971) and Eigen (1984) can carry out the replication of polyribonucleotides more than 100 units long but, so far, they both require protein enzymes. The replication-first hypothesis has intellectual attraction and receives, in part, logical support from the dogma of molecular biology, but in its present form still faces what appears to be unsurmountable obstacles.

The discovery of catalytic properties of RNA molecules has led Lazcano (1986) and Gilbert (1986) to suggest that protein synthesis and the use of DNA as an informational macromolecule are the result of natural selection in cells in which both metabolism and reproduction depended on RNA molecules. Thus, the emergence of primordial cells in which both metabolism and reproduction depended basically on the catalytic properties of RNA molecules, preceded the origin of the translation apparatus (genetic code, tRNA, ribosomes, etc.) and its products, i.e., proteins. These were initially selected for by neo-Darwinian processes to increase the catalytic potential of RNA cells. The appearance of double-stranded DNA molecules as informational macromolecules in RNA-cells is explained on the basis of selective pressures that favored an enhanced stability of the genetic information (Lazcano et al., 1987b). This hypothesis, however, does not imply that the first forms of life were naked catalytic and replicative molecules. As shown in Figure 2b, the hypothesis that both metabolism and reproduction originally depended on RNA molecules -- or on nucleic acid-like molecules such as those studied by Schwartz and Orgel (1985) -- and other chemical species of prebiotic origin requires the existence of membranes in order to achieve the differential reproduction necessary for natural selection to take place.

METABOLISM-FIRST HYPOTHESIS

The metabolism-first hypothesis which by implication is also considered as the protein-first or the enzyme-first hypothesis, was originally proposed by Oparin (1924). Although it is sometimes claimed that it never included the discoveries of modern genetics, detailed historical analysis has shown that Oparin was able to incorporate into his ideas on the origins of life the essential concepts of molecular biology (Kamminga, 1980). More recently, Dyson (1985) has given reasons for preferring the metabolism-first hypothesis to the problem of the origin of life rather than the replication first hypothesis. Dyson considers that the existence of organisms is possible in the absence of the precise replication of nucleic acids. These organisms, he suggests, were primarily composed of proteins. He prefers to start with proteins rather than nucleic acids because his mathematical toy model of the Oparin theory (Dyson, 1982) works well with ten or more (it would also work with twenty) species of monomers (amino acids), but would work badly with only four species of monomers (nucleotides).

It should be added that this hypothesis is reasonable in the light of experimental work on prebiotic chemistry. Indeed, in Miller's prebiotic model of organic synthesis by means of electrical discharges, amino acids are formed in substantial amounts (Miller, 1953), but purines and pyrimidines are not usually generated, or formed only in amounts about 100 times smaller (Oro, 1961). Thus, the ten amino acids required by Dyson's (1982) model are much easier to synthesize than the four nucleotides required by the nucleic acid-first hypothesis discussed above.

There is a third argument which is borrowed from the endosymbiotic theory of the origin of eukaryotic cells. Following the concepts elaborated by Margulis (1981) on the significance of symbiosis and parasitism as driving forces of early cellular and biochemical evolution, Dyson (1985) puts nucleic acids into the role of primeval parasites. The parasites come into play when the protein catalysts enclosed within coacervate droplets have developed their homeostatic metabolic pathways. Accordingly, some of these pathways give rise to the formation of ATP and other nucleoside triphosphates from which RNA-like molecules may be synthesized later.

Prior to the presence of an RNA-type replicating molecule, the metabolizing cell (lacking any kind of nucleic acid molecules) would be able to undergo only spontaneous division as a consequence of the tendency of the cell surface to weaken as the cells expands in volume by assimilating fresh monomers from the external environment (Dyson, 1985). According to this model, the first cells passed "genetic information" to their offspring in the form of enzymes which were probably proteins. It is assumed that these enzymes were directly involved in the synthesis of other enzyme molecules. Perhaps the only experimental evidence along this line is the biochemical formation of glutathione and antibiotic peptides directly by enzymes (which are of course encoded in DNA genomes), a process which Lipmann (1972) considered to have preceded in early evolution the synthesis of proteins by the ribosomal machinery of present day cells. In this sense, according to Dyson (1985), there is no logical reason why a population of primeval enzymes mutually catalyzing each other's synthesis should not serve as carrier of genetic information.

Summarizing the fundamental concepts of his model, Dyson (1985) has said that the distinguishing features of these primordial cells are : 1) metastasis, rather than replication; 2) error tolerance, rather than precision in information transfer; and 3) flexibility of the "protein genome," rather than the tyranny of the gene. Finally, the cells would evolve by random drift as described by the concepts of neutral evolution (Kimura, 1983), rather than by natural selection according to Darwinian evolution. Unfortunately for Dyson's theory, no protein has been shown to replicate itself.

THE MEMBRANE-FIRST HYPOTHESIS

Cells have a genotype and a phenotype. Even though self-splicing and elongation of nucleic acids could be considered pheonotypic functions (Cech, 1986), by themselves they are unable to generate a cellular phenotype in the absence of a membrane and other key biochemical molecules (Figure 2). On the other hand, in spite of claims to the contrary, proteins can not replicate themselves and act as the genome which defines the genotype of the cell. Even the enigmatic prions (Prusiner et al., 1985) have their amino acid sequence encoded in the nucleic acid of the cell genome (Robertson et al., 1985), indicating that they are synthesized like all the other encoded proteins. Thus, on the basis of what we know today, it appears practically impossible to conceive cellular life without the replication of nucleic acids or related molecules such as those suggested by Joyce et al. (1987), and the metabolism which depends on enzymes of a proteinic character.

Our own ideas on the emergence and the nature of the first living systems are somewhat different. On the one hand, Oro (1980) has argued elsewhere that the cooperative association of prebiologically synthesized oligomers leading to the emergence of the first self-duplicating entity must have required the non-enzymatic chemical synthesis of 1) an informational self-replicating molecule (proto-DNA or proto-RNA); 2) of a code translating molecule (proto-tRNA); 3) a peptide-synthesizing complex (proto-ribosome); and 4) the self-assembly of a semipermeable liposome (proto-membrane). This model is summarized in Figure 2A. On the other hand, Lazcano (1986) has suggested that the reproduction and metabolism of the first cells depended on the replication and catalytic properties of RNA molecules -- or comparable nucleic acid-molecules with similar properties, such as those suggested by Joyce et al. (1987) -- encapsulated within the boundaries of prebiotic liposomes (Figure 2b).

However, the two of us agree that no theory on the origin of life is complete unless it explains how catalysis and replication were unified into a single integrated process with positive feedback in which both nucleic acids and proteins participate. Therefore, it appears to us that an essential part of the elucidation of biopoiesis has to be sought in the molecular cooperation processes which led to the synergistic interaction between nucleic acids and proteins within a membrane. According to this idea, molecular interaction has to occur within a distinct boundary with selected permeability or discriminating properties, loosely represented by a coacervate droplet (Oparin, 1972), but better by a liposomic structure (Oro, 1980; Deamer, 1986). This prebiotic interphasic structure is the equivalent of a cell membrane. The significance of this protocellular membrane is ordinarily assumed, but is very seldom explicitly recognized. Its prebiotic role has been discussed elsewhere (Deamer and Oro, 1980; Deamer, 1986), so that we need to point out only its three most important features below.

In the first place, the protomembrane provides the separation between the internal microenvironment and the external surroundings which is necessary to maintain higher reactions rates inside the primeval cell than outside. This has been clearly demonstrated by the enzyme-mediated synthesis of poly (A) within the boundaries of coacervate drops (Oparin, 1971). Equivalent experiments with more realistic models of precellular membranes are lacking. Secondly, the protomembrane acts as a filter, or discriminator (a biochemical

Maxwellian demon if you wish) causing the selective retention of some of the synthesized oligomeric molecules within the internal microenvironment. This, potentially, causes growth and eventual reproduction (only if replicative nucleic acids of nucleic acid-molecules are present), two of the driving forces of life. Thirdly, it provides for the primeval cell its own individual identity from an overall (functional and structural) phenotypical point of view, in relation to the external world so that it can compete with other primeval cells and be subject to natural selection.

Recognizing that it is impossible to know the historical sequence of major events and of the molecular cooperation steps leading to the emergence of life, we can only make some suggestions which may eventually form the frame of a working hypothesis. The usefulness of such suggestions will depend on the extent to which they may be testable by experiment. The following is a brief summary of some of the sequential steps of chemical and precellular evolution which need to be described before a working hypothesis can be suggested.

From a structural point of view, one of the most important precellular processes was the self-assembly of phospholipids and other amphiphilic lipids with hydrophobic peptides into liposome structures which can act as semipermeable membranes as well as molecular-size discriminators. The data available from prebiotic as well as other studies, show how readily these structures can be spontaneously formed (Deamer and Barchfeld, 1982; Cullis and Hope, 1985). The relevance to the question of the origin of life of the pioneering studies of Oparin on coacervates (Oparin, 1938 et seq.; Haldane, 1954) and others on semipermeable membranes is well known. However, these studies were not revived until the prebiotic synthesis of amphiphilic lipids was accomplished in Deamer's (Hargreaves _et al._, 1977) and our own (Epps _et al._, 1978; Rao _et al._, 1982) laboratories. Starting from simple precursors and following a sequence of step by step reactions, phosphatidylcholine and phosphatidylethanolamine were obtained in good yields (Rao _et al._, 1982). Furthermore, these phospholipids can readily form bilayered liposomic structures which behave like the semipermeable membranes of cells (Figure 3).

The Fischer-Tropsch type reactions leading to the non-enzymatic synthesis of fatty acids (Figure 3) probably occurred in the later stages of the nebular phase of the Solar System (Miller _et al._, 1976), but are somewhat difficult to imagine taking place in the early terrestrial environment. As noted by Ferris and Usher (1983), "the only obstacle remaining to a convincing prebiotic phospholipid synthesis is the lack of a plausible abiotic synthesis of linear C12 to C18 fatty acids." Prebiotic liposomes may have been formed by even simpler molecules than those shown in Figure 3, such as ionic, single-chain fatty acids (Hargreaves and Deamer, 1978). Moreover, since non-polar molecules extracted from samples of the Murchison meteorite have amphiphilic properties and are capable of forming boundary structures (Deamer, 1985, 1986), clearly the potential for the occurrence of molecules capable to form liposomes, exists in the absence of life.

It is at this point that the processes of prebiotic synthesis begin to connect with the ones that generate boundary structures and other cooperative molecular interactions. These are the organization steps of precellular evolution which eventually gave rise to the emergence of protocells. As noted by Tanford (1978), biological organization may be viewed as consisting of the biosynthetic and the assembly stages. Accordingly, the appearance of precellular systems can be divided into 1) the non-enzymatic synthesis of biochemical monomers and oligomers; and 2) their assembly into polymolecular systems bounded by lipidic membranes which resulted from the position of lowest chemical potential of their amphiphilic components. Neither the size (Baeza et al., 1987b), nor the functional properties (Mayhew et. al., 1977; Dimitriadis, 1978 a, b, c) of the polyribonucleotides are modified due to their encapsulation within liposomes (Table 1). However, it is reasonable to assume that, when this process took place in the prebiotic environment, new physicochemical properties could appear as a result of the interactions between the components of these polymolecular systems. As shown by Hammes and Schullery (1970), the interactions between liposomes and different water-soluble polypeptides leads to major changes in the morphology and permeability of liposomes of phosphatidyl-L-serine, and the dramatic transition of poly-L-lysine from a random coil into an alpha helix. It is likely that in the prebiotic environment comparable to interactions between phospholipids. oligonucleotides and polypeptides of non-enzymatic origin may have led to important changes in the stability and catalytic properties of the precellular systems.

STRUCTURAL ORGANIZATION AND INTERACTION STEPS

ENCAPSULATION OF OLIGOMERS AND MONOMERS

Once liposomes were formed by the self-assembly of phospholipids and other amphiphilic lipids in the presence of a large variety of biochemical monomers and oligomers, an important process which probably occurred is the encapsulation of these chemical species by liposomes. The formation of liposomes can take place by evaporation of mixtures of amphiphilic lipids and subsequent wetting, or by other simple means (Deamer and Oro, 1980). The hydrophobic bonding between phosphatidyl-L-serine and poly-L-serine molecules (Hammes and Schullery, 1970) suggests that the prebiotic liposomes may have contained some peptides of hydrophobic amino acids of non-enzymatic origin in their matrix.

Experiments on the encapsulation of DNA (Deamer and Barchfeld, 1982) and polyribonucleotides by liposomes (Table 1) show that this process occurs very readily and can take place in the presence of ions and prebiotic condensing agents (Baeza et al., 1987, a, b). As shown in Table 1, encapsulation of polyribonucleotides does not require cholesterol, a molecule that very likely was absent in the prebiotic environment. The encapsulation of polyribonucleotides is enhanced by Zn++ (Figure 4), which has been extensively used in non-enzymatic template-directed activated nucleotide polymerization reactions (Inoue and Orgel, 1981, 1982). Histidine, an amino acid of considerable prebiotic catalytic significance (Shen et al., 1986), has been used as a component of the buffer solutions employed during the encapsulation of polyribonucleotides. Evidence is also available on the encapsulation of peptides, proteins and other compounds. The encapsulation of oligonucleotides and catalytic peptides by a protocellular membrane could initiate some of the first cooperative processes. Thus, one can visualize the encapsulation of small RNA-like molecules, together with some short aa-tRNAs selected initially for their ability to bind to amino acids, and of a few small catalytic oligopeptides within the microenvironment of the liposome. Under the appropriate environmental conditions this mixture could be organized into a structurally interacting molecular matrix capable of starting the RNA-instructed synthesis of an oligopeptide.

PROTO-RNA/PROTO-tRNA-aa INTERACTIONS

As a specific example of the above molecular matrix we can consider a short polyribonucleotide and several small (approx. 5 nucleotides) tRNAs. They would assemble into an orderly matrix by the hydrogen bonding of their complementary bases, as in the biochemical codon-anticodon interactions. The binding would occur at environmental conditions (low temperatures, for instance) that would allow the formation of a dynamically stable structure between these charged proto-tRNAs and a linear RNA molecule.

But how did these proto-tRNA molecules appear and accumulate within the boundaries of the liposomes? Why and how where they selected for? As noted by Dayhoff et al. (1972), it is extremely unlikely that a cell (or a protocellular system) could independently evolve even two different tRNAs that would function smoothly in the coding process. Thus, it is likely that all tRNAs are derived from an ancestral form that appeared and diversified before the evolution of the specificity of the code (Dayhoff and McLaughlin, 1969). If a tRNA-like molecule appeared prior to protein synthesis, then the gene coding for it must have been initially transcribed in a polyribonucleotide replication-like fashion by catalytic RNA molecules (Lazcano, 1986), or by catalytic oligopeptides of non-enzymatic origin.

Amino acids were among the most abundant products of non-enzymatic chemical synthesis in the prebiotic environment (Miller, 1953; Miller et al., 1976; Lazcano et al., 1983), and their coexistence with both catalytic and non-catalytic oligonucleotides must have led to many types of interaction between them inside and outside the boundaries of liposomes (Lazcano, 1986). In the absence of enzymes, these proto-tRNA molecules must have become charged with amino acids either by catalyzing themselves the aminoacylation reaction (Smithies, in Crick, 1968; Lazcano, 1986), or with the assistance of oligopeptides of prebiotic origin (Oro, 1980).

The aa-tRNAs which did not bind to the proto-RNA or were discriminated against, would not form part of the matrix and would be eventually degraded or diffused out of the protocellular membrane environment without being involved in the subsequent chemical reactions. Models for these types of associations have been suggested by several authors. For instance, Kuhn and his co-workers (cf. Lehmann and Kuhn, 1984) have visualized that both, the proto-RNA and the aa-tRNAs form helical hairpin structures which after they are assembled protect each other. A modified version of this model is shown in Figure 5. The relative sizes for this proto-RNA and the aa-tRNA are 30 and 20 respectively, although smaller aa-tRNA of about 5 nucleotides have also been considered. This is precisely the size that Balasubramenian (1985) has suggested on the basis of X-ray data and other modelling studies. It is interesting to point out that the development of homochirality within each of the two types of biological polymers (D-pentoses of nucleic acids and L-amino acids in proteins) may have ben derived from the selective chiral interactions in the process of translation involving the above aminoacyl-tRNA as suggested by some authors (Usher and Needels, 1984; Balasubramenian, 1985). The liposomic internal environment, in addition to these oligonucleotide molecules, would contain oligopeptides and other compounds that could facilitate further reactions.

PROTO-RNA INSTRUCTED OLIGOPEPTIDE SYNTHESIS

One of the most difficult and less understood steps in the origin of life is the one in which the transfer of information from oligonucleotides, forming the earliest genomes (proto-RNA) to oligopeptides takes place. This is, of course, the step in which the operation of the genetic code is initially started. Once the proto-RNA and several aminoacyl-tRNA are self-assembled into the above matrix, the different aminoacyl groups would be in close proximity and in linear sequence, so that the carboxyl group of the first amino acid could react with the second acid; the process would continue until all the amino acids would be combined into an oligopeptide chain. The free-energy change of this reaction is about zero, since the amino acid were already activated when they formed the aminoacyl-adenylates and the aminoacyl-tRNA. Initially the energy may have been provided from prebiotically synthesized ATP, or other prebiotic equivalent.

It has been suggested that the earliest translational molecular apparatus required only mRNA and a few primitive tRNAs (Woese, 1972). It has been also proposed that protein synthesis could have begun in the absence of a protoribosome, if the primitive tRNA could bind to the template with five base-pairs instead of the present three (Crick et al., 1976). However, in the absence of ribosomes, the errors introduced would have been too large, and the translation device would be inoperative or highly inefficient at best (Oro and Lazcano, 1984). Whether one accepts that the first organisms were capable of synthesing oligopeptides (Oro, 1980), or that protein synthesis is in itself a result of natural selection (Lazcano, 1986; Gilbert, 1986), we hypothesize that an oligonucleotide-instructed oligopeptide synthesis could not begin in the absence of a proto-ribosome. It is possible that this structure initally lacked proteins (Crick, 1968; Crothers, 1982). The similarities in base sequence between many prokaryotic and eukaryotic tRNAs with their respective rRNAs suggest that these two types of structures may have had a common origin (Bloch et al., 1983, 1985; Nazarea et al., 1985).

Among the oligopeptides synthesized from the information of prebiotically formed proto-RNA, there may have been a considerable variety, including some that probably contained reactive histidyl, lysil, arginyl, or other amino groups which eventually were selected for because of their catalytic role in the formation of phosphodiester or peptide bonds. In this way, oligonucleotides and peptides would be formed with increased efficiency.

PROTOENZYME-MEDIATED SYNTHESIS OF OLIGONUCLEOTIDES

In the process just mentioned of nucleic acid-instructed synthesis of catalytic peptides, it can be expected that some of these may have been selected for because of their ability to catalyze the formation of some of the linear genomic molecules. In this case, the synthesis of specific proto-RNA would be increased, leading in turn to a faster production of the specific catalytic peptides involved in the synthesis of proto-RNA as long as the necessary precursor monomers were available. It is at this point that the unique cyclic process of positive feedback or synergistic catalysis begins between a specific proto-RNA and a specific enzyme. The latter would be a rudimentary RNA-instructed polymerase, or replicase (Lazcano *et al.*, 1987a). From here onwards it is a matter, first, of specifically improving by Darwinian selection the efficiency of the synthetic catalytic activity of the protoenzyme and, second, in a more general way, of finding alternate synthetic reactions which either this or other protoenzymes may catalyze.

EMERGENCE OF METABOLISM

From a genetic and structural point of view, two of the most important bonds are the phosphodiester and peptide bonds, since the first is synonymous with nucleic acids and the second with proteins. From a metabolic point of view, the most important bonds are pyrophosphate and the acylphosphate, since they represent the two prevailing forms of high-energy compounds involved in the chemical activation of either anabolic or catabolic reactions.

The cyclic repetition of the above synergistic catalytic processes can be expected to lead to larger molecular weight products among nucleic acids and peptides, and thus increase the complexity of the system, as long as the precursors are available from the prebiotic soup (Oparin 1924). Different schemes for facilitating the transport of amino acids, nitrogen bases and other ionic molecules have been discussed elsewhere (Stillwell, 1980; Deamer and Oro, 1980; Ferris and Usher, 1983; Deamer, 1986). Low-selective pores must have also existed in the lipidic membranes of the early cells, since it is known that these structures develop both in mixed lipid bilayers (Robertson, 1983) and when a molar residue excess of poly-L-serine is added to liposomes (Hammes and Schullery, 1970). When the different precursors became exhausted from the prebiotic soup as a consequence of a decrease in their steady state formation, metabolic pathways would develop and expand (Margulis, 1981).

SUMMARY

The hypotheses that life emerged as the result either from 1) naked catalytic nucleic acid molecules capable of self-replication; or 2) catalytic oligopeptides which induced metabolic reactions prior to the existence of life itself have been critically examined and found to be necessary, but insufficient for biopoiesis in the light of our present knowledge of the fundamental characteristics of cells. Although we two have somewhat different opinions regarding the nature of the first living systems and their immediate predecessors, we both agree that the first cells resulted from the evolution of polymolecular systems whose identity was maintained due to the presence of lipidic vesicles of non-enzymatic origin. Although some authors have argued that the formation of membranes resulted after the development of a replicating system,the lack of compartment would not only limit severely the possible cooperative interaction between the different molecules which formed the replicating and catalytic apparatus, but would also lead eventually to their dispersal. Membranes are essential to life not only because they allow cells to maintain an internal microenvironment substantially different from the exterior, but also because any self-replicating or catalytic system lacking them would be unable to undergo preferential accumulation and, eventually, differential replication.

ACKNOWLEDGMENTS

We are indebted to Drs. Lynn Margulis and David D. Deamer for several useful discussions on the significance of membranes on biological and prebiological evolution. We are also grateful to Mlle. Celia Ramirez and Dr. Ricardo Amils for several discussions on the origin and early evolution of ribosomes, and to J. Fastag and V. Llaca for their help in preparing the Figures and Table 1. Partial support for this work was provided by NASA Grant NGR 44-005-002 to J.O.

REFERENCES

Baeza, I., Ibanez, M., Lazcano, A., Santiago, C., Arguello, C., Wong, C. and Oro, J., 1987a: Origins of Life (in press).

Baeza, I., Ibanez, M., Santiago, C., Wong, C., Lazcano, A. and Oro, J., 1987b: Adv. Space Res. (in press).

Balasubramenian, R., 1985: Proc. Int. Symp. Biomol. Interactions, Suppl. J. Biosci., 8, 823-835.

Biebricher, C. K., 1983: In M. K. Hecht, B. Wallace and G. T. Prance (eds.), Evolutionary Biology, 16, 1-52.

Bloch, D. P., McArthur, B., Widdosowson, R., Spector, D., Guimaraes, R. C. and Smith, J., 1983: J. Mol. Evol., 19, 420-428.

Bloch, D. P., McArthur, B. and Mirrop, S., 1985: BioSystems, 17, 209-225.

Cech, T. R., 1986: Proc. Natl. Acad. Sci., 83, 4360-4363.

Crick, F. H. C., 1968: J. Mol. Biol., 38, 367-391.

Crick, F. H. C., Brenner, S., Klug, A. and Pieczenik, G., 1976: Origins of Life, 7, 389-397.

Crothers, D. M., 1982: J. Mol. Biol., 162, 379-391.

Cullis, P. R. and Hope, M. J., 1985: In D. E. Vance and J. E. Vance (eds.), Biochemistry of Lipids and Membranes, Benjamin-Cummings Publ., Menlo Park, 25-72.

Dayhoff, M. O. and McLaughlin, P. J., 1969: In M. O. Dayhoff (ed.), Atlas of Protein Sequence and Structure 1969, 4, 89-94.

Dayhoff, M. O., McLaughlin, P. J., Barker, W. and Hunt, L., 1972: In D. L. Rohlfing and A. I. Oparin (eds.), Molecular Evolution: Prebiological and Biological, Plenum Press, New York, 297-317.

Deamer, D. W., 1985: Nature, 317, 792-794.

Deamer, D. W., 1986: Origins of Life, 17, 3-25.

Deamer, D. W. and Barchfeld, G. L., 1982: J. Mol. Evol., 18, 203-206.

Deamer, D. W. and Oro, J., 1980: BioSystems, 12, 167-175.

de Llorens, R., Cuchillo, C. M. and Oro, J., Abstracts of the Fifth ISSOL Meeting on the Origin of Life, Berkeley, California, July 21-25, 1986, 137-138.

Dimitriadis, G. J., 1978a: FEBS Lett., 86, 289-293.

Dimitriadis, G. J., 1978b: Nature, 274, 923-924.

Dimitriadis, G. J., 1978c: Nucleic Acid Res., 5, 1383-1386.

Dyson, F. J., 1982: J. Mol. Evol., 18, 344-350.

Dyson, F. J., 1985: Origins of Life, Cambridge University Press, Cambridge.

Eigen, M., 1971: Naturwissenschaften, 58, 465-523.

Eigen, M., 1984: In G. Nicolis (ed.), Aspects of Chemical Evolution, John Wiley and Sons, New York, 119-138.

Eigen, M. and Schuster, P., 1977: Naturwissenschaften, 64, 541-565.

Epps, D. E., Sherwood, E., Eichberg, J. and Oro, J., 1978: J. Mol. Evol., 11, 279-292.

Ferris, J. P. and Usher, D. A., 1983: Origins of Life, G. Zubay (ed.), Biochemistry, Addison Wesley, Reading, Massachusetts, 1191-1241.

Fleischaker, G. R. and Margulis, L., 1987: Adv. Space Res. (in press).

Gilbert, W., 1986: Nature, 319, 618.

Haldane, J. B. S., 1929: The Rationalist Annual, 148, 3-10.

Haldane, J. B. S., 1954: New Biology, 16, 12-27.

Hammes, G. G. and Schullery, S. E., 1970: Biochemistry, 9, 2555-2563.

Hargreaves, W. and Deamer, D. W., 1978: Biochemistry, 17, 3759-3768.

Hargreaves, W., Mulvihill, S. and Deamer, D. W., 1977: Nature, 266, 78-80.

Hill, D. and Blumenthal, T., 1983: Nature, 301, 350-352.

Inoue, T. and Orgel, L. E., 1981: J. Am. Chem. Soc., 103, 7666-7667.

Inoue, T. and Orgel, L. E., 1982: J. Mol. Biol., 162, 201.

Irvine, W. M. and Hjalmarson, A., 1984: Origins of Life, 14, 15-23.

Joyce, G. F., Inoue, T. and Orgel, L. E., 1984: J. Mol. Biol., 176, 279-283.

Joyce, G. F., Schwartz, A. W., Miller, S. L. and Orgel, L. E., 1987: Proc. Natl. Acad. Sci. USA (in press).

Kamminga, H., 1980: "Studies in the History of Ideas on the Origin of Life from 1860," Ph.D. Thesis, University of London.

Kamminga, H., 1987: Origins of Life (in press).

Kimura, M., 1983: The Neutral Theory of Molecular Evolution, Cambridge University Press, New York.

Kulpa, C. F. and Tinghitella, T. J., 1976: Life Sci., 19, 1879-1888.

Kuhn, H., 1976: Naturwissenschaften, 63, 68-75.

Lavelle, D., Paxton, W. B., Blaustein, D. I., Ostro, M. J. and Giacomoni, D., 1982: Arch. Biochem. Biophys., 215, 486-497.

Lazcano, A., 1986: In L. Margulis, R. Guerrero and A. Lazcano (eds.), "Origin of Life and the Evolution of Cells," Treballs Soc. Catalana Biol., 39 (in press).

Lazcano, A. and Oro, J., 1981: In C. Ponnamperuma (ed.), Comets and the Origin of Life, Reidel, Dordrecht, 191-225.

Lazcano, A., Oro, J. and Miller, S. L., 1983: Precambrian Res., 20, 259-282.

Lazcano, A., Fastag, J., Gariglio, P., Ramirez, C. and Oro, J., 1987a: Jour. Mol. Evol. (submitted).

Lazcano, A., Guerrero, R., Margulis, L. and Oro, J., 1987b: Biosystems (submitted).

Lehmann, U. and Kuhn, H., 1984: Adv. Space Res., 4, 153-161.

Lipmann, F., 1972: In D. L. Rohlfing and A. I Oparin (eds.), Molecular Evolution: Prebiological and Biological, Plenum Press, New York, 261-269.

Magee, W. E., 1980: In G. Grigoriadis and A. C. Allison (eds.), Liposomes in Biological Research, John Wiley and Sons, New York, 249-263.

Margulis, L., 1981: Symbiosis in Cell Evolution, Freeman Co., San Francisco.

Mayhew, E., Papahadjopoulos, D., O'Malley, J. A., Carter, W. A. and Vail, W. J., 1977: Mol. Pharmacol., 13, 488-495.

Miller, S. L., 1953: Science, 117, 528.

Miller, S. L., Urey, H. C. and Oro, J., 1976: J. Mol. Evol., 9, 59-72.

Mills, D. R., Peterson, R. L. and Spiegelman, S., 1967: Proc. Natl. Acad. Sci. USA, 58, 217-224.

Muller, H. J., 1929: Proc. Int. Cong. Plant. Sci., 1, 897-921.

Muller, H. J., 1955: Life. Science, 121, 1-9.

Nazarea, A. D., Bloch, D. P. and Semrau, A. C., 1985: Proc. Natl. Acad. Sci. USA, 82, 5337-5341.

Oparin, A. I., 1924: Proiskhodenie Zhizni (Moscovsky Rabotchii).

Oparin, A. I., 1938: Origin of Life, Macmillan, New York.

Oparin, A. I., 1971: In A. P. Kimball and J. Oro (eds.), Prebiotic and Biochemical Evolution, North Holland, Amsterdam, 1-7.

Oparin, A. I., 1972: In C. Ponnamperuma (ed.) Exobiology, North Holland, Amsterdam, 1-15.

Oro, J., 1961: Nature, 191, 1193.

Oro,J., 1980: In H. C. Halvorson and K. E. van Holde (eds.) the Origins of Life and Evolution, Alan R. Liss, New York, 47-63.

Oro, J. and Lazcano, A., 1981: In B. Vennesland, E. E. Conn, C. J. Knowles, J. Westley and F. Wissing (eds.), Cyanide in Biology, Academic Press, London, 517-541.

Oro, J. and Lazcano, A., 1984: Adv. Space Res., 4, 167-176.

Ostro, M. J., Giacomoni, G., Lavelle, D., Paxton, W. and Dray, S., 1978: Nature, 274, 921-923.

Prusiner, S. B., Groth, D. F., Bolton, D. C., Bowman, K. A., Cochran, S. P., Bendheim, P. E., and McKinley, M. P., 1985: In K. Maramorosch and J. J. McKelvey, Jr. (eds.), Subviral Pathogens of Plants and Animals: Viroids and Prions, Academic Press, New York, 337-391.

Rao, M., Eichberg, J. and Oro, J., 1982: J. Mol. Evol. 18, 196-202.

Rao, M., Eichberg, J. and Oro, J., 1962: J. Mol. Evol., 18, 196-202.

Robertson, H. D., Branch, A. D. and Dahlberg, J. E., 1985: Cell, 40, 725-727.

Robertson, R. N., 1983: The Lively Membranes, Cambridge University Press, Cambridge.

Shapiro, R., 1986: Origins: A Skeptic's Guide to the Creation of Life on Earth, Summit Books, New York.

Shen, C., Yang, L., Miller, S. L., Oro, J., 1986: Abstracts of the Fifth ISSOL Meeting on the Origin of Life, Berkeley, California, July 21-25, 1986, 99-100.

Schuster, P., 1984: Adv. Space Res., 4, 143-151.

Schwartz, A. W., 1987: Adv. Space. Res. (in press).

Schwartz, A. W. and Orgel, L. E., 1985: Science, 228, 585-587.

Spiegelman, S., 1971: Quater. Rev. Biophys., 4, 215-231.

Stillwell, W., 1980: Origins of Life, 10, 277-292.

Straub, S. X., Garry, R. F. and Magee, W. E., 1974: Infect. Immunity, 10, 783-792.

Szoka, F. and Papahadjopoulos, D., 1978: Proc. Natl. Acad. Sci. USA, 75, 4194-4198.

Tanford, C., 1978: Science, 200, 1012-1018.

Troland, L. T., 1914: The Monist, 24, 92-133.

Usher, D. A. and Needels, M. C., 1984: Adv. Space Res., 4, 163-166.

Varela, F. J., Maturana, H. R. and Uribe, R., 1974: BioSystems, 5, 187-196.

Wilson, T., Papahadjopoulos, D. and Taber, R., 1979: Cell, 17, 77-84.

Woese, C. R., 1972: In C. Ponnamperuma (ed.), Exobiology, North Holland, Amsterdam, 301-341.

Zaug, A. J. and Cech, T. R., 1986: Science, 231, 470-475.

ADDENDUM

Deamer, D. W., and Pashley, R. M., 1989: Origins of Life (in press).

Joyce, G. F., 1989: Nature, 338, 217-224.

Lazcano, A., Guerrero, R., Margulis, L., and Oro, J., 1988: J. Mol. Evol., 27, 283-290.

Morowitz, H. J., Heinz, B. and Deamer, D. W., 1988: Origins of Life, 18, 281-287.

Orgel, L. E., 1987: Cold Spring Harbor Symp. Quant. Biol., LII, 9-16.

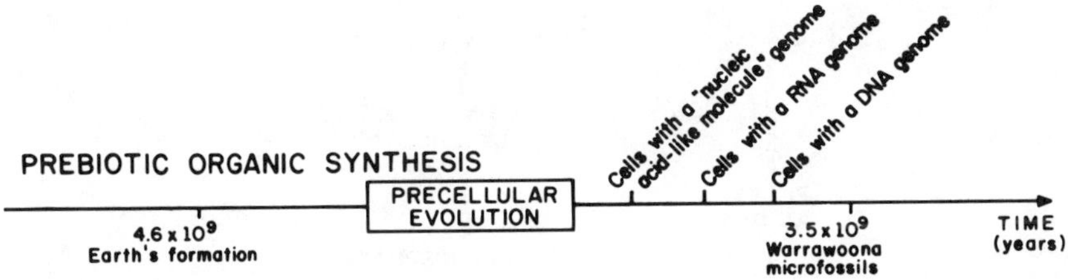

Figure 1. Precellular evolution as a transitional phase between chemical and biological evolution.

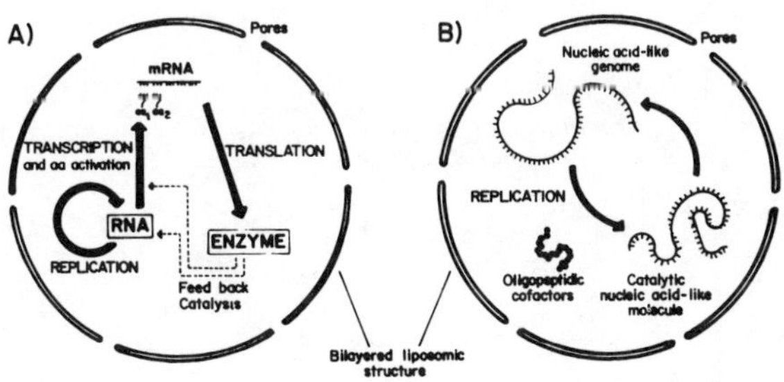

Figure 2. A) A model of a protocell (modified after Oro, 1980). A positive feedback is represented among the replicative proto-RNA and a catalytic proto-enzyme. B) A model of the first biological systems according (modified after Lazcano, 1986). Reproduction and metabolism depend on the catalytic properties of replicative RNA molecules (or other nucleic acid-like molecules with comparable properties), while amino acids, oligopeptides and other chemical species of non-enzymatic origin are cofactors. Note, that in the two models membranes are essential.

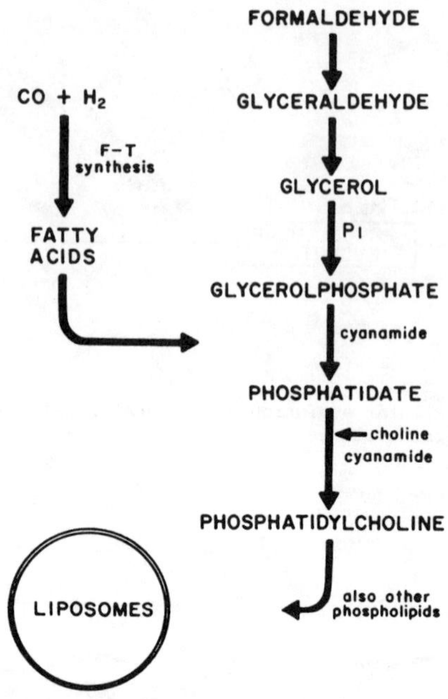

Figure 3. Prebiotic formation of phospholipids and liposomes.

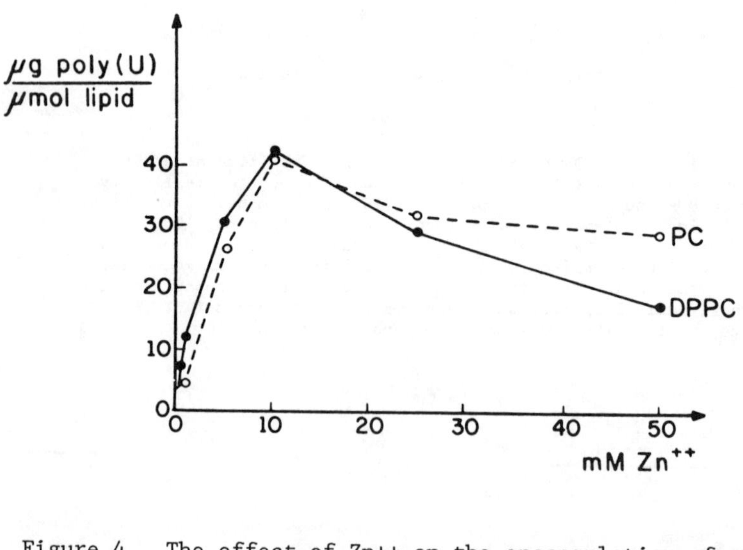

Figure 4. The effect of Zn++ on the encapsulation of poly(U). PC, phosphatidylcholine; DPPC, dipalmitoyl phosphatylcholine.

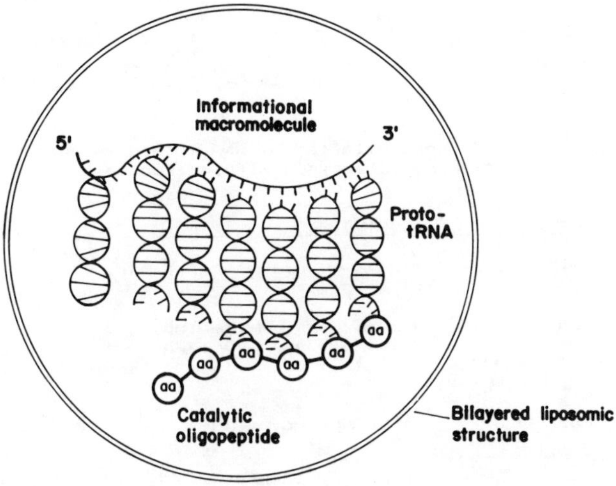

Figure 5. Emergence of a translation apparatus in the absence of ribosomes and within the boundaries of a liposome. The formation of a polypeptide due to the aggregation of proto-tRNA hairpins is shown (modified after Kuhn, 1976).

Table 1. Encapsulation of RNA molecules and polyribonucleotides within liposomes

POLYNUCLEOTIDES	LIPIDS*	COMMENTS	REFERENCE
poly(I) : poly(C) or poly(I)	DPPE; SPM + LEC; SPM	liposomes were prepared under a N_2 atmosphere	Straub *et al.*, 1974
poly(U)	DPPC + CHOL; DPPC + CHOL + SA; DPPC + CHOL + DCP; Egg LEC + CHOL + SA	entrapped poly(U) is RNase-resistant; liposomes were prepared under a N_2 atmosphere	Kulpa and Tinghitella, 1976
poly(I) : poly(C) or poly(I) or poly(C)	beef brain PS	the NaCl buffer included histidine	Mayhew *et al.*, 1977
globin mRNA	beef brain PS	the NaCl buffer included histidine; mRNA remains bilogically active	Dimitriadis, 1978 a, b
reticulocyte rRNA (28s, 18s, 7s, 5s); globin mRNA; yeast tRNA	beef brain PS	the NaCl buffer included histidine; liposomes were prepared under a N_2 atmosphere	Dimitriadis, 1978 c
poly(A); 25s RNA	PC + PG + CHOL	liposomes were prepared under a N_2 atmosphere	Szoka and Papahadjopoulos, 1978
globin mRNA	LEC + DCP	mRNA remains biologically active	Ostro *et al.*, 1978
poliovirus RNA	PS + DPPC	the NaCl buffer included histidine; liposomes were prepared under a N_2 atmosphere	Wilson *et al.*, 1979
poly(I) : poly(C)	SMP; LEC; PE; PE + LEC + SMP + CHOL	entrapped poly(I) : poly(C) is RNase-resistant	Magee, 1980
E. coli RNA + murine rRNA	beef brain PS + PC; DCP + PS	the NaCl buffer included histidine	Lavelle *et al.*, 1982
poly(C); poly(U)	DPPC + CHOL; DPPC; PC	the buffer included urea; encapsulation was enhanced by Zn^{++}	Baeza *et al.*, 1987 a, b

* Symbols: phosphatidylethanolamine, PE; dipalmitoylphosphatidylethanolamine, DPPE; dipalmitoylphosphatidylcholine, DPPC; phosphatidylcholine, PC; phosphatidylserine, PS; phosphatidylglycerol, PG; lecithin, LEC; cholesterol, CHOL; stearylamine, SA; dycetylphosphate, DCP; sphingomyelin, SPM.

PHOTOCHEMICAL ORIGINS OF BIOMOLECULAR CHIRALITY

William A. Bonner and Edward Rubenstein
School of Medicine
Stanford University
Stanford, California 94305

ABSTRACT

Recent theoretical and experimental investigations into the origin of biomolecular chirality are reviewed briefly. Biotic and abiotic theories are evaluated critically with the conclusion that asymmetric photochemical processes with circulary polarized light (CPL) constitute the most likely mechanisms. The three principal CPL processes are examined, and it is concluded that asymmetric photolysis with CPL is by far the most effective mechanism. Solar sources of terrestrial CPL appear too random and too weak to be effective. We suggest an alternative CPL source, namely, the circulary polarized synchrotron radiation from the remnants of supernova explosions. It is suggested that such chiral synchrotron radiation could asymmetrically process racemic compounds in the organic mantles of the dust grains in dense interstellar clouds, and that the enantiomeric excesses of the resulting chiral molecules could be transferred to Earth by cold accretion as the solar system periodically traverses these interstellar clouds.

INTRODUCTION

It was over 125 years ago that Pasteur first pointed out the crucial connection between optical activity and living matter, a connection which today is familiar to all of us. In the intervening years it was subsequently discovered that terrestrial optical activity also encompasses a unique chiral homogeneity in the monomer units of the crucial biopolymers associated with living organisms (L-amino acids in proteins, D-sugars in DNA, RNA and the plant polysaccarides, etc.), and it is commonly accepted today that such chiral homogeneity is a prerequisite for life, and that life as we know it would be impossible without it (Avetisov et al., 1985). Since the chemistry of contemporary life involves highly stereoselective, 3-dimensional interactions between enantiomerically pure molecular species, it is perhaps reasonable to speculate that optically active molecules of homogeneous chirality, and possibly even substantial enantiomeric purity, must have been available on

ISBN 0-937194-18-2

the primitive Earth before life could begin, and that the origin of molecular chirality (or optical activity) might thus be looked upon as a critical early stage in the origin of life. Fundamental questions thus arise: when, where and by what mechanism(s) did optical activity originate, or arrive, on the early Earth, and how did it subsequently develop its absolute enantiomeric homogeneity? Such questions have intrigued scientists for over a century.

Theories for the origin of chirality fall into broad categories, biotic and abiotic. Biotic theories presuppose that life began on a racemic basis and then that living forms, utilizing chiral secondary and tertiary molecular structures, selected and enhanced a particular molecular chirality (Wald, 1957), thus permitting evolution into more complex organisms. Thus chirality and enantiomeric purity developed as life itself developed, and the latter originated independently of the former (Ageno, 1972; Harrison, 1973; Morozov, 1979; Ferracin, 1982; Balasubramanian, 1983; Cairns-Smith, 1986). Such theories are frequently imprecise as to details, are quite unamenable to experimental verification, and have recently been the subject of several fundamental criticisms (Avetisov et al., 1985). Abiotic theories assume that optically active molecules of a predominant chirality originated and became enantiomerically enriched on the primitive Earth independently of living matter by one or more physical processes involving either chance or determinate mechanisms. Such theories suggest that the prebiotic Earth already had available chiral molecules of substantial enantiomeric purity which could permit subsequent stereoselective molecular evolution to the stereochemically intricate processes characteristic of the present biosphere. Chance mechanisms postulate physical processes for symmetry breaking on the molecular level which have an equal probability of producing an excess of either one enantiomer or its antipode. Determinate mechanisms assume that any one of several known asymmetric physical forces in nature may have acted on racemic or prochiral substrates during pertinent prebiotic epochs in such a way as to cause the persistent formation of an excess of one particular enantiomer. Since the times of Pasteur, Van't Hoff and Le Bel, both chance mechanisms and determinate mechanisms for the origin of optical activity have been the subjects of numerous theoretical and experimental investigations - with a tremendous upsurge in both during the past several decades.

Experimental investigations of abiotic chance mechanisms for the origin of optical activity have involved, among others, spontaneous resolution during crystallization (Pincock et al., 1971; Jacques et al., 1981; Bernal, 1985), lattice controlled reactions in chiral crystals (Green et al., 1979), and asymmetric adsorption on the surface of optically active quartz crystals (Kavasmaneck and Bonner, 1977). Such experiments have consistently afforded substantial enantiomeric excesses of one antipode from the racemic substrates under investigation. All arguments for the chance origin of optical activity depend, however, on the assumption of only a very limited number of terrestrial origin sites, a fact which Mann and Primakoff (1983) have developed into a convincing general refutation of chance mechanisms for the origin of the homochirality of our biosphere.

Determinate abiotic mechanisms for the origin of optical activity have received the most experimental and theoretical attention during the past several decades. Such mechanisms have involved parity violation during β-decay (Bonner, 1984), weak neutral current interactions (Kondepudi and Nelson, 1984; Tranter, 1985), magnetic and gravitational field effects (Dougherty, 1981), and the effects of circulary polarized light (Balavoine et al., 1974). Details of earlier, as well as more recent investigations of biotic and abiotic theories for the origin of molecular chirality have recently been systematically reviewed (Bonner, 1972, 1988).

With one exception, all of the determinate mechanisms for the origin of molecular chirality which so far have been investigated have led to experimental results which could not be duplicated in other laboratories, or to the theoretical prediction of effects which are well below the threshold of experimental observation. The single exception is the mechanism of the action of circularly polarized light, where the experimental effects observed have proved to be not only large and unambiguous, but also totally reproducible. This mechanism, and its extrapolation onto a cosmic scale, is the subject of the present communication.

Circularly polarized light (CPL) consists of an electromagnetic wave the electric vector of which spirals clockwise (right-;RCLP), or counterclockwise (left-;LCPL) along its propagation axis. CPL thus constitutes a physical force having "true" chirality (Barron, 1986) and accordingly has the capability of inducing stereoselective diastereomeric photochemical interactions with enantiomeric or prochiral molecules. This was appreciated well before the 1900s by Le Bel (1874), and Van't Hoff (1894), who suggested that optically active substances in nature might have been formed by transformations induced by R- or LCPL, and by Byk (1904) who argued that RCPL, which "predominated on the Earth's surface in reflected sunlight," constituted "an asymmetrical force of solar photochemical energy" available during "countless ages of evolution" (Ritchie, 1933). Such ideas became the impetus for numerous experiments in the early 1900s to generate optically active products by CPL-mediated asymmetric syntheses or degradations. Such experiments were uniformly unsuccessful for three decades, until Jaeger (1930) pointed out that in order for asymmetric reactions with CPL to be successful, the reactions must necessarily be photochemically mediated, and until Kuhn and coworkers performed the first successful asymmetric photolyses of racemic ethyl α-bromopropionate, $CH_3CH(Br)COOEt$ (Kuhn and Braun, 1929), and of α-azidopropionic acid dimethylamide, $CH_3CH(N_3)CONH(CH_3)_3$ (Kuhn and Knopf, 1930). In the latter experiments, after ca. 40% photolysis the residual substrate had optical rotations of -1.04° and +0.78°, depending on the chirality of the CPL employed. In the next several decades a number of other successful asymmetric reactions induced by CPL were reported in the literature (Cf. Bonner, 1972).

The first investigator to demonstrate that R- and LCPL interacted differently with enantiomers was A. Cotton (1896) who discovered the phenomenon of circular dichroism, the unequal absorption of R- and LCPL by an optically active compound as the wavelength of the CPL approached that of a characteristic absorption band of the molecule. It is now recognized that all CPL-mediated processes are dependent on the phenomenon of circular dichroism, that is, on the fact that the absorption coefficients for R- and LCPL characteristic of one enantiomer are different, i.e. $\epsilon_R \neq \epsilon_L$. This inequality requires also that the two enantiomers of a chiral molecule likewise have different ϵ-values for CPL of a particular handedness. Kuhn (Kuhn and Knopf, 1930) defined circular dichroism as $\Delta\epsilon = \epsilon_L - \epsilon_R$, and also defined g, the "anisotropy factor," as $g = \Delta\epsilon/\epsilon = \Delta\epsilon/0.5(\epsilon_L + \epsilon_R)$, where ϵ is the average extinction coefficient for R- and LCPL (or that for unpolarized light). If one plots circular dichroism *versus* wavelength (λ), $\Delta\epsilon$ (which may be (+) or (-) depending upon whether ϵ_L is greater or less than ϵ_R) approaches a maximum (if $\Delta\epsilon > 0$) or (for the enantiomer) a minimum ($\Delta\epsilon < 0$) as λ approaches λ_{max}, the wavelength characteristic of maximum absorption by the pertinent chromophore in the molecule. Thus CPL-mediated reactions are best carried out using wavelengths close to λ_{max}, where $|\Delta\epsilon|$ is at its maximum. Circular dichroism has an important general consequence in CPL-induced reactions. Since in photochemical reactions the reaction rates are proportional to the amount of light absorbed (i.e. to ϵ), D- and L-enantiomers (or prochiral conformers) will, due to circular dichroism, react at different rates with CPL of a given handedness, thus leading to the possibility of an overall asymmetric reaction. The anisotropy factor, g, is important, as seen below, in that it determines the maximum optical yields obtainable in various CPL-induced reactions.

Today there are three general types of reactions known to proceed photochemically in an asymmetric manner under the influenced of CPL. These are 1) asymmetric photoequilibration, 2) asymmetric photochemical synthesis, and 3) asymmetric photolysis. Each of these has been well studied, and each has consistently afforded readily observable and reproducible yields of optically active products from optically inactive precursors. We shall consider each of these processes briefly, citing only limited examples, and examining the potential efficacy of each as a mechanism for the prebiotic origin of enantiomeric excesses in Nature. Additional examples can be found in the reviews by Bonner (1972, 1988).

Asymmetric photoequilibration (also called photoenantiomerization, optical activation or partial photoresolution) consists of the simple equilibration of two enantiomers using CPL (Eq. (1)).

Each interconversion

$$D \underset{k_L}{\overset{k_D}{\rightleftharpoons}} L \qquad (1)$$

follows a simple first order rate equation, so that at equilibrium $[D]/[L] = k_L/k_D$. With ordinary light, $k_D = k_L$, so that $[D]/[L] = 1.0$. With CPL, however, $k_D = \epsilon_D$ and $k_L = \epsilon_L$, so that at equilibrium $[D]/[L] = \epsilon_L/\epsilon_D$ and, since $\epsilon_D \neq \epsilon_L$, $[D]/[L] \neq 1.0$. The illumination of a racemate with CPL of a suitable wavelength and a particular handedness converts it into an optically active $[D] > [L]$ mixture whose composition is determined by the ϵ_L/ϵ_D ratio, while illumination with CPL of opposite handedness converts the racemate to a $[L] > [D]$ mixture of equal and opposite composition and optical rotation. The optical yield after such an asymmetric photoequilibration is given by
$([D] - [L])/[D] + [L]) = |\epsilon_L - \epsilon_D| \;/\; |\epsilon_L + \epsilon_D| = \Delta\epsilon/2\epsilon = g/2$. Since for most optically active compounds $g<\approx 0.02$, the optical yield is limited to the low value of ca. 1%, or so. Such asymmetric photoequilibrations have been extensively studied by Stephenson and coworkers (1972, 1976) for octahedral transition-metal complexes with bidentate ligands, such as $Cr(ox)_3{}^{3-}$.

Asymmetric photochemical synthesis involves the photochemical conversion of an optically inactive prochiral reactant into an optically active product by the action of CPL. The most thoroughly studied cases of such reactions (Buchardt, 1974; Moradpour *et al.*, 1975) involve the photochemical ring closures (followed by aromatization) of various diarylethylenes into polyhelicenes, which are chiral by virtue of their right- or left-handed screw structure. A prototypical example is the conversion of 1-(2-naphthyl)-2-(3-phenanthryl)ethylene (1) into hexahelicene (2) using ultraviolet CPL. With RCPL the 2 product had $[\alpha]^{234}{}_{36} = -30.0°$, while with LCPL the 2 had a rotation of +30.5°. The mechanism of such asymmetric ring closures is well understood, and it can be shown that the optical yields in such reactions are again limited to g/2, or ≈ 1%.

Asymmetric photolysis involves the differential photodestruction of two enantiomers in a racemate by CPL (Eq. (2)). Again, because of circular dichroism, $k_D \neq k_L$, and the two enantiomers photolyze at different rates.

$$\text{D-Product(s)} \xleftarrow{k_D} \text{DL-Substrate} \xrightarrow{k_L} \text{L-Product(s)} \qquad (2)$$

Thus, if the reaction is terminated before photolysis is complete, the *unphotolyzed* residue will be optically active, i.e. $[D] \neq [L]$.

Kagan and coworkers (1974) have shown quantitatively, both theoretically and experimentally, how the optical purity of the unphotolyzed residue depends on both the extent of photolysis and the anisotropy factor, g. They show that the greater the extent of degradation and the higher the g-value for the substrate, the greater will be the optical purity of the unphotolyzed residue, and that for sufficiently high values for completion of reaction and g, the optical purity of the residue can approach 100%. Kagan _et al._ (1974) obtained an optical purity of ≈ 20% for the residual camphor after photolyzing DL-camphor to the extent of 99%, while Flores _et al._ (1977) obtained a 2.5% optical yield for D-leucine on photolyzing DL-leucine in aqueous solution to only 75% completion, using 2128 Å LCPL Thus, asymmetric photolysis with CPL is capable of providing enantiomeric enrichments far exceeding the mere ≈ 1% characteristic of asymmetric photoequilibrations and photochemical syntheses. Furthermore, asymmetric photolysis is not limited to the rather esoteric systems used in the other studies (asymmetric photoequilibrations and photochemical syntheses have never been demonstrated in molecules associated with life), but is in principle applicable to any organic molecule having absorption bands at photochemically pertinent wavelengths. Thus, of the three CPL-mediated processes, asymmetric photolysis would seem to be the most likely mechanistic candidate for the prebiotic origin of optical activity.

For asymmetric photolysis to have engendered enantiomeric excesses in the chiral molecules on the primitive Earth, a preponderance of CPL of one particular handedness must have been persistently available during long periods of prebiotic time. Furthermore, such CPL must have been of sufficient intensity and suitable wavelength to promote asymmetric photolysis at such rates and in such locations as to permit, for subsequent chemical evolution, local accumulations of substantial quantities of molecules having the favored enantiomeric excesses. Early authors tacitly assumed all of the above, but only vague suggestions were made as to specific mechanisms for the natural production of CPL (cf. Bonner, 1972), and no estimates were made of the intensity of the CPL, or the magnitude of its net polarization. These remain unanswered questions which are now only beginning to be addressed.

The ultimate radiation source for all photochemical processes on Earth is, of course, the Sun, whose spectrum is rich in all of the photochemically pertinent ultraviolet and visible wavelengths. How can such solar radiation be converted to terrestrially available CPL? It is well known that sunlight becomes linearly polarized by reflection, or by Raleigh and/or Mie scattering from atmospheric molecules and/or aerosols, and that elliptically and circularly polarized light can be produced by suitable additional reflections or multiple scatterings (Hecht and Zajac, 1974). That circular polarization in specific directions in the sky is not random has been shown by two recent quantitative studies. Angel and coworkers (1972) found that the 7800-8800 Å sunlight at the eastern horizon

40 minutes before sunrise had a 0.25% excess of LCPL, while that in the west just after sunset had a similar excess of RCPL. More recently Wolstencroft (1985) reported on the circular polarization of both 5825 Å visible and 3500 Å ultraviolet skylight at a fixed direction (34° altitude,due south from Sutherland, South Africa) at various time of the day. The polarization at each wavelength was 0% at sunrise, noon, and sunset, but peaked to 0.1% (5825 Å) and 0.48% (3500 Å) with LCPL in the early morning hours, and peaked similarly with RCPL in the late afternoon. Thus for an LCPL excess in the AM, there was a cancelling RCPL excess in the PM. Wolstencroft showed, moreover, that if Earth were a smooth sphere the overall net chirality of the CPL from all directions of the sky at any specific location was 0%. If there were slopes or mountains in the local terrain, however, such that certain areas of the sky were obscured, there could be a net preponderance of RCPL and LCPL in such localities. Wolstencroft estimated that the overall value for net circular polarization due to such obscuration might vary between 0.001% and 0.05% in a given location, depending upon the terrain and the atmospheric aerosol concentration. It is difficult to see, however, how a net terrestrial preponderance of CPL could arise by such a mechanism when integrating such topographical and meterological irregularities over the entire surface of Earth. Moreover, due to the equivalent diurnal production of AM LCPL and PM RCPL by the aerosol scattering mechanism, any asymmetric effects produced by the morning LCPL should be cancelled by the afternoon RCPL. The only diurnal asymmetry potentially operative over the entire Earth would be the fact that subsequent temperature dependent non-photochemical reactions involving the initial chiral photochemical products would occur at greater rates in the warmer afternoon environments, thus favoring the chemical evolution of RCPL products. In this connection, it is interesting but probably coincidental that 2128 Å RCPL favored formation of "natural" L-leucine on asymmetric photolysis of DL-leucine (Flores et al., 1977).

Developing ideas originally proposed by Van't Hoff (1894) and Byk (1904), Mörtberg (1971) has suggested an alternative genesis for a net chirality of terrestrial CPL. Because of the Faraday effect,the RCPL and LCPL components of linearly polarized skylight suffer unequal refraction on descending through the increasing refractive index gradient of Earth's atmosphere in the presence of the Earth's magnetic field. Thus, one of the CPL components is preferentially refracted toward Earth and reaches the surface in slight excess. Mörtberg then demonstrates that if the Earth has a tilted (not perpendicular) axis with respect to the incoming sunlight and travels in an elliptical (not circular) orbit about the Sun, there would result a net difference in the intensity of the RCPL and LCPL reaching the two hemispheres, and this difference would not cancel at any geographical location on Earth during its yearly orbit. However, there are two important factors which could nullify Mörtberg's mechanism, namely, the periodic reversals of the Earth's magnetic field which have occurred frequently since Precambrian times (McIlhinny, 1971), and the precession of the equinoxes, which symmetrically reverses the tilt of Earth's axis every 13000 years. Such factors might be vitiated, however, by the overall global asymmetry of Earth's land mass, which would insure that there be no precise cancellation of the effects of Mörtberg's mechanism, either by geomagnetic reversals or precession of the equinoxes (Bonner and Rubenstein, 1987). The weakest point in Mörtberg's hypothesis is the admitted smallness of the anticipated effect. While he makes no attempt at quantitative estimation, he stresses that the small Verdet constant for gases, the shallow refractive index gradient of the atmosphere, the relatively small tilt to Earth's axis, the trivial ellipticity of Earth's orbit, and the weakness of Earth's magnetic field (≈ 0.5 G) would combine to result in a very small excess indeed of terrestrial CPL of a preferred chirality.

Therefore, asymmetric photolysis with CPL is a demonstrably efficient abiotic mechanisms for generating enantiomeric excesses, but, since terrestrial sources of CPL, as outlined above, are probably too weak and/or too random to be effective, Rubenstein, Bonner and coworkers (1983, 1987) have recently suggested an alternative, cosmic source of CPL, namely, the circularly polarized synchrotron radiation emanating from the neutron star remnants of supernovae.

The energy releasing nuclear fusions which occur as a star heats up during gravitational collapse are 1) H → He, 2) He → C, 3) C → N, O, Si and 4) Si → Fe. Further fusion absorbs energy rather than releasing it, so fusion stops at a Fe core. When a star has finally exhausted all of its nuclear "fuel," such energy releasing transformations no longer balance the gravitational attraction of the star's mass and the star collapses on itself. If the star is of sufficient size (≈ 8-10 solar masses), its interior collapses in a matter of seconds and "as one piece" to a core of nuclear density. But since nuclear matter is incompressible, a shock wave rebounds outward from the core, turning the stellar collapse into a supernova, a gigantic explosion which blows off the outer mantle of the star, leaving a neutron star core and releasing vast amounts of energy (10^{51} J) (Bethe and Brown, 1985; Clark and Stephenson, 1977). The latter produces shock waves which compress the interstellar medium, forming new stars (Herbst and Assousa, 1979), while the expanding nebula of ejecta cools and condenses into the microscopic grains of refractory dust which populate the interstellar medium (Clayton, 1982). Supernovae are relatively frequent, three being recorded in our galaxy during the past 1000 years (Bethe and Brown, 1985). At this rate, some 14 million supernovae would have occurred in our galaxy alone since the formation of our solar system, which itself is believed the consequence of a supernova some 4.7 billion years ago (Cameron and Truran, 1977; Schramm and Clayton, 1978). Because of their intrinsic brightness, 10 or more supernovae per year are sighted in distant galaxies (Bethe and Brown, 1985).

The neutron star remnant from a supernova is thought to be a small object (≈ 10 km diameter), incredibly dense (10^{14} g/cc) Clark and Stephenson, 1977), rotating extremely rapidly (milliseconds per revolution) (Ashworth et al., 1983) and possessing an intense magnetic field (≈ 10^{12} G) (Lyne, 1984). This rotating magnetic field induces colossal electric fields around the neutron star, which in turn act to accelerate plasma electrons and protons around the core to near relativistic velocities. These charged particles are constrained by the intense magnetic field, however, to move in circular paths orbiting the neutron star, and in doing so the orbiting electrons produce synchrotron radiation photons, as suggested originally by Shklovskii for the spectrally continuous light emitted by the Crab Nebula (Cf. Oort, 1957). These processes are illustrated schematically in Fig. 1.

Synchrotron radiation has several experimentally established characteristics (Winick, 1980) which are potentially relevant for the origin of optical activity. First, it comprises a continuous spectrum of wavelengths from radio frequency to x-rays, and thus has the broad capability of inducing photochemical reactions. Second, in the plane of the orbiting electrons synchrotron radiation is linearly polarized, with the plane of polarization perpendicular to the magnetic field. Above the orbit plane and depending on the angle, the radiation becomes elliptically and ultimately circularly polarized with a specific handedness, while below the plan one finds elliptically and circularly polarized radiation having the opposite handedness. Thus above the plane of orbit of the electrons accelerating around a neutron star, there will exist vast domains of space bathed by synchrotron radiation having elliptical and circular polarization of one chirality, while below the orbit plane there will be vast regions of space having synchrotron photons of the opposite chirality, as depicted schematically in Fig. 2. How might such synchrotron radiation from supernova remnants influence the chirality of molecules on Earth?

Any direct interaction of synchrotron CPL from a neutron star with molecules on the surface of Earth appears rather unlikely because of the small angle in space which would be subtended by Earth and the consequent low intensity of the synchrotron CPL which would reach its surface. It seems more probable that the asymmetric effects of such synchrotron CPL might instead have been transferred to Earth over vast periods of time via the dust grains in interstellar molecular clouds.

Interstellar matter is comprised of H and He atoms along with vastly smaller quantities of H_2 and other small inorganic and organic molecules, as well as 0.02-0.3 μm dust grains of refractory silicates and metal oxides synthesized by thermal fusion in stars and blown into space during supernova events. As with water droplets in our atmosphere, interstellar matter is not uniformly distributed throughout space, but is bunched into clouds of varying densities -- from "diffuse" clouds to dense "molecular" clouds of densities up to 10^6 particles/cc (Whittet, 1981).

The surface composition of interstellar dust grains is thought to depend on the type of cloud in which they occur and the radiation to which they have been subjected (Greenberg, 1984). When the grains find themselves in a dense molecular cloud at very low temperature (10-100° K) they accrete an "ice" mantle consisting of frozen H_2O, NH_3, CH_4, CO, CO_2, CH_2O and other small organic molecules in the cloud. These "ices," then, become irradiated with ultraviolet and ionizing radiation and free radicals are formed, frozen into the ices. On subsequent slight warming, the trapped radicals interact to produce higher molecular weight organic substances, including polymers, which form refractory organic mantles on the dust grains. As further accretion of smaller molecules and radiation processing continues,more complex mantles form on the grains. These may subsequently be "exploded" back into space by the heating of the mantle-laden grains during collisions, leaving barren grains now ripe for "recycling" by similar accretion and radiation processes. Greenberg (1983, 1984), who estimates that from 1-10% of the simple molecules in dense molecular clouds are processed, as above, during the 10^7 year mean lifetime of such clouds, has not only described such mantle development sequences in detail, but has also conducted laboratory simulations of the condensation-irradiation-warming stages of grain mantle evolution. Mixtures of H_2O, NH_3 and CO, typical interstellar molecules, were condensed under high vacuum on a cold (10°K) metal surface simulating an interstellar grain, whereupon the condensate was irradiated with ultraviolet light and warmed slightly to induce recombination of the trapped free radicals. Subsequent investigation of such irradiated "mantles" by gas chromatography and mass spectrometry revealed that the refractory polymeric material ("yellow stuff") which was formed also contained a number of smaller organic molecules, including such racemic and potentially optically active compounds as lactic acid, glyceric acid and glyceramide (Agerwall et al., 1986).

We have recently suggested (Bonner and Rubenstein, 1987) that the circularly polarized ultraviolet synchrotron radiation from the neutron star remnants of supernovae could asymmetrically photolyze the racemic organic constituents in the mantles of interstellar grains in molecular clouds, thus producing optically active molecules of a specific chirality on the interstellar grains themselves.

At approximately 110-million year intervals our solar system, revolving around the center of our galaxy, passes through dust clouds at the inner edge of the spiral arms of the galaxy. The Earth has thus traversed such dust regions some 40 times since the formation of the solar system and some 5 or 6 times prior to the earliest life on Earth, ≈ 3.8 billion years ago. Greenberg (1983) has estimated that at each such pass-through, the Earth accretes some 10^8 - 10^{10} metric tons of matter, a quantity $\approx 10^{-10}$ to 10^{-12} of its total mass and far greater than its current biomass. The small size of the interstellar grains insures that they don't heat appreciably on traversing Earth's atmosphere, so that the organic constituents of the grain mantles can be captured unpyrolyzed at the surface of the Earth. Greenberg has suggested that such cold accretion must have provided some of the complex organic molecules from which life on Earth began. We have suggested further that the optically active components of the grain mantles, formed as above, and protected from racemization by the cold interstellar temperatures, could have been transported to Earth with their enantiomeric excesses intact.

The above hypothesis for the extraterrestrial origin of optically active molecules on Earth, summarized in Fig. 3, allows for two presently unverifiable predictions. Due to the relative frequency of supernovae in our, and other, galaxies and to the prevalence of molecular clouds containing dust grains with organic mantles, life based on chiral organic molecules should occur inevitably and relatively frequently in the Universe. Secondly, because of the overall symmetry inherent in the circularly polarized synchrotron radiation in the vast regions of space around a neutron star, the chirality of the biomolecules associated with extraterrestrial life need not be similar to that found on Earth, but might vary randomly throughout the universe.

REFERENCES

Agarwal, V. K., Schutte, W., Greenberg, J. M., Ferris, J. P., Briggs, R., Conner, S., van de Bult, C.E.P.M. and Baas, F., 1986: Origins of Life, 16, 21-40.

Ageno, M., 1972: J. Theor. Biol., 37, 187-192.

Angel, J. R. P., Illing, R. and Martin, P. G., 1972: Nature, 238, 389-390.

Ashworth, M., Lyne, A. G. and Smith, F. G., 1983: Nature, 301, 313-314.

Avetisov, V. A., Anikin, S. A., Gol'danskii, V. I. and Kuz'min, V. V., 1985: Doklady AN SSSR-Biophysics, 282, 115-117.

Avetisov, V. A., Kuz'min, V. V. and Anikin, S. A., 1987: Chem. Phys. 112 (2), 179-187.

Balasubramanian, R., 1983: Origins of Life, 13, 109-112.

Balavoine, G., Moradpour, A. and Kagan, H. B., 1974: J. Am. Chem. Soc., 96, 5152-5158.

Barron, L. D., 1986: Chem. Soc. Rev., 17, 189-123.

Bernal, I., 1985: Inorg. Chim. Acta, 96, 99-110.

Bethe, H. A. and Brown, G., 1985: Sci Am. 252 (5), 60-68.

Bonner, W. A., 1972: "Origins of Molecular Chirality," in Exobiology, C. Ponnamperuma (ed.), North Holland, Amsterdam, 170-234.

Bonner, W. A., 1984: Origins of Life, 14, 383-390.

Bonner, W. A., 1987: "Origins of Chiral Homogeneity in Nature," in Topics in Stereochemistry, Vol. XVIII, E. Eliel and S. Wilen (eds.), Interscience, Wiley, New York, 1-96.

Bonner, W. A. and Rubenstein, E., 1987: Biosystems, 20, 99-111.

Buchardt, O., 1974: Angew. Chem. Int. Ed., 13, 179-185.

Byk, A., 1904: Z. Physik, Chem., 49, 641-687.

Cairns-Smith, A. G., 1986: Chem. in Britain (June), 550-561.

Cameron, A. G. W. and Truran, J. W., 1977: Icarus, 30, 447-461.

Clark, D. H. and Stephenson, F. R., 1977: The Historical Supernovae, Pergamon, New York, 1-13.

Clayton, D. D., 1982: "Supernovae and the Origin of the Solar System," in Supernovae: A Survey of Current Research, M. J. Rees and R. J. Stoneham (eds.), Reidel, Boston, 535-564.

Cotton, A., 1896: Ann. Chim. Phys., 8 (7), 317.

Dougherty, R. C., 1981: Origins of Life, 11, 71-84.

Ferracin, A., 1982: J. Theor. Biol., 94, 517-518.

Flores, J. J., Bonner, W. A. and Massey, G. A., 1977: J. Am. Chem. Soc., 99, 3622-3625.

Goldanskii, V. I. and Kuz'min, V. V., 1988: Z. Phys. Chem. (Leipzig) 269 (2), 216-274.

Green, B. S., Lahav, M. and Rabinovich, D., 1979: Acct. Chem. Res., 12, 191-197.

Greenberg, J. M., 1983: "The Largest Molecules in Space: Interstellar Dust," in Cosmochemistry and the Origin of Life, C. Ponnamperuma (ed.), Reidel, Boston, 71-112.

Greenberg, J. M., 1984: Sci. Am., 250 (6), 125-135.

Greenberg, J. M., 1986: Astrophys. Space Sci. 128 (1), 17-31.

Harrison, L. G., 1973: J. Theor. Biol., 39, 333-341.

Hecht, E. and Zajac, A., 1974: Optics, Addison-Wesley, Menlo Park, California, 239-253.

Herbst, W. and Assousa, G. E., 1979: Sci. Am., 241 (2), 138-145.

Jacques, J., Collet, A. and Wilen, S., 1981: Enantiomers, Racemates and Resolutions, John Wiley & Sons, New York.

Jaeger, F. M., 1930: Optical Activity and High Temperature Measurement, McGraw Hill, New York, 55 76, 203-214.

Kagan, H. B., Balavoine, G. and Moradpour, A., 1974: J. Mol. Evol., 4, 41-48.

Kavasmaneck, P. R. and Bonner, W. A., 1977: J. Am. Chem. Soc., 99, 44-50.

Kondepudi, D. K. and Nelson, G. W., 1984: Physica, 125A, 465-496.

Kuhn, W. and Braun, E., 1929: Naturwiss., 17, 227-228.

Kuhn, W. and Knopf, E., 1930: Z. Physik. Chem., 7B, 292-310.

Le Bel, J. A., 1874: Bull Soc. Chem. Fr., 22, 334-347.

Lyne, A. G., 1984: Nature, 308, 605-606.

McIlhinny, M. W., 1971: Science, 172, 157-159.

Mann, A. K. and Primakoff, H., 1983: Origins of Life, 13, 113-119.

Moradpour, A., Kagan, H., Baes, M., Morren, G. and Martin, R. H., 1975: Tetrahedron, 31, 2139-2143.

Morozov, L., 1979: Origins of Life, 9, 187-217.

Mörtberg, L., 1971: Nature, 232, 105-107.

Oort, J. H., 1957: Sci. Am., 196 (3), 52-60.

Pincock, R. E., Perkins, R. R., Ma, A. S. and Wilson, K. R., 1971: Science, 174, 1018-1020.

Ritchie, P. D., 1933: Asymmetric Synthesis and Asymmetric Induction, Oxford, London, 44 ff.

Rubenstein, E., Bonner, W. A., Noyes, H. P. and Brown, G. S., 1983: Nature, 306, 118.

Schramm, D. N. and Clayton, R. N, 1978: Sci. Am., 239 (4), 124-139.

Schutte, W. and Greenberg, J. M., 1986: Astrophys. Space Sci. Lib. 124 (Light Dark Matter), 229-232.

Stephenson, K. L., 1972: J. Am. Chem. Soc., 94, 6652-6654.

Stephenson, K. L. and Baker, R. L., 1976: Inorg. Chem., 15, 1086-1088.

Tranter, G. E., 1985: Mol. Phys., 56, 825-838.

Van't Hoff, J. H., 1894: The Arrangement of Atoms in Space, 2nd Ed., 30; 3rd Ed. (1908), 8.

Wald, G., 1957: Ann. N. Y. Acad. Sci., 69, 353-368.

Whittet, D. C. B., 1981: Q. J. R. Astron. Soc., 22, 3-21.

Winick, H., 1980: "Properties of Synchrotron Radiation," in Synchrotron Radiation Research, H. Winick and S. Doniach (eds.), Plenum, New York, 11-20.

Wolstencroft, R. D., 1985: "Astronomical Sources of Circularly Polarized Light and their Role in determining Molecular Chirality on Earth," in The Search for Extraterrestrial Life: Recent Developments, M. D. Papagiannis (ed.), Symp. Int. Astron. Union, 171-175.

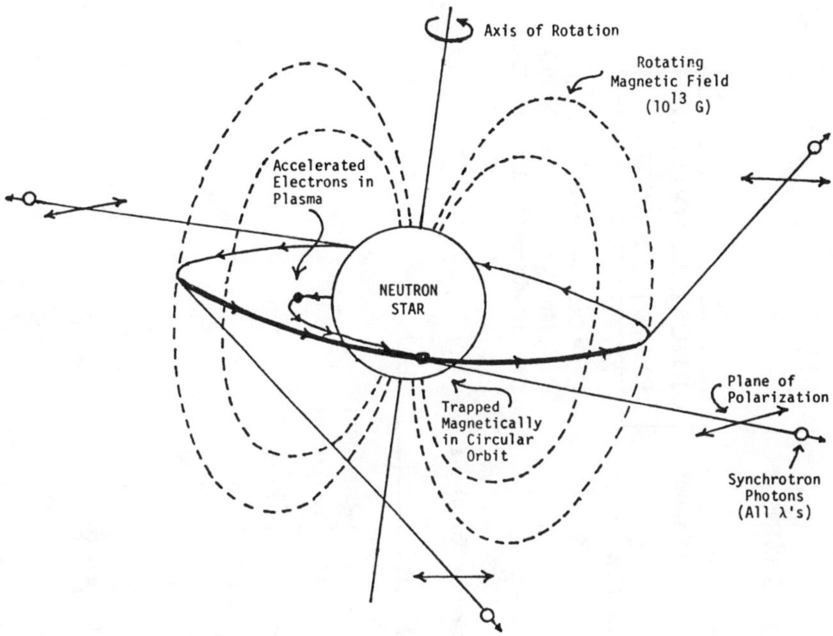

Fig. 1. Production of synchrotron radiation from supernova remnant. (From Bonner and Rubenstein, *BIOSYSTEMS*, 1987)

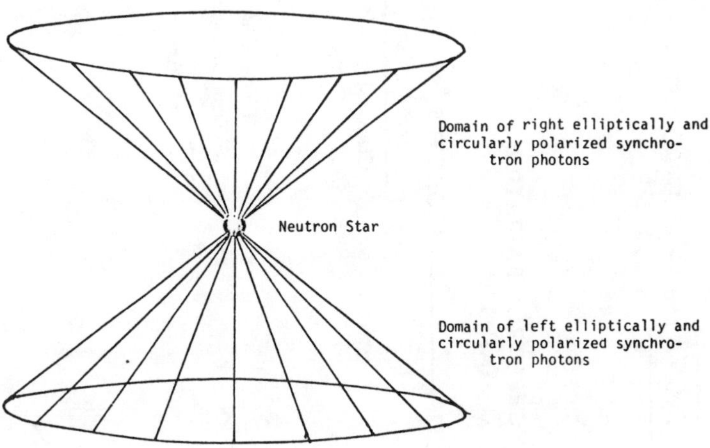

Fig. 2. Domains of polarization of synchrotron radiation around neutron star.

(From Bonner and Rubenstein, *BIOSYSTEMS*, 1987)

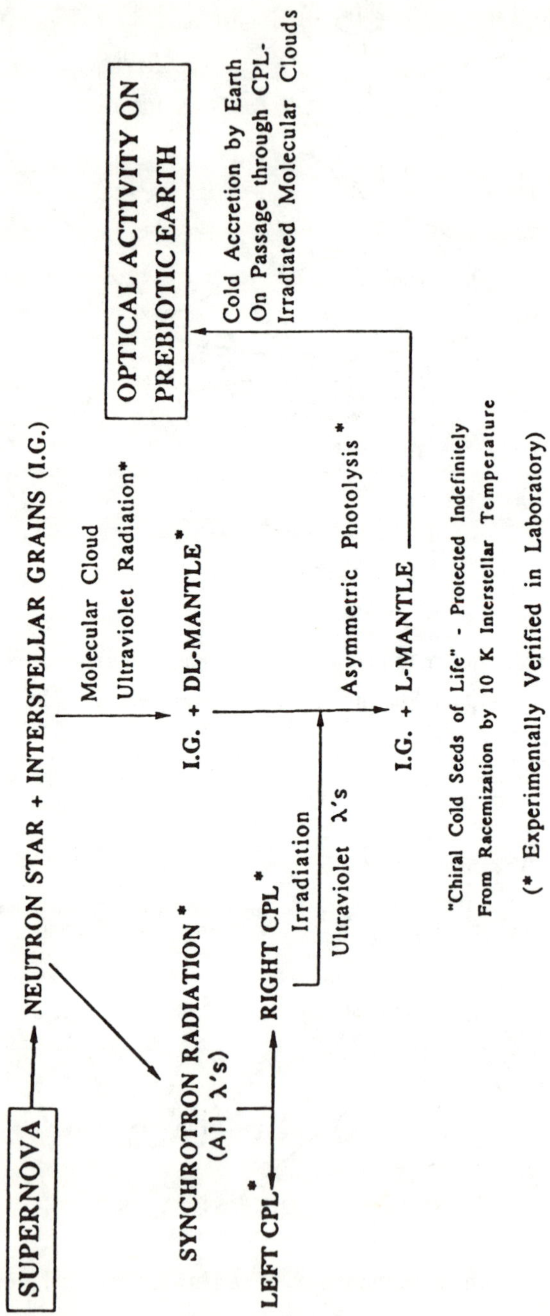

Fig. 3 The Cosmic Connection

ENTROPY AND THE EVOLUTION OF COMPLEXITY AND INDIVIDUALITY

Jerome Rothstein
The Ohio State University
Columbus, Ohio 43210

ABSTRACT

The thermodynamic steady state is characterized by a minimum entropy production rate (Onsager). Approach to steady state is a generalization of approach to equilibrium. Extended to small amplitude "periodic steady state" this implies thermodynamic evolution of a mutable system toward formation of heat engines, with Carnot cycles as the ideal culmination (zero entropy production), and with storage of the work done (if not stored it dissipates). Only systems complex enough to admit richly varied metastable structures can evolve much beyond ordinary uniform steady states.

The cooling rotating earth eventually reached a temperature at which metastable configurations could form, change, and often survive the small amplitude diurnal thermal cycling between day (solar heat source) and night (space heat sink). Phase changes, chemical reactions and polymerization provide rich configurational structures, while quantum stationary states provide basic stability. Pauli principle, excluded volume, and vanishing transition probabilities for proper symmetry conditions yield possibilities for metastable energy storage, while catalysts can unlock the stored energy. Thermodynamic evolution is then able to push toward the formation of complex macromolecular heat engines.

Individuality co-evolves with complexity in the sense that the number of stable or metastable isomers with essentially the same energy increases so rapidly as a function of the number of sub-units that eventually procedures capable of producing isomer mixtures, with roughly comparable probabilities of forming each one, will almost never produce two identical isomers. From this point of view we are isomers of the homo sapiens macromolecule, but since we are always changing, our quantum theoretical description is something between a variable superposition and a variable mixture of such molecular pictures. However, such systems of interest are always open subsystems of a larger system, making non-equilibrium thermodynamics and systems engineering important for

ISBN 0-937194-18-2

constructing tractable models. Communication and computation play a crucial role in them, as do informationally-enriched entropy and its relation to computational complexity.

INTRODUCTION

If one believes that life evolved after a complex prebiotic chemical evolution occurred, and that relevant complex molecular species were preceded by simple precursors, then one must explain how complex organization can emerge from a collection of disorganized simple building-blocks without violating well-established physical laws. The most important of these, in our opinion, is the second law of thermodynamics, for it implies that the natural direction for spontaneous change is toward disorganization, not its opposite. Entropy increases monotonically, reaching its maximum at equilibrium. Thereafter, no further spontaneous changes, except fluctuations, can occur. But ordinary actions of living matter like growth and reproduction, which organize constituents of the environment into parts of an organism or replicas of the parents, seem to contradict the second law. At death of course, the entropy of the corpse increases when it rots, leading to occasional past evocation of mysterious disentropic vital forces which become inoperative at death.

Contradiction with the second law is, of course, only apparent. Life can be maintained indefinitely only in an open system. Like most familiar open systems, e.g., heat engines and refrigerators, living things store energy or reduce entropy locally, with compensatory entropy increases exported to the environment. The same can be argued for the evolution of life in the open sun (energy source) -earth-space (sink) system (Figures 1 and 2). This was appreciated by Blum (1955), Schroedinger (1944), and Oparin (1953), but they did not have the concept of an informationally enriched entropy, which will often be referred to as generalized entropy in the sequel, to carry the argument into the areas of selective behavior, of stored information controlling sequences of physicochemical events at a molecular level, or of precisely locating, within thermodynamics, a rationale for Darwinian evolution. This was done in a series of papers which the present one continues. The earlier work is reviewed herein and integrated into new arguments that complexity will evolve cumulatively, and that increasingly complex systems will increasingly display individuality. Both complexity and individuality, like information and organization, are defined in terms of entropy. The importance of metastability, which has been somewhat overlooked in the rush to exploit new kinetic approaches to non-equilibrium phenomena and open systems, is stressed. It permits the evolution of such things as DNA-programs, so that evolving complex systems can justly be called computers. In this coupling of selective behavior with computation one can also sense the origin of mind. Darwinian selection among competing survival strategies would then link the evolution of intelligence smoothly to biological and chemical evolution.

Many hesitate to accept so fundamental a role for thermodynamics as the one sketched above. They prefer to search for life's explanation in the quantum mechanics of complex systems. They can point to the fact that the triumphal march of molecular biology, biophysics and biochemistry in unraveling the mysteries of heredity, virus, protein synthesis, muscle and nerve ultrastructure, photosynthesis, vitamin and hormone action, etc., etc., is being daily extended, with the end nowhere in sight. Furthermore, it is only through quantum mechanics that the molecular structures and reaction mechanisms on which the build up of complexity or metabolism depend, can even begin to be understood in fundamental terms. It has also been said that thermodynamics is only derivative in the sense that its concepts are probabilistically explained or derived on the basis of the underlying fundamental dynamics of microscopic constituents, and further that the accuracy of the laws of thermodynamics is, au fond, only that of the laws of large numbers applied to the molecules, or ions comprising a macroscopic system. This is, after all, Boltzmann's immortal contribution to physics. Modern kinetic theory, statistical mechanics and statistical thermodynamics can be said to begin with it. But the Newtonian dynamics he used, we now know is not correct at the molecular level. Quantum mechanics is correct, and when the statistical approach (including the important contributions of Gibbs) is coupled to it we get both thermodynamics and deep insight into the physics and chemistry of matter in bulk unobtainable by the methods of classical thermodynamics. Remembering also that a single molecule of DNA or RNA, not a statistical ensemble of them, suffices to program the development and growth of a living creature, how can one possibly avoid the conclusion that quantum mechanics, with a little help from statistics, is the proper place to search for an explanation of life?

Arguments along the above lines have been considered so convincing that even the doubts of so eminent a group as Bohr (1958), Elsasser (1961), and Wigner (1965), who question or even deny the power of quantum mechanics to explain life, have been viewed uncomfortably, by many scientists, as a kind of neo-vitalism best set aside and neglected. It must also be admitted that classical thermodynamics generally, and entropy specifically, do indeed seem to be insufficiently fundamental or powerful to serve as an adequate foundation for biology. We consider these questions further below, and argue that the operational answer to the question "How much can we know, and what unavoidable cost must we pay to know it?" is as fundamental an epistemological question as one could hope to ask, that the cost in question is the entropy price of measurement (pace, Heisenberg), that quantum theory presupposes, as one of its undefined primitives, the concept of measurement, that measurements yield information, and that the generalized entropy concept to which these ideas lead provides exactly what is lacking in quantum mechanics and is needed to explain life.

The fundamentality of entropy can also be argued in another way. The phenomenal success of the Newtonian world-machine, which so impressed Spinoza, Kant, and the scientists of the eighteenth and nineteenth centuries, ultimately led to discovering its limitations and thus to modifying it. Newtonian dynamics had to be changed for very high velocities (special relativity), very small scale phenomena (quantum mechanics), and very large scale phenomena (general relativity). But Newtonian mechanics and the three successor theories to it just mentioned are all reversible, as is electrodynamics, while real systems all display irreversibility (monotonically increasing entropy). This is a real limitation on the detailed applicability of Newtonian physics to actual systems and of its successor theories as well. It was not viewed as such because the dynamical laws were assumed valid in their application to the fundamental constituents. The problem was that there were so many of them that their motions were complex and thus apparently so disorganized that they could not be described in detail, forcing reliance on statistical methods. Note that the revolutionary upheavals in dynamics mentioned above left thermodynamics completely unscathed. Future revolutions (say discovery of a satisfactory unified field theory) are not expected by anyone to topple the laws of thermodynamics, though they may supplement or complete them. Indeed one can say that special relativity, via the mass-energy relation, evaluates the undetermined energy constant implicit in the first law, while quantum mechanics evaluates the undetermined entropy constant of the second (or third) law. Surely "revolution-proof" laws must be at least as fundamental as laws which can be toppled! We can not resist quoting Einstein's view of the second law as a fitting conclusion to this section (Schilpp, 1957): "It is the only theory of universal content concerning which I am convinced that, within the framework of the applicability of its basic concepts, it will never be overthrown (for the special attention of those who are skeptics on principle)." Thermodynamics even motivated his search for universal principles, for speaking of the Lorentz transformation, he says: "This is a restricting principle for natural laws, comparable to the restricting principle of the non-existence of the perpetuum mobile which underlies thermodynamics."

ON THE NEED TO SUPPLEMENT QUANTUM MECHANICS IN CONSTRUCTING A THEORETICAL FOUNDATION FOR BIOLOGY

The purpose of this section is to show that a generalization of the entropy concept is needed, in addition to quantum mechanics, to build an adequate theoretical foundation for biology. Those willing to accept this conclusion and not interested in the history of attempts to explain biology quantum mechanically, can proceed immediately to the next section. The arguments demand little in the way of quantum mechanical or statistical mechanical formalisms, but they do presuppose some familiarity with the fundamental concepts of both. It is necessary to include this section to disarm two oft-repeated arguments. The first is that any fundamental understanding of matter, including living matter, requires a full quantum discussion of all properties in question, and once such a discussion is given it is final in the sense that everything else must follow from it. The second is that thermodynamics and statistical mechanics are derived theories, not fundamental like quantum mechanics, and are either inherently unable to explain or irrelevant to vital mechanisms. Our position is that both are needed. No structure is understandable in a fundamental sense without quantum mechanics, but a thermodynamics based on generalized entropy is needed to explain why complex structures can evolve.

The importance of quantum mechanics for theoretical foundations of biology is beyond dispute even though we believe it to be an inherently incomplete foundation. Bohr's brilliant intuition of stationary states for the hydrogen atom, later given a general theoretical basis in quantum mechanics removed the chief difficulty in the way of an electrical theory of matter. That theory had become unavoidable after J. J. Thompson's discovery of the electron, Rutherford's scattering experiments forcing the concept of the nuclear atom, Lorentz's extension of Maxwell's electromagnetic theory to ponderable matter and a host of other experimental and theoretical studies. The difficulty was the instability of systems of electric charges, for stable static systems are prohibited by Earnshaw's theorem, while dynamic systems (like the Bohr-Rutherford miniature solar system model of atomic electrons in orbits about a massive positively charged nucleus) were unstable because accelerated charges radiate energy and thus electrons would ultimately spiral into the nucleus.

With quantum mechanics stationary states generally and the ground state (the one of lowest energy) in particular, became stable, while Pauli's principle (no two electrons can be in the same state or orbital, in a quantum system) led to a deep understanding of structural build-up. This included the structure of atoms and the periodic table, the shapes and structures of molecules and the structure of solids. The interactions of matter and radiation were explained. Not only were stationary states derived as eigenfunction solutions of Schroedinger's time-independent equation: $H\Psi = E\Psi$, but non-stationary states were described by his time-dependent equation: $H\Psi = -(h/2\pi i)\ \partial\Psi/\partial t$. Here H is the Hamiltonian (energy) operator, Ψ the wave-function and E an energy eigenvalue. The enormous success of quantum mechanics for all cases in which the equations could be solved, exactly or by approximate methods (e.g., perturbation theory) is still being extended in ever more complex cases with modern high-speed computers. The time-dependent equation, essentially, permits understanding of dynamic processes in principle, from radioactivity and nuclear reactions to chemical kinetics, while wave-function symmetry properties and symmetry breaking even enable us to explain metastability and catalysis.

The foregoing argues strongly in favor of a quantum explanation of life, for if the structure and reactions of all matter are explainable by it, why should living matter be an exception? We need to consider both the difficulties raised by Bohr (1958) and Elsasser (1961) on the one hand, which say, in effect that it is pointless to consider a living system to be a gigantic molecule in a "living state," and that of Wigner (1961), who says, in essence, that quantum mechanics has no reasonable way to describe reproduction. We will then show that the first difficulty reduces to the complexity limitation, still so often unrecognized, but present already at the birth of thermodynamics, while the second involves either implicit preparation procedures not covered by Schroedinger's equation, or another form of the complexity limitation. Obviously one can hardly hope to explain the origin and evolution of life by quantum theory if no quantal description of even ordinary properties of living matter can be given. In the end, Elsasser feels a need for special "biotonic laws," consistent with, but independent of, quantum mechanics, while Wigner is led to seek a more general theory which ascribes a fundamental role, even in quantum mechanics (and thus in all physical science), to consciousness. We see no need for such assumptions in the sense that measurement and preparation of systems by other systems do not require human consciousness but only behavioral modes possible for physical systems, while "biotonic laws," we believe, merely reflect selective behavior of systems resulting from the internal structure and organization (constraints, relaxed and imposed over time). They are thus more like design parameters, compartmental structures and boundary or initial conditions than laws. We will argue that these kinds of system behavior and their origin can both be explained by a thermodynamics of open systems using a generalized entropy concept broad enough to include measurement, information, and organization.

Our view is thus that generalized entropy supplies what Bohr, Elsasser, and Wigner needed in principle to close the gap between quantum mechanics and biology. We do not consider the existence of that gap to be a shortcoming of quantum theory in the sense that we expect it to persist in any future fundamental theory not incorporating something like generalized entropy *ab initio*. Furthermore, the gap itself is really old, for quantum mechanics does not explain irreversibility, nor has it yet arrived at a consistent theory of measurement. All of those problems, we propose, plus evolution, complexity and individuality can be treated in a methodologically unified manner with the help of generalized entropy, and without introducing extraneous or exotic hypotheses.

A short history of attempts at quantum explanations of life, with a useful bibliography, has been given by Mehra (1973). Bohr (1958) stressed, early on, that the drastic interventions constituting quantum measurements of parameters of all particles comprising a living system (of interest to predict later states of the system) would literally explode the system. He then suggested a "generalized complementarity principle" applied to biology, namely that the very measurements needed to distinguish whether the system was in one of a large number of "living states," or in one of a presumably even larger number of non-living states were incompatible with the continued existence of such living states. This is a strong argument, in our opinion, against regarding a living system as a supermolecule, with eigenfunctions capable of being classified as living or non-living. There is simply no way (so far, at least) to define a quantum observable corresponding to making that distinction. As a practical matter, physicians and biologists are able to make that distinction fairly easily, most of the time, by very mild non-lethal procedures. They correspond to classical observables, which commute, are macroscopic, and are frequently closer in concept to thermodynamic than to mechanical variables. See Rothstein (1979, 1982) for further discussion.

Bohr later acknowledged the importance of complexity (1963) for biology, but lacking the generalized entropy concept he did not carry the analysis very far. In our opinion, complexity puts the quietus on all attempts to achieve a total reduction of life to quantum mechanics. We believe life is not exceptional in this regard, for we do not believe the properties of any other real macroscopic system can be totally reduced to quantum mechanics either. Nobody even knows how to begin constructing a Hamiltonian operator for a particular piece of glass which will permit prediction of whether it will crack when chilled. Even if we could, it probably would not help at all, for acquiring the detailed information needed for the prediction would explode the glass a'la generalized complementarity (and even if this were not so, the complexity of the resulting Schroedinger equation would be at least as intractable as solving Newton's equations of motion for all the molecules of a turbulent gas!). But "non-lethal" measurements, e.g. stress analysis using polarized light, might well permit just such a prediction. These are macroscopic measurements. To invoke quantum mechanics here is as pointless as invoking kinetic theory and statistical mechanics to explain the actions and properties of sewing machines (the latter bon mot, if memory serves, is Boltzmann's).

This is not to deny the key role of quantum mechanics in explaining the origins and roles of homopolar and ionic bonding in the vitreous state of matter, the transparency of glass and so on.

Quantum theory can or will explain the vitreous state, but the myriad structure-sensitive properties of any ordinary piece of glass are simply too complex for quantum mechanics. When great care is taken in specimen preparation, as with germanium single crystals, it is possible to achieve much more detailed correspondences between properties predicted by quantum mechanics and what is observed than was formerly the case. But there is virtually no end to the residue of characteristics still eluding detailed quantum description, from "speckles" in reflected or transmitted coherent light to ultrasonic attenuation, from spatial variations in the impurity distribution to the pattern of dislocations quenched in during crystal growth, from effects due to handling the specimen subsequent to growth to aging effects and crystal damage due to cosmic rays or natural radioactive background. No real macroscopic system, we believe, can ever be completely described. Theoretical descriptions cover either an ensemble of different systems, namely all systems compatible with the operations performed in preparing them to meet the theoretical specifications, or apply only to idealized cases (no such real system can be actually prepared and possibly is not even preparable). Macroscopic systems can be individually identified if we take the trouble to devise sufficiently mild, albeit complex, procedures. Quantum objects have no individuality in the sense that similarly prepared systems are indistinguishable.

Elsasser (1937) realized the crucial importance of statistical mechanics in closing the gap between quantum mechanics and biology, following it up in later publications (Elsasser, 1961, 1958, 1965). His writings contain many valuable insights, but we feel uncomfortable with some of his assumptions and conclusions. He clearly grasped the problem implicit in the facts that the basic concepts of atomic physics are microscopic while biological systems are macroscopic, which he tries to solve by making two assumptions about the classes (ensembles) representing biological systems. These are that the usual assumptions of homogeneity and infinite class membership do not apply to biological classes. Inapplicability of the homogeneity assumption seems to be the same as saying that the conventional ensemble of similar systems is replaced by one in which the individual systems can be distinguished from one another somehow,i.e., they are dissimilar. The second assumption seems to say that one simply cannot prepare an indefinitely large number of similar systems. The two taken together, thus appear to be equivalent to saying that the systems have individuality. I agree with that conclusion, but fearing that the ensemble picture he introduces virtually eliminates the utility of the Gibbs (or any other) ensemble, thus making statistical mechanics itself inapplicable, I prefer a different way to relate individuality to phase space. Individuality reflects the organization of basic raw material entailed by a characteristic set of internal constraints. See Rothstein (1952, 1957, 1962, 1971), for discussions of the connections between organization and entropy. The representative point for each individual then traverses a characteristic region of phase space, and distinguishability of individuals means that their regions have no points in common.

The more complex the individuals, the higher the dimensionality of their associated phase space, the larger the number of possible individuals which can be assembled from the same constituents (isomers) and thus the larger the number of non-ergodic components into which the phase space is partitioned. Specifying a particular individual is then analogous to choosing an initial or boundary condition. This last point was intuited by Polanyi (1968) and proposed independently by Rothstein (1958, 1979, 1987), who based his arguments on generalized entropy. The assumption of special biotonic laws becomes unnecessary in his development.

Wigner (1961) argues against quantum explanations of reproduction in the sense of there being a quantum state

$$\phi = \nu \times w$$

describing the state of the system organism-environment (ν is organism, w is environment, and ϕ is their Kronecker product), which can evolve into state $\Psi = \nu \times \nu \times r$, i.e., the Kronecker product of parent, offspring and (depleted) environment. He shows that the conditions on the "collision matrix", S, needed to permit such a process make its existence so overwhelmingly improbable that one is invoking miracles rather than providing a rationale for reproduction when one hypothesizes such a matrix. He does not specify the S ensemble otherwise than to say that S "has no particular properties but is a random matrix." In that ensemble the probability of finding one with the desired property could well be comparable to that of the kettle of water freezing when put over a fire.

How, in the world of thermodynamics and chemistry, should one describe a situation in which a complex molecule, say a polymer, is introduced into a solution containing the monomeric species from which it is constructed, and catalyzes the formation of another such molecule? This problem was considered (Rothstein, 1963) without taking the molecule as a linear polymer necessarily (perhaps it could be a sheet or three dimensional stacked structure), asking if some simple property were conducive to its having this self-replicative ability. Assuming ordinary nearest-neighbor kinds of interaction gives rise to a template kind of assembly. This rules out a solid structure in favor of sheets or chains (the "outside" would not "know" what the "inside" was doing), and demands exothermic or catalyzed assembling reactions to overcome the entropy of solution of the constituents. The resulting dimer must then dissociate, for otherwise the process would be better called dimerization or incipient crystallization than reproduction. This dissociated state clearly has higher entropy. If this added entropy suffices to make the change in free energy, F = E - TS, negative, where E is energy, T absolute temperature and S entropy, i.e., if the entropy of solution can overpower the binding energy of template to replica, the dissociation is thermodynamically "pushed" to completion. The template has then catalyzed the formation of its replica. A simple calculation comparing the increase in configuration entropy when a two-strand ribbon of 2N units dissociates into two chains of length N with that occurring when a double layer with 2N units dissociates into two sheets, each with N units, then led to the conclusion that the dissociation into two chains was overwhelmingly more favorable than dissociation into two sheets, (for comparably strong two-unit interactions between units of the model and units of the replica). A similar discussion can be given for double strand cases like DNA.

The results of linear template reactions, seen to be so improbable from the quantum viewpoint, become natural and even probable, from the viewpoint of statistical thermodynamics. Nature's preference for chain molecules for reproductive purposes (DNA, RNA), or more generally, for catalyzing many kinds of complex biosyntheses (proteins, hormones, etc.), also becomes understandable. Furthermore, one can even sense that the statistical thermodynamic approach may be sounder in principle than the quantum approach, heretical as it might sound. For there are no operations one can perform on the solution, unless one were at least as powerful as a quantum version of Maxwell's demon, to determine the wave-functions written down in Wigner's discussion. It is not even clear that the wave functions, ν, w, r, above are meaningful because of interaction terms, ditto for the use of Kronecker products. It is surely wrong to write the wave-function of a hydrogen atom as the Kronecker product of the wave-functions of an electron and of a proton unless the atom is ionized and both particles have high energies compared to the ionization potential. The Coulomb interaction is essential to deriving the properties of the hydrogen atom. Bohr's notion of the "wholeness" of the system is clearly important here (also his generalized complementarity). The complex assembly process discussed thermodynamically above should be viewed as a sequence of acts of preparing the intermediate stages (Rothstein, 1969). The quantum discussion then requires wave-packet reduction at each stage, which is irreversible and not described by Schroedinger's equation, as in discussions of measurements in quantum mechanics. The programmed catalysis needed for reproduction is encoded in the chain units. Different units represent different "letters" of the "alphabet" (nucleotides or amino acid residues in heredity and proteins respectively) in terms of which the "program" is "written."

Another virtue of the thermodynamic approach is the clarity with which a single macromolecule can be seen to affect phenomena at a macroscopic level. Even before the gene was shown to be essentially a molecular structure, Jordan (1941) tried to build a quantum biology. He concluded that because single microscopic objects can cause biological changes (e.g. radiation effects on Drosophila) a macroscopic explanation of life must be ruled out,i.e., that living things can not be explained as machines. This objection to mechanism denies machine status to Geiger counters and much else. We feel that quantum mechanics is needed to ensure the stability of biological structure just as it is needed for that purpose in all matter, and it is particularly needed, as emphasized by Schroedinger (1944), for the stability of the hereditary code-script. But reading and executing catalytic programs is like measurement and preparation of systems, requiring free energy sources, amplification, and irreversibility. If the notion of machine is broad enough to admit heat engines, measuring equipment, amplifiers and computers as subsystems, then rejection of machine-like explanations of life in principle seems premature. A similar caveat, we feel, applies to a paper by Wigner (1969). The machines of greatest interest here, of course, are chemical machines. They deserve much study.

We now turn to generalized entropy as the key element, not provided by quantum theory, needed to understand biology in principle. Included in that understanding are evolution, complexity and individuality, as well as reproduction and other ordinary properties or attributes of living matter. The phrase "in principle" is essential, for the explanation of detailed mechanisms and structures needs quantum mechanics, chemistry and physics as much as ever.

GENERALIZED ENTROPY, MEASUREMENT, INFORMATION AND ORGANIZATION

Boltzmann's statistical entropy became important in the frame of communication theory with the publications of Wiener (1950) and Shannon (1948). Entropy was shown to be a quantitative measure of the amount of information, not dependent on the meaning or value of that information. There has since been an explosive growth in engineering applications of information theory (thousands of papers), as well as extensive theoretical developments and attempts to apply the concepts to many other fields (linguistics, psychology, biology, and cryptography, to name a few). The initial periods of inflated expectations and subsequent disappointment have been succeeded by an era of continuing maturation and sophisticated application of the now classical concepts and methods.

Information is obtained by making selections from a set of alternatives, and its quantitative measure is exactly the same as the prior measure of uncertainty about which ones will be selected. For the simple case of n alternatives, of given probabilities p_1,

p_2, ..., p_n, the average information I, conveyed when it becomes known which one has been selected, is given by: $I = -\Sigma p_i \log p_i$. It is customary to take the logarithms to base 2 in information theory and computer science for discrete cases. The unit of information, one bit, is then exactly the amount conveyed by specifying the result of tossing an honest coin. In the continuous case, the base, e, of the natural logarithms is sometimes chosen, as in physics. Information is then given by the integral $I = -\int p(x) \log x \, dx$ and Boltzmann's statistical entropy, S, becomes: $S = kI$, where k is a constant (named Boltzmann's constant by Planck).

Unfortunately, Shannon used the letter H, calling it entropy, for the quantity I, above, while Boltzmann used H for -I, writing: $S = -kH$. This led to much confusion in the literature. In statistical mechanics, S measures uncertainty with respect to what the detailed microscopic state of the system is, and thus is proportional to the information conveyed by specifying that state. As $I = 0$ for the case where all $p_i = 0$, except p_j which equals 1, the case of complete information ($S = 0$) corresponds to a negative entropy, -S, compared to the (macroscopic) state of entropy S. Statements in the literature of information theory to the effect that maximum information corresponds to maximum entropy, uncritically combined with standard physical formulae for entropy, have thus often led to confusion and absurdity in talking about things like the thermodynamics of systems with stored information (say in DNA).

Many thought that thermodynamic and information entropies were related only by mathematical analogy, but Rothstein (1951) realized that the analogy was a logical identity in the following sense. From an operational viewpoint the cells of phase space correspond to distinguishable results of measurement, which are the operationally based alternatives in terms of which a system of interest is described. With the standard assumption of equal a'priori probabilities for equal volumes of phase space, a physical state of entropy S, is rigorously characterized in terms of information by $S = kI$. There is thus a thorough-going analogy between communication systems and measurement systems, shown in Figure 3, taken from (Rothstein, 1951), in which Shannon's diagrammatic representation of communication also represents measurement. As quantum mechanics deals with the results of measurements on quantum systems, this implies that the wave-function can be viewed as a kind of generating function for the various kinds of information obtained by different measuring procedures. This leads immediately to a very simple way to consider the Einstein-Podolsky-Rosen paradox (Rothstein, 1951) (later referred to as EPR) and Bohr's response to it (de Baere, 1986). This informational interpretation of the wave-function includes the ensemble interpretation, so we disagree with Ballentine's (1970) objection that the informational interpretation is subjective, a view shared to some extent by Heisenberg

(1958), and also by Wigner (1965) who was thereby led to ascribe a physical role to consciousness. But in physics, information is what is obtained by measurement. Its use is thus no more open to the charge of subjectivity than the most objective things to be found anywhere in physics. Macroscopic thermodynamic entropy (as opposed to statistical entropy) was then examined to see if the measurement-information connection entailed an informational interpretation of it (Rothstein, 1952). This turned out to be the case, ensuring that no inconsistency could ever arise from taking physical entropy as a measure of the operationally defined physical information provided by an actual or contemplated measurement procedure.

The germs of entropy as a measure of ignorance, disorder or disorganization go back at least to Boltzmann's day. Motivated by the idea that biological organization can be viewed as information encoded somehow in structure (briefly noted in Rothstein (1951) and foreshadowed as "negative entropy" in Schroedinger (1944)), let us define an amount of organization quantitatively by a suitably defined entropy (Rothstein, 1952). Consider a set of elements (they will be later identified as the subunits of the organization) each associated with its own set of alternatives, for each of which an entropy is defined. Defining a set of complexions as a "vector" of choices, one for each set of alternatives, permits us to define a complexion entropy which, for independent elements (uncorrelated choices), is simple the sum, over all elements, of the entropies of their sets of alternatives. Call this independent element case the state of zero organization of those elements. When the elements become "organized" this can only mean that they become interdependent. The correlations, constraints or couplings between the sets of alternatives which characterize any such organized situation must reduce the complexion entropy in comparison to what it would be if the elements were independent. We define this entropy reduction as the amount of organization. Organization is maximum, if there is only one possible complexion (zero entropy). Clearly, organization is information coded in structure if the alternatives are structural, as in metastability. Later papers (Rothstein, 1957, 1962, 1971, 1979, 1982) explore further aspects of the organization-information-entropy connection. Organization can be dynamic, or behavioral, as well as structural, just as information can be either transmitted by a propagating wave or be stored in a computer memory, printed page, or other medium.

In quantum mechanics, measurement and preparation of a system are described in the same formal terms. Preparation is an example of manufacturing, which can be viewed generally as impressing some desired organization on inchoate raw materials, or encoding the information contained in a specification in attributes of the product. More detail is given in (Rothstein, 1956) from which Figure

4 is taken. It shows the analogy between communication and manufacturing, which also amounts to a formal identity.

In a biological context, reproduction can be described as the preparation or manufacture of an offspring. This requires free energy expenditure and irreversibility (entropy increase), in agreement with Wigner's argument against quantal descriptions of reproduction. A very similar situation obtains in computers, for whenever information is stored, the state of some portion of its memory is appropriately prepared, and whenever instructions are read out of memory, or status flags and condition codes are set or consulted, additional acts of measurement or preparation are carried out. We therefore believe that constructing a thermodynamically reversible general purpose computer meets the same difficulties in principle as reversibly carrying out reproduction. For more details see Rothstein (1986), also on how information storage in computer memories, or DNA, exploits the metastability of the imprinted substrate.

As pointed out, (Rothstein, 1971) and above, measuring a system and <u>finding</u> it to be in some state, <u>preparing</u> it to be in that state, <u>specifying</u> it to be in that state, or <u>constraining</u> it so that it can be in no other state, are formally equivalent in the sense that the end result is described in the same way. They share the same admissable phase space regions and correspond to the same entropy (zero if the state is pure, positive if the state is mixed). Important methodological points now crop up, EPR being only one. Others arise from other paradoxes, such as those of Loschmidt, Zermelo and Gibbs, from Maxwell's demon, and from Laplace's superior being (who knows the initial data for all particles in the universe and the forces acting on them and thus knows its state for all time).

An important point for methodology is that the operational viewpoint, which necessarily assumes the legitimacy of measurement and acts of preparation, where both imply coupling (organization) of systems of interest and means to measure and prepare it, is clearly committed to information and organization as part of its very language (Rothstein, 1962). The notion of a physical law is also inseparable from the concepts of information and organization (Rothstein, 1957, 1971). Consider a simple case like Ohm's law, $V = IR$. In an experimental set-up in which a voltmeter and an ammeter provide the sets of possible values for voltage V and current I, say $\{V_i\}$ and $\{I_j\}$, the law asserts that not all pairs (V_i, I_j) can be observed and says which ones can be. The law thus restricts the permissable pairs to a subset of the measurable pairs and, therefore, can be viewed as organizing the elements V and I, or as providing information. Given a value V_i for V, the uncertainty about what value of I occurs is drastically reduced by the law.

This observation is developed in (Rothstein, 1957) to yield measures of quality for theories on the basis of which choices can be made between rival theories. It is also shown that all the reasons one normally uses, intuitively or formally, to make such choices, can be reduced to providing maximal information from (or inducing maximum organization in) experimental data. Now Einstein showed that geometry, to the extent that it is used in physics, becomes thereby a physical theory. Geometry is "true" insofar as it does not refer to reality. As part of physics it has to be potentially falsifiable by experiment (Einstein, 1947). Feeling this had to be true for any formalism used in physics (or science generally), Rothstein examined logic from this point of view (Rothstein, 1956) (doing the same later for space-time (Rothstein, 1962)), finding it to inhere implicitly in the operational viewpoint, and more obviously in the information-organization approach. It also became clear that this operational generalized entropy provided a philosophically neutral language (for example one is not forced to take sides in many issues like the "reality" debate between Bohr and Einstein) which did simultaneous justice to the empirical and theoretical sides of the scientific enterprise.

In the example of Ohm's law, the same discussion holds whether the law be viewed as empirically based or derived mathematically within some broader theory. With exact laws and data subject to errors, the latter are generally handled by use of statistical techniques, and do not affect the situation because they are taken into account when element or complexion entropies are calculated. Hidden variables have no effect either. They merely add an unknown constant to an entropy (corresponding to a hidden splitting of some alternatives into sets of subalternatives). "Black boxes" are analogous to hidden variables. In Ohm's law we can view the resistor R as a black box (or as a computer) which, given an input (stimulus) V, provides (computes) an output (response) I. The internal structural complexity of the black box, or the complexity of the theory connecting input and output variables (or equivalently, the complexity of the computation a computer performs to generate its output from the programs and data constituting its input), can be taken as "hidden" at the level of input-output analysis. For thermodynamic (macroscopic) systems, microscopic states are hidden. Statistical mechanics has the task of explaining the input-output relations of macroscopic systems in terms of statistics and the dynamics of microscopic states.

Integration of the dynamical equations of physics (equations of motion) permits prediction, or retrodiction, of the states of a system from any initially specified state. The equations thus organize the set of pairs of states, just as Ohm's law organizes the set of (V, I) pairs. The initial state specification is approximate, corresponding to locating a representative point in phase space for the system in phase volume, $\Delta\Omega$, of the total phase space Ω. The system as specified represents a choice from $\Omega/\Delta\Omega$ equally probable alternatives, so $k\ln(\Omega/\Delta\Omega)$ is the entropy difference between the system as specified and the equilibrium state corresponding to phase volume, Ω. The unavoidable finiteness of $\Delta\Omega$ is a consequence of the operational impossibility of paying an infinite entropy cost to prepare the system in a perfectly sharp state. The equations of motion convert any phase space point into a phase space trajectory, hence $\Delta\Omega$ is point-wise transformed into $\Delta\Omega'$, say. The phase volumes $\Delta\Omega$ and $\Delta\Omega'$ are equal by Liouville's theorem, whence the "microscopic entropy" remains rigorously constant. But the <u>operational</u> thermodynamic entropy is specified in terms of dissecting Ω into a collection of cells like $\Delta\Omega$, corresponding to the operationally distinguishable alternatives allowed by the measurement and preparation procedure. This dissection does not change relative to Ω. It is fixed initially and can only change if the operational language in terms of which the system is prepared and described is changed. Such a change is a new preparation procedure which is generally incompatible with allowing the undisturbed system to follow its equations of motion. It is tantamount to a new boundary condition, or new initial condition (Rothstein, 1979).

The volume $\Delta\Omega'$, though unchanged in magnitude, changes in shape. The phase space coordinates comprise equal numbers of position and momentum coordinates. Any spread in the latter (like a spread in velocity), entails a monotonically increasing spreading in the former. The initial $\Delta\Omega$ evolves into an increasingly filamentary shape threading through a monotonically increasing number of cells of the original phase space dissection. A repetition of the application of the same measuring scheme would thus assign the representative point probabilistically to any of those cells. The evolving situation is therefore described as increasing entropy (increasing uncertainty as to which cell will be found to contain the representative point). This discussion shows how increasing thermodynamic entropy and irreversibility are compatible with constant microscopic entropy and reversible dynamical equations of motion for the microscopic constituents of the system, for they are not the same thing. For more detailed discussion, including resolution of the paradoxes of Loschmidt and Zermelo and the significance of spin echo experiments (as showing how even dynamical information storage suspends a corresponding amount of entropy increase) see (Rothstein, 1957). For analysis of operational versus microscopic languages and demonstration that the above paradoxes can only arise by confusing the two, see Rothstein (1974).

Another famous paradox, due to Gibbs, is also easily resolved by examining the operational-informational context. Suppose two equal volumes V, separated by a partition, contain two perfect gases A and B, at pressure P, which can mix without reacting. Removing the partition gives the well-known increase in entropy (entropy of mixing): $2k \ln V$. Now let the nature of one gas, say B, be gradually altered so that it becomes more and more like A.

The entropy change is unaffected. But when B becomes A, the change is zero, unlike when it is arbitrarily close to being A.

Either operational means to distinguish A and B in principle exist, or they do not. If they do, there is entropy of mixing, otherwise there is not. This forcefully illustrates, once again, how proper attention to the operational-informational situation avoids paradox. One can not admit, as bearers of legitimate physical information, choices from alternatives which can not be distinguished from each other. One must consider two such cases as a single alternative, and if the language used makes such distinctions, then mathematical theory must discard those distinctions.

Symmetry numbers in molecular physics, and symmetrization of wave functions in quantum mechanics do such discarding. In effect, the latter deprives fundamental particles of individuality. In a wave function for a system containing n electrons, say, with the ith one having coordinate x_i, it is wrong in principle to work with a wave function $\Psi(x_1, x_2, \ldots x_n)$, whose value changes, if any pair of x_i are interchanged. It must be symmetrized over all permutations of $x_1, \ldots x_n$, i.e. the correct wave function φ must be such that $|\varphi|^2$ is not changed in value by such a permutation. This gives rise to the two kinds of quantum statistical mechanics, Bose-Einstein, and Fermi-Dirac. In the first case, φ is a normalized sum of all the Ψ's, and in the second, a sum with coefficients ± 1, where the minus sign goes with odd permutations, plus with even ones. In both cases, complex similar systems containing many particles (no upper limit on their number) will lack individuality, for they can be interchanged by a sequence of exchanges of the indistinguishable particles from which they are constructed. This lends interest to the problem of explaining the origin of individuality, so characteristic of living things. As we shall see, individuality, complexity and metastability are associated and individuality is inherently a macroscopic concept.

PREBIOTIC EVOLUTION AND THE ORIGIN OF LIFE

Three important approaches to biogenesis and molecular biology, which have developed with less mutual cross-fertilization and support than one would have hoped for, are integrated into the foundation of the generalized entropy approach. The first is based on non-equilibrium thermodynamics, primarily open systems (e.g., the work of Prigogine, Eigen, and their schools later referred to as PE). The second is more chemical (or biochemical) and seeks specific reactions, precursors, intermediates, mechanisms, synthesis routes and so on, to form a bridge from the abiotic primitive earth to the biosphere. The current phase is often considered to begin with the work of Miller and Urey (1959), Fox and his school (1960), and others (Blum, 1961). A veritable army of investigators is now tacking the unending variety of problems in this area. The third approach is more formal and mathematical, with systems theory, cybernetics, information theory, automata theory and computer science, rather than traditional theoretical physics and chemistry as its chief tools. While the three often overlap, and many researchers use elements of more than one approach, there is something like a cultural gap between them in that the first is most strongly based in theoretical physics (and chemical physics), the second in chemistry, (mostly organic and biochemical), while the third has a strong engineering flavor.

A readable survey by Prigogine et al. (1972) which summarizes the thrust of Eigen's school as well, carries the (non-equilibrium) thermodynamics of evolution to where new assumptions, not grounded explicitly in thermodynamics, need to be frequently invoked. While this may distress theorists who would like to emulate the mathematicians and state all the axioms at the outset, it does not surprise thermodynamicists. They well know that statistical and quantum mechanics must provide all rate coefficients and other needed kinetic, chemical or structural detail and reaction mechanisms. But thermodynamics has always been able to give much insight, both quantitative and qualitative, into the thermodynamical course of events, like the approach to equilibrium or the steady state, despite its inability to predict the time scale, or provide mechanisms. The PE school (and others) have significantly extended thermodynamics into kinetics, but an extension in the direction of evolving structural complexity and individuality is also needed for evolution. This requires metastability and generalized entropy, for no stable memory is possible without them. Dissipative structures (linear and non-linear) are important in growth, metabolism and reproduction, and also play a role in evolution, but only metastable information storage permits evolution to be cumulative. It is this quality of increasing organization over time that evades the kinetics of dissipative systems, but is captured by generalized entropy. We discuss this combination of dissipative and metastable below, but first the chemical and engineering approaches must be further analyzed.

It is hard to imagine a time when the chemical approach ceases being central to understanding the details of the origin and operation of the machinery of life. In biology, quantum mechanics is used mostly to elucidate biochemistry, while the task of physiology is essentially to explain vital functions in terms of chemistry, physics and system concepts. Most of the papers at this conference (and others) are devoted to the chemical approach and properly so. We can thus largely restrict our remarks to the other approaches, confirming that it is mainly through chemistry that the force of quantum effects becomes manifest in biology. Chemical substrates are the vehicles of generalized thermodynamics in life as we know it. Chemical interactions and effects provide the main channels for the communication and control actions which are the very being of biological organization. Chemical structure has the stability to permit information storage and the existence of individuality, and also the flexibility to permit the stored information to be changed, i.e., for evolution to occur. There is no contradiction in this apparent conjunction of stability and instability, for stability over times comparable to, or even greatly exceeding, the life span of an individual is compatible with the variability over the eons needed for organic evolution. Metastability carries these connotations of short-term stability and long-term variability, and will frequently be invoked, though stable-unstable terminology will also be used. This is partly traditional (we regard steel and glass as stable even though they are metastable) and partly a convenience. A jug of water with a spigot is metastable in the sense that although a small force, properly applied, causes the water to run out, one can ordinarily regard this system as stable, for even fairly large disturbances generally will not change the water level. This simple situation is similar to reactions which occur at a negligeable rate in the absence of a catalyst, to actions of digital control or switching circuits, to amplifiers and to many other cases.

The last paragraph has been deliberately phrased in terms making some of the engineering approach more compatible with chemistry. Automata theory, for example, deals with the following kind of "black box." An automaton has a set of states, a set of inputs, a set of outputs, and a pair of mappings (or relations). The last two determine the next state and the next output as functions of the current state and current input. This is the deterministic case. Weakening the mappings to relations, so that the next state can be any of some specified subset of states, while the output can be any of some specified subset of outputs, defines a nondeterministic automaton. If all three sets are finite, we have finite state automata (FSA's). It is a theorem that deterministic and non-deterministic FSA's are equivalent in the sense that given one, a procedure exists to construct an example of the other with the same input-output behavior. It is clear that automata are abstract models of stimulus-response behavior, that automata can be cascaded, with the outputs of one serving as inputs for the next, and that this can be generalized to networks of automata and to "cellular automata" (CA's). In these, the next state and the output depend on a set of inputs from other automata (called the set of neighbors in CA's) and the outputs also affect a set of other (neighbor) automata. Networking is important in both computer science and neural modeling, and it and cellular automata capture the abstract essence of compartmental models and system concepts. The interconnections organize the set of cells or automata in the sense earlier defined. Measurement and communication systems are special cases of automata.

If an FSA is augmented by a "tape" from which it can "read" its inputs and on which it can "write" its outputs, as well as "move" from one "tape square" (containing one symbol or a blank) to another, then it has been given an effectively unbounded memory. Equivalently, it has become an "infinite state" machine, called a Turing machine (TM). Its transition function is now a table of quintuples, for given the state of the FSA and the symbol in the currently scanned square, it specifies the new symbol to be written there, the move (left or right) of the FSA to an adjacent square, and the new state the FSA will be in when it gets there. It is now known that TM's have complete generality in the sense that any symbolic procedure, or algorithm, can be executed by an appropriate TM. This includes computations, derivations within formal systems (any logical or axiomatic system) and data processing. One can construct a universal TM (UTM) which, given a properly encoded description of an arbitrary TM and the data which the latter is to process, will process the description and data to generate properly encoded descriptions of what the arbitrary TM does. The UTM is a theoretical model of a general purpose computer. It is now believed that no more general model is possible. Unbounded cellular automata can be TM's and UTM's, for the state sets of the cells can be chosen large enough to correspond to the state-symbol pairs of the TM FSA and the tape symbols, while the move is absorbed into the cell transition function (the cell states must be augmented by an "off" or "quiescent" state if one is not already present). An advantage of CA's is their parallel computation capability, which becomes even greater in "bus automata" (BA's). In the latter, modular interconnection (switching) networks, with each module "belonging" to a cell which controls it, enables the set of effective neighbors of any cell to vary controllably from zero to the entire BA. BA's have proved their worth as models for general computation with parallel computing capability, and thus as computable theoretical models for complex organizations in general (Rothstein, 1976, 1977; Moshell et al., 1979; Rothstein, 1978; Rothstein et al., 1979; Champion et al., 1987; Rothstein, 1987) and we suspect them to be the most general such model possible. They combine distributed computation with non-local communication, and constitute a hopeful approach to modeling vision and brain function (Rothstein, 1977, 1982, 1986, 1978), and thus "artificial intelligence" as well. The comparison between BA and CA reminds one of that between animals with well-developed nervous systems and metazoans without them. We believe the BA to be the most general model of a biological system possible (or of complex systems generally) in the same sense that TM's are thought to be the most general models possible of what can in principle be computed.

Research is very active in computational models for biology (Pattee, 1966), but unless they relate to chemistry and thermodynamics, they are of only formal interest as far as the origin of life and individuality are concerned. We must therefore show how generalized thermodynamics "pushes" the evolution of pre-biotic organization, and then indicate how chemistry can support the evolution of complexity, information storage, and computation.

The prebiotic earth in relation to the sun and space, is shown in Figures 1 and 2, (Rothstein, 1985). We concentrate on Figure 2, as it presents the least favorable case for the evolution of organization, e.g., metastable energy storage. When the earth has cooled enough to allow the formation of permanent bodies of water and the persistence of many organic compounds (should they be formed), the earth's rotation will have subjected most regions to daily alternating exposures to sun and space. On an evolutionary time scale a day is infinitesimal, so one can think of an average steady state of the globe. Every day, an average quantity of heat Q_2 arrives from the sun and every night a quantity Q_1 is radiated to space. For the steady state, $Q_2 - Q_1 = 0$. Describe Q_2 as supplied from a heat source at an effective temperature T_2, and Q_1 as supplied to a heat sink at temperature $T_1 < T_2$. Then the entropy increase per day, ΔS, is given by: $\Delta S = Q_1/T_1 - Q_2/T_2 \geq 0$.

As time goes on, the global steady state is expected to change for the following reason. Onsager (1931) showed that the steady state of dissipative system is one in which the rate of entropy production is a minimum. Rayleigh's theorem of minimum dissipation in mechanics, of which the familiar "sleeping" spinning top is an example, is a special case of Onsager's theorem. If the earth were a heat engine, doing work W per day, we would have, by the first law of thermodynamics, $W = Q_2 - Q_1 > 0$, and thus Q_1 would no longer equal Q_2, but $Q_2 - W$, Fig. 5. The resulting ΔS is then: $\Delta S = (Q_2 - W)/T_1 - Q_2/T_2$, i.e., the rate of entropy production is diminished. Onsager's theorem therefore says, in effect, that a system will "try" to "evolve" to the state of minimum dissipation by doing work. In the case of the spinning top, the work involved is simply to raise the center of gravity as high as possible. Similar things would happen to the earth, like the establishment of global wind patterns and tides. This kind of evolution to the steady state, for dissipative systems, is fundamental to much of the work of the schools of Prigogine and Eigen. But work need not only be stored as mechanical energy, kinetic or potential. There are chemical engines, electric cells, and many other thermodynamic systems to which the same argument applies. We are, of course, most interested in chemical, photochemical, electrochemical and similar systems for which the work W could be stored as free energy. The theoretical possibility for exotic life forms to evolve based on other than organic chemistry is examined, for example, in Rothstein (1979).

In the case of the spinning top, once the center of gravity is as high as possible, its frictional contact with the supporting surface has become minimal, and no further reduction in the rate of entropy production is possible. The top "goes to sleep," i.e., spins in one place, until its kinetic energy is drained away and it "dies." It would spin indefinitely, if its kinetic energy could be replenished (i.e., if the system were open).

In chemical systems more stable forms of free energy storage are available than in the case of the top, particularly in metastable chemical states. Dynamical systems in a steady state, e.g., patterns of vortices, quickly die like the top (except in special cases like superfluids or astronomical systems) soon after they are decoupled from the energy source. Chemical systems can run on stored free energy, using e.g., food as fuel. They can thus "live" through longer periods of adversity. We can call chemical systems "plants" for which the sun is the immediate energy source. Those which exploit the free energy stored in plants are "animals" (Rothstein, 1985), Fig. 6.

Before the top goes to sleep, it obviously dissipates energy at more than the minimum rate, and this is true for any system while it is evolving toward the steady state. When a steady state system is decoupled from its source, it goes to equilibrium, i.e. its entropy increases. The amount of that increase represents the amount of organization that can be dynamically stored in the system by the source. For metastable free energy storage (any compound of a free energy of formation > 0), the source "manufactures" the metastable system, in the beginning. But what of later on? Is there a thermodynamic rationale for reproduction? There is.

By our earlier argument one would expect many local heat engines, whose efficiency is near zero, initially to be "spontaneously generated." The more there are, and the more efficient they are, the closer will be the global approach to the steady state. If a chemical engine could catalyze formation of a replica, even under conditions and at locations where it could not originate spontaneously, then its multiplication and spread into such regions would enhance the global approach to the steady state. From this point of view both increase in efficiency and ability to reproduce, favor the heat engines so endowed over those less well endowed. This thermodynamic argument supports a kind of Darwinian evolution as well as the evolution of prebiotic chemical organization and reproductive ability.

The same argument may also shed light on dissipation rates at various stages in a life cycle. If organization must be paid for by a compensating entropy increase elsewhere, the earlier the stage in embryogenesis, the higher would one expect the specific rate of entropy production to be. Manufacture plus metabolism should burn more fuel than metabolism alone. Prigogine (1972) cites Zotin and Zotina _et al._ (1967) who found 6 kcal/day/gm of heat production for chicken eggs at the fourth day of development, which dropped to about 1 kcal/day/gm by the sixteenth day. He speculates that this might be connected with the behavior of nonlinear systems driven far from equilibrium and subject to a succession of unstable transitions leading to spatial order and increasing entropy production. We do not understand exactly what is meant by this and prefer the following explanation. Cell division is most rapid in the early stages of development. This means most rapid manufacture of highly complex chemical species, like DNA and protein, from simpler constituents and thus the most rapid rate of compensatory entropy production. The free energy of food in the yolk sac is harnessed to drive the synthetic machinery.

A heat engine argument for an evolution of symbiosis can also be given. If heat engine E_2 operates between T_3 and $T_2 < T_3$, -- engine E_1 between T_2 and $T_1 < T_2$, with E_1 using the heat Q_2 rejected by E_2, then the overall entropy production is decreased compared to when E_1 operates alone. This can be generalized. It is interesting that the evolution of cooperation between engines, and that of the engines themselves, are based on the same thermodynamic argument.

The end-point for thermodynamic evolution to the steady state occurs when the engine performs according to a perfect Carnot cycle, for then the rate of entropy production is exactly zero. It is thus of considerable interest to note that many biochemical reactions take place under close to reversible conditions. Such apocalyptic extrapolations should not be pushed too far, however, because the driving system also changes (and the steady state, like all scientific concepts, is anyhow an approximate description of an idealized reality). Earth and sun also evolve. Indeed, life has such profound effects on the earth that growing numbers of people urge viewing earth and biosphere as a single co-evolving system. For example, the atmosphere seems to have evolved to its present composition primarily because of photosynthetic oxygen production by plants. Though most would hesitate to go so far as to apply the appellation "living" to the entire earth (Lovelock's Gaia concept (1979)), this may be due more to linguistic habit than to any logical or scientific defect in the concept itself (see Rothstein (1985) for more detailed analysis). In the end it comes down to the question of how life, or the living state, should be defined (this is treated in more detail in a forthcoming companion paper (Rothstein, 1986).

Living systems are thermodynamically open to energy flow and exchange of matter, so that the exact placement of a boundary between the living system and its environment has much arbitrariness. At a molecular level this problem becomes unmanageable. The situation is very reminiscent of von Neumann's discussion of quantum measurement or preparation (1983), in which the arbitrary placement of the boundary between the system of interest and the means used to measure or prepare it, is emphasized. This is no accident, for his discussion of irreversibility of measurement went almost far enough to reach the concept of generalized entropy. As von Neumann cites Szilard (1929), I am tempted to believe he missed it because he viewed measurement in quantum mechanics too much from the microscopic side, mindful of the uncertainty principle, at the expense of its inherently operational-macroscopic aspect. It is now obvious that at a molecular level this problem of boundary arbitrariness becomes unmanageable in principle (generalized complementarity), i.e. our definition of "living system," at a molecular level is inherently fuzzy.

Consider Figure 6, which implicitly indicates that living things are open to inflows of information about their environments, as well as to energy and chemical fluxes. They are also sufficiently complex to compute appropriate responses to that information. The heat engine is now a "well-informed heat engine" (WHE). We will discuss this concept further in connection with complexity and individuality, but note now how measurement and preparation are involved.

The information influx can be viewed as a result of measurement processes performed by WHE on whatever part of the environment we take to be the information source. Alternatively, since receipt of any information necessarily changes the state of the receiver (if not, then the information has not registered), it is permissable to regard an act of information transfer as preparation of some state of the receiver by the source. For molecular WHE's, this double interpretation is to be expected from our previous discussion, from EPR and from quantum mechanical discussions of measurement. Consideration of photosynthesis, molecular aspects of vision or any other sense, the mechanochemistry of polyelectrolytes applied to the polymer fiber engine constructed by Katchalsky _et al._ (1960) and numerous other examples show that molecular WHE's exist. Some may find this _macroscopic_ coalescence in concept of measurement and preparation surprising. However, when work, W, is done on any system capable of storing the corresponding energy, say by lifting a weight or compressing a spring, that system has been "prepared" so as to be in a higher energy state. This too is true (or should we say a truism) for molecular WHE's.

For macroscopic systems, the distinction between measurement and manufacture (equivalent to preparation) is usually meaningful and fairly unambiguous, but a transition between them, in a loose way, may well be associated with the origin of life. When solar radiation starts to brew complex compounds in the prebiotic sea-soup, it seems more natural to view the brew as the result of a kind of preparation procedure in a solar cooker, than one in which the disorganized soup makes measurements on the sun and becomes thereby organized. But when a plant displays a positive phototropism, it seems more natural to view the process as a response by the plant to the stimulus provided by the solar radiation, i.e. the information about the source of free energy (supplied by measurement) is used to compute and execute the appropriate response. Customary usage generally reserves terms like preparation, manufacture and processing for cases where drastic changes occur to the starting materials, e.g. reduction of ore to metal, keeping the term measurement for cases in which changes in the measured system can be neglected (non-perturbing).

Assuming life had to evolve from non-living matter, we face the problem of defining attributes or criteria whose absence means non-life, and whose presence means life. We believe that operationally sharp definitions of life, devoid of arbitrariness in placement of the boundary between life and non-life, are no more possible in principle than that sharp distinctions between means of observation and system observed can be made free of arbitrariness. But it is permissable to seek criteria, attributes and operational definitions of a macroscopic thermodynamic nature whose collective presence or satisfaction means life, whose violation or absence means non-life, and whose partial presence or satisfaction gives ambiguous cases about which disputes can rage, or which might be resolved by future research. A related approach seeks _necessary_ conditions for life and for non-life, aiming for sets of them which are _sufficient_ as well. This general kind of search is even older than science, for it has been pursued via philosophy, religion, law and medicine, by healers and witch doctors to the present day. The sea-soup and plant discussion of the previous paragraph leads to the following thermodynamic approach to this problem, which is an elaboration of its opening sentence.

Few would deny the necessity of a large thermodynamically open system (call it the global system) for the origin of life. This paper, and earlier ones in the series, assert the necessity for the global system to have a structure in which heat engines can evolve as subsystems of the global system. We now assert the necessity that some of them be WHE's, with information as well as energy inputs, which _prepare_, as well as _measure_, portions of their global environment. This is an example of (computed) selective behavior, which is further analyzed in Rothstein (1982). Feedback loops are involved via the information input needed at various stages of the preparation procedure. Besides the input means, WHE's organization includes the preparation programs, the computer to execute them, and effectors to carry them out. In effect, processors have evolved within the global environment which was at first processed only by the primary energy source (sun). Later on, WHE's processed the global reservoir of materials in many different ways. Storage of free energy (e.g., manufacture of carbohydrates by plants), synthesis of complex compounds, digestion, digging of burrows, refining of ores, construction of skyscrapers, regeneration, healing, growth and reproduction are but a few randomly chosen examples of such processing.

It has been proposed that "Any dynamic pattern in a non-equilibrium system capable of replicating itself will be said to exhibit generalized life-like behavior, with the proviso that the elements of the pattern are part of a higher entropy configuration before combination than after" (Rothstein, 1979). It was further suggested that "A system exhibiting life-like behavior will be said to exhibit intelligent life-like behavior if it can play survival games. More explicitly, it can gather information (by measurement) about its environment and compute a response which preserves it, whereas without such an appropriately computed response, it could be destroyed" (Rothstein, 1979). These were the most general and least arbitrary definitions I could think of at the time, and they clearly satisfy the necessary conditions stated earlier. But general as they are, there is no assurance that the class of systems falling within their scope could not be broadened still further. Only if no further generalization is possible, could satisfaction of these (or any other) definite conditions be considered as necessary and sufficient for life.

Here are some of those generalizations. Though the point was overlooked at the time, it may well be false that lysis of a cell by an attacking virus, making many viral copies, yields a lower overall entropy than virus plus intact cell. Surely the total (generalized) free energy decreased, the cell organization decreased, and the viral organization increased by a factor one less than the final number of virus particles. It seems quite possible that, if all the entropies (of all kinds) and energies involved were properly taken into account, the overall entropy would increase, for utilization and partial degradation of free energy originally stored in the cell, provides the thermodynamic drive for viral replication. One can argue that this situation is not different in principle from that in which animals prey on plants. There too, one kind of organization is cannibalized by another. In both cases the metastably stored energy is unlocked by enzyme or catalyst keys. Does this pose difficulties for our picture of evolution?

There is obviously no problem in all such cases with the second law, but generalization of Onsager's principle, or perhaps even generalization of the Carnot engine may be required to treat them adequately. Referring to Figures 1 and 2, one notices that there is no compelling physical necessity, as distinguished from possible practical necessity to simplify models of reality to the level of theoretical tractability, to replace Figure 1 by Figure 2 <u>in toto</u>. High grade energy can be utilized directly without first being converted into heat. To use an electric power system analogy,one can generate power hydroelectrically, as well as thermally. The problem is that in the cases contemplated water and coal are in distinguishable, so to speak. In the electric analogy one can envision a strange hybrid generator driven by both water and steam simultaneously. In the chemistry-plus-radiation situation of interest here, the complexities can be enormous, as research in photosynthesis shows.

Perhaps this difficulty is not all that great at a thermodynamic level. Taking a hint from the electric analogy, we divide the system's energy input into a work part and a heat part, say W_2 and Q_2, the output into work $W_1 + W$ and heat Q_1. Here W_1 is W_2 diminished by the losses in the mechanical conversion part of the engine, and W is that part of $Q_2 - Q_1$ which has been converted into work by the thermal part of the engine. The net entropy production is $Q_1/T_1 + (W_2 - W_1)/T_c - Q_2/T_2$. Here, T_c is some effective temperature appropriate to the losses in the mechanical conversion process and will presumably always satisfy $T_2 > T_c$, and probably would be close to T_1 in cases of interest here. One can argue that this contribution is likely to be small, will be constant if the mechanical conversion part doesn't evolve and, even if it does, will probably tend to a minimum anyhow by Rayleigh's principle. The essentials of the previous discussion of evolution are thus likely to be preserved.

An important point now arises, namely that interest has shifted from the evolution of WHE's in a global environment to a further evolution within a subsystem which is already highly evolved. We can take that subsystem as a new global environment (presumably close to steady state), and note that evolution within it can lead to WHE's within WHE's and to new possibilities not accessible directly from the original global system. If the subsystem is largely autonomous, i.e. loosely coupled to the global environment, this second evolution will now be to a local minimum of dissipation. It should not be confused with a global minimum, which is the smallest of all the local minima. Because of metastability, the global environment can be expected to split up into many regions within which evolution can occur to different minima at different rates. Just as polycrystalline ingots never reach the perfect crystal state of lowest energy because of metastable quenching-in of non-equilibrium defect concentrations and crystallite boundaries, so will there be virtually no chance of ever evolving to the true global minimum.

The above suggests that something like viruses could have evolved within living systems. Some might have evolved within metazoans, others when only unicellular life forms existed, or they may have co-evolved with cells, from precursors of modern DNA and RNA competing in the sea-soup. Cancer and aging thus might have something to do with thermodynamic evolution within cells and tissues (Rothstein, 1963).

A very different extension of the definition of generalized life results from broadening the class of patterns. For chemically based life, patterns would consist of complex configurations of atomic constituents, an involved structure within which one might discern complex macromolecules. It is hard to see how a continued existence of a WHE could occur in the face of chemical, thermal, and radiation perturbations without something like a fairly stable cell wall to protect it. From the facts of stereochemistry and crystallography, geometric patterns (which need not be regular) would likely be the dominant type. But as noted in Rothstein (1979), quantum mechanics allows complex patterns in momentum space also, and superfluids like helium II display them. There is thus no a priori reason to exclude the possibility of evolving patterns in momentum space which could be called life-like. The wave functions in momentum space are Fourier transforms of those in position space. Information storage in such a system would smear over position space in a holographic sort of way. But this barely begins the potentialities for pattern. The many-particle Schroedinger equations lead to patterns in configuration space of enormously high dimensionality. Position and momentum operators do not exhaust the repertoire of possible Hermitean operators. All of these have their corresponding eigenfunction expansions (which generalize Fourier expansions) and their conjugate operators (defining corresponding generalized Fourier transforms). There are patterning possibilities, apparently, for an uncountably infinite class of cases (say the totality of bounded self-adjoint transformations in Hilbert space). These are wild speculations, but they might become of interest as we explore the cosmos further. We now return to the only life we know of, chemically based life, to see what we can discern about the evolution of chemically based computation.

The main subsystems of a computer are its memory, the central processing unit (CPU), and the means for giving it its data input, or for obtaining its data output (I/O). The control unit can be taken as part of the CPU. It also contains a number of storage registers, which we can consider to be memory, and the arithmetical-logic unit (ALU). The ALU consists of combinational logical circuitry, namely switches and "gates," which perform basic logical operations individually and, in properly designed combinations, compute as output any desired logical functions of the inputs.

At the molecular level, inputs can be of many kinds. For example, a photon, an incident particle providing energy via inelastic collisions, a phonon or other means of providing energy can excite the system to a higher energy state (preparation) in which it displays different behavior than when in the ground state. Donors or acceptors can change the electron distribution and change the reactivity of the complex. Many reagents can be considered as inputs, which cause reactions to occur, the products being regarded as prepared systems or waste products. A frequently occurring scheme of this kind may be written: A + B <-> C + D, where the left side can be taken to mean that stimulus or preparation means A, is applied to system B, to form the prepared system C and the waste product D. Omitted is the activated complex which is formed as an intermediate stage. (Note that, if we identify B with C, likewise A and D, we have the EPR situation, in essence.) Removing the waste product pushes the act of preparation toward completion.

Chains of such acts of preparation can be envisioned, with characteristic amounts of free energy degraded per stage. When the stages are polymerization steps whose units are choices from a set of distinguishable alternative units (say amino acid residues for proteins, or nucleotides for DNA and RNA), we have information storage in the polymer,i.e. memory. Clearly, the entropy price of this kind of organization is paid in part by the free energy of the starting constituents. Specific memory configurations correspond to specific catalysts.

This leaves the molecular logical circuits to be provided for. Early steps in this direction were made (Rothstein, 1959) by reasoning that quantum arguments explaining why solids were metals, semiconductors, or insulators would also apply to large (and even not so large) molecules. Metal-like "hook-up wire" could be provided by systems of conjugated single and double bonds, whereby a donor (or acceptor) at one end of a chain could cause charges to shift to (or from) the other end. This electron, or hole, type of conduction also had an ionic analogue involving hydroxyl or hydrogen ions (and other groups). Resonance, tautomerism, and isomerism provided a rich spectrum of possibilities from labile to stable, with exquisite control possibilities and amplification, as well as logical functions. These logical functions (Boolean algebraic expressions) are implemented by molecular analogues of the switches and logic gates used by computer circuit engineers. In more recent times, molecular and quantum computation have begun to receive much attention, and are sure to lead in many intriguing directions. Though the subject is still in its infancy, it now seems reasonable to believe in the possibility, in principle, of molecular computation, particularly for molecules of biological interest. On molecular computing see (Conrad, 1986), and for attempts to realize molecular devices see (Carter, 1982, 1987). Quantum computers, so called, deserve more discussion than we can give in this paper. See (Benioff, 1980, 1982; Deutsch, 1985; Feynman, 1985; Peres, 1985). Quantum computation and reversible computation are both highly controversial. For a pro-reversibility view see (Bennett, 1982). For a critical view of both quantum and reversible computing see (Rothstein, 1988).

Chemical evolution entails the evolution of molecular computers for exactly the same thermodynamic reasons as those that urge the evolution and replication of WHE's of increasing efficiency! Computation at a molecular level, which means computation of "appropriate," rather than "inappropriate" behavior in particular circumstances, clearly confers thermodynamic-Darwinian advantages. This encompasses activities like access to and efficient utilization of free energy sources, and also reproduction, as discussed earlier. There is obviously a large world of chemical algorithms to be explored. Virtually all of biochemistry can be viewed as a chronicle of such explorations. An interesting difference between ordinary computation and chemical computation is the obvious need, in the latter, to manufacture intermediate substances to store both energy and information. Conventional computers usually have all devices operating off a common power supply, but chemical computers may well operate with a distributed power supply (local fuel cells burning nutrient).

Conventional computers generally operate synchronously, i.e. with almost all operations controlled by a common clock. Asynchronous operation involves extra communication overhead, e.g. in "handshaking" between source and destination, the use of buffer registers to smooth out information flows between slow and fast devices, strobe lines and so on. In chemical computation asynchronous operation is much more prevalent, with intermediates often serving as buffers and as carriers of the analogues of the ubiquitous identifying codes in computers. Catalytic selectivity is often the chemical analogue of access codes in computers. Vitamins, hormones, etc., perform related functions. But common clocks are not absent in biology, existing, for example, in the cerebellar coordination of vertebrates, and the ubiquitous diurnal rhythms common to most living things.

THE EVOLUTION OF COMPLEXITY AND INDIVIDUALITY

Much has been said previously about complexity, but there are important semantic and mathematical points to be made which go beyond obvious consequences of the generalized entropy considerations already given. They point to and beyond individuality, through computation, to the possibility of the evolution of intelligence as a further outgrowth of thermodynamic evolution.

The evolution of heat engines bears an analogy to phase changes in the sense that separation into two "phases," engine (active) and environment (passive), occurs. Many engines are likely to evolve more or less simultaneously, like crystal embryos forming crystallites from melt or solution. The complexity (in an intuitive sense) of the situation increases, for the macroscopic description of masses of tiny crystals and mother liquor seems much more complex than that of the homogeneous phase from which they are formed. The same is obviously true for the spontaneous generation of embryonic heat engines in the prebiotic sea-soup. As development toward WHE's continues, there occurs a further differentiation of parts of the heat engines. They perform the various functions, already discussed, which give thermodynamic advantages like higher efficiency, etc. This consideration can be carried further, but enough has been said already to justify the view that the thrust of evolution is toward systems of increasing complexity. This means more highly organized systems, with more elements becoming part of them, which thus require more complex descriptions and more complex programs to construct them.

Workers in the field of computer science have developed many measures of computational complexity, of which the Kolmogoroff complexity (K-complexity) is of most immediate interest here (Kolmogoroff, 1965; Chaitin, 1975, 1977; Abu-Mostafa, 1986). Very briefly, the K-complexity of a string of symbols is the shortest program which, when executed, generates that string. The specification "shortest" is needed because, given any program at all, one can always write a longer program which computes the same thing. It turns out that K-complexity is essentially information, for the program "selects" the specific string it computes from an ensemble of possible strings. This idea is so fundamental that it has even been used to provide a new mathematical foundation for probability and statistics (Martin-Lof, 1966; Kolmogoroff, 1968). In the context of this paper we can say that thermodynamics pushes the evolution of systems (WHE's) whose specifications are of increasing K-complexity. Since DNA, or RNA, are generally agreed to embody specifications of living things, i.e. to program their development, this fits in nicely with the empirical observation that the more highly evolved organisms have longer functional nucleotide sequences in their genetic codes than the more primitive ones.

Higher complexity, e.g. as measured by the K-complexity of DNA, is associated with lower thermodynamic entropy than equilibrium. It measures metastably stored information and indicates the current level of organization which the thermodynamic evolution has reached. When there is no information storage, the K-complexity can concur with thermodynamic entropy. This occurs for the K-complexity of a microscopic description of the state of a gas, which is discussed in more detail below. Evolution requires changes in <u>stored</u> information, i.e. in K-complexity. These are conveniently handled by using conditional K-complexity, which is analogous to using conditional probability in place of absolute probability. The conditional K-complexity of string A with respect to string B is the length of the shortest program permitting A to be computed, if B is given. Applied to DNA, it can be taken as a measure of genetic or evolutionary relatedness, to help establish phylogenetic trees, consanguinity, and so on. That part of the total entropy which corresponds to K-complexity is clearly of great interest. To illuminate the relation between K-complexity and thermodynamic entropy, let us look again at crystallites in equilibrium with their mother liquor.

There is no doubt that entropy is larger for a mass of liquid than for the same mass in crystalline form. The difference is the heat of fusion divided by the absolute temperature, L_f/T. It is very large compared to the macroscopic informational entropy in which a collection of crystallites are viewed as alternatives, for it contains a factor of the order of Avogadro's number, N_o (6.02 X 10^{23}). The entropy e.g., for a choice between two equiprobable alternatives is k ln 2, versus N_o k ln 2 for such a choice for each molecule of a mole of gas. Even for 10^{15} crystallites for one mole of substance, where each one is regarded as a macroscopically distinguishable alternative, we are dealing with an entropy of only 50 bits, not 10^{23}. It will now be apparent that introducing macroscopic distinctions is barely noticeable, entropy-wise, compared to the magnitude of the entropy changes involved in phase changes, cooperative effects, sol-gel transformations, solvation, or polymerization reactions. The rate of paying the entropy price for the maintenance and increase of WHE's K-complexity is easily paid by the reduction in the rate of entropy production which Onsager's theorem entails.

A simple kinetic theory example will now be given, in which there is no information storage, and where the increasing K-complexity of the description of an evolving state is reached for cases of the approach to equilibrium. The maximum K-complexity of the state description, multiplied by Boltzmann's k, becomes the equilibrium entropy. One must not fall into the semantic trap of identifying this K-complexity, which increases monotonically along with entropy, with the increasing K-complexity of evolving metastable configurations, which correspond to <u>decreasing</u> entropy (<u>stored</u> information).

Consider a pendulum, slowly damped by gas in a sealed container. For simplicity assume the only losses to be via the viscosity of the gas, which consists of perfectly elastic billiard-ball molecules initially bouncing back and forth between and perpendicular to walls parallel to the oscillation plane of the pendulum. Add four walls far away to make a big flat box. Let the N molecules each have mass m, the bob mass M, with the initial total energy: $E = 1/2\ Nmv_o^2 + Mgh$, where h is the initial elevation of the bob above its minimum position and v_o the common initial velocity of all gas molecules. Let the gas have low enough density that there will be only a few collisions per swing between molecules and bob. We know what will happen. The highly ordered molecular motion will gradually become randomized (thermalized), ending with the bob in Brownian motion about its lowest position. The initial situation is easily fully described, and one can at first follow details of collisions, thereby completely describing later situations. But the descriptions become increasingly complex in detail, so that for $N = 10^{15}$, say, some few dozens of generations of avalanching numbers of collisions already take us to calculations beyond the capabilities of the fastest computers, and the world becomes too small to hold a detailed description of the state of the system. So, simple descriptions go with the initial state of low entropy, and the maximum complexity of the description goes with the subsequent state of highest entropy (disorder, equilibrium). But the course of evolution led to evolved states with increasingly complex descriptions, i.e. highly organized and of a lower entropy, thus characterized in a seemingly contradictory way to what was just asserted about the pendulum.

The paradox is resolved the same way as in earlier discussions of the paradoxes of Loschmidt and Zermelo (Rothstein, 1957; Rothstein, 1974), namely by not confusing microscopic non-operational descriptions with macroscopic operational ones. The latter are simple because complicated detail is averaged out. In the example from the kinetic theory, the average is ultimately taken over the entire phase space Ω, all of whose positions have become available to the representative point of the system. Initially, most of Ω was incompatible with the way the system was specified. Growing ignorance of microscopic detail is measured quantitatively by Boltzmann's statistical entropy, which increases until it reaches the equilibrium value. The evolution of macroscopic complexity in a system, on the other hand, corresponds to decreasing thermodynamic entropy as measured by the increasing amount of information stored in it (metastably for the most part, though dynamical storage is also possible as in spin echo experiments (Rothstein, 1957)). The entropy price for the evolution of macroscopic complexity is paid by the larger system of which the evolving one is a subsystem. In principle, both the information stored metastably as K-complexity and the energy stored in a dynamical steady state, constitute free energy which can be tapped directly or be made available catalytically. In the language of microstates, the evolution of metastable complexity corresponds to an increasing restriction of the portion of Ω available to the representative point for the system. The more highly evolved a system becomes, the farther it departs from equilibrium (the lower its thermodynamic entropy). The foregoing says, in essence, that the tendency for complexity and life to evolve from simple substrates is more easily understood thermodynamically than molecularly. By this we do not mean to deny that chemical and molecular mechanisms are the basis for the machinery of life, but only to emphasize once again the crucial role of non-equilibrium thermodynamics, based on generalized entropy, in selecting with <u>high</u> probability, the seeming <u>improbable</u> class of hard-to-define molecular configurations which we call living.

There exists an interesting connection of another kind between evolution, computer science and statistical physics. Evolution of life toward great complexity, in our picture, needed daily thermal cycling of the sea-soup, required to allow evolution of life as a succession of approximations to solving an "optimum problem" (Carnot engines have optimal efficiency; maximum efficiency means minimum rate of entropy production). The configurations "quenched in" at night are "annealed" by day, permitting successively better approximations to an optimum. "Simulated annealing" has recently been found to be a good approach to solving many complicated problems in combinatorial optimization (Kirkpatrick et al., 1983) which are too big and too complicated to pick out the optimum analytically or by exhaustively testing all cases. This method gives something reasonably close to optimum in usefully short computation times. Apparently Nature discovered this ages ago!

A few more remarks on phase change should be made before discussing the evolution of individuality, for the notion of a "living phase" is probably just as valid a simplification of reality as is the notion of a "living molecule." Melting and evaporation are but two of a large number of transitions in bulk properties of matter as a function of the macroscopic thermodynamic parameters. Magnetism, for example, arises from the interactions of electron spins localized at sites of the crystal lattice. When the magnetic moments of all spins point in the same direction (Ising model), the magnetized state is produced as a "cooperative effect." There are many other kinds of cooperative effects, including helix-chain transitions in DNA. Volkenstein suggested that cooperative effects might play a fundamental role in biophysics, but his discussion is based on equilibrium statistical mechanics (1962). Realization of the importance of non-equilibrium led to the idea of steady-state phases (examples are flow birefringence and the plasma state of a gas carrying electric current between two electrodes) as more fitting candidates (Rothstein, 1969). Progress in developing rigorous techniques in the statistical mechanics of non-equilibrium and phase change may eventually make this a fundamental area of theoretical biophysics.

The evolution of individuality, as a macroscopic property which occurs in sufficiently complex systems, must be shown to evade the difficulty that quantum mechanics creates for complex systems, in that it denies individuality to fundamental particles, and thus render two like systems built out of such particles indistinguishable in principle. Our argument (Rothstein, 1987) notes first the enormous difference between microscopic and macroscopic individuality, illustrated by the observation that two buildings, built from the same kinds and numbers of indistinguishable bricks, are easily distinguished by their architectural variations. Note that the distinction depends on essentially non-perturbing observations. Also, no amount of study of the structure of bricks and the means for bonding them into larger structures yields much insight into principles of architecture. General constraints, like the inadvisability of attempting to build a brick structure five miles in height, can be inferred, of course. But esthetics, utilitarian, economic, topographic, cultural and similar considerations actually determine the architecture and individuality of dwellings.

Similarly, the molecular and supermolecular architecture or organization of living beings broadly depend, like the structure of all matter, on quantum mechanics, but their specific structure is not specified by quantum mechanics. Specific structures mirror the myriad of accidents of evolution and history, and very different ones could presumably have evolved. Only macrosystems can evolve, microsystems being essentially chemical systems. The latter can be synthesized or destroyed, but two examples of one such system are indistinguishable. At some point, molecular systems become large enough to be considered macroscopic objects, but the transition is fuzzy. Similarly, at some point the evolution of complexity reaches a level at which no two systems can be prepared which cannot be distinguished by procedures mild enough to avoid disruption of the metastable structures on which their uniqueness (equivalent to individuality and distinguishability) depends.

Every individual system can be represented as an information source, whose messages are elicited by mild procedures. The stochastic nature of this message source distinguishes it from other message sources using the same alphabet (different alphabets represent different measurement schemes and/or experimental conditions), and a theory of the system's behavior predicts the statistical parameters of the source. Note that the concepts of measurement, recognition, and identification coalesce here, that identification of an individual can range from statistical discrimination tests to reading a name tag, and that the theory can be deterministic (causal) as in the extreme case of 0 or 1 for various probabilities (usually conditional). However, as quantum mechanics tells us, getting information about a microscopic system (from measurement) changes its state, i.e., a different information source is needed to represent it. Generally, large numbers of identical microsystems have to be measured, as in studies of spectral lines emitted by a gas plasma.

What can be meant by a population or ensemble of systems endowed with individuality which is not untenable quantum mechanically? It must mean that carrying out the operations which distinguish that population from the rest of the universe always produces members of that population which can be distinguished from each other by mild procedures. In biology, a typical mild procedure is to observe the behavior of an animal or plant over a period of time. To understand why evolving chemical complexity should spawn such systems, we first consider a family of molecules of increasing complexity, the series of alkanes, methane, ethane, etc. Their composition is given by the formula C_nH_{2n+2}, where n is the number of carbon atoms in the "skeleton." Different isomers, for given n, become possible when n is 4 or more. Two isomers are distinguishable from each other by mild procedures, though two molecules of the same isomer are not distinguishable by any procedure. The number of isomers increases very rapidly with n, e.g. for n = 10, 20, 30, 40, the respective numbers of isomers are 75; 366,319; 4,111,846,763 and 62,491,178,805.831. The number of isomeric monosubstituted alkanes is enormously greater, disubstituted enormously greater than that, and so on. Concerning nucleic acids, there can be 4^n nucleotide chains of n units, each of which can be one of the four nucleotide bases. For n = 40 their number exceeds 10^{24}. It is clear that for molecules like DNA, even for chain lengths like those of the smallest virus, the number of isomers for a single n completely dwarfs numbers like 10^{24}. If we had a procedure for making some number of these giant molecules, say for $n = 10^6$, and could produce some fraction of the total number of isomers with comparable probabilities, then the probability of ever preparing the same isomer twice becomes negligeable, i.e. such isomers have become individuals.

It may well be that reality is even more tilted in favor of individuality than the above indicates. Recalling the possibility of molecular switching and computation discussed earlier, consider the number of functions of n Boolean variables (they take on values 0 or 1 only). There are 2^n possibilities for the various input conditions, and a Boolean function is assigned by specifying whether its value is 0 or 1 for each input condition. The number of such functions $|B_n|$ is

$$|B_n| = 2^{2^n}$$

a number which becomes enormous for moderate n; e.g. for n = 100 it is about 2407 followed by 10^{30} zeros.

Any of these functions can be realized by connecting n switches in the appropriate series-parallel circuit. If one were to construct at random one series-parallel circuit of 100 switches every 10^{-12} second for 100 billion years, the probability of getting two which calculate the same Boolean function is of the order of 10^{-60}. Remembering that our brains contain some 10^{10} or 10^{11} neurons, each of complexity far greater than a switch (one neuron of the associative cortex is connective to five to ten thousand others), with some large but unknown fraction of the neurons modifiable by experience, our individuality is not at all surprising. At this level of complexity, perhaps our intelligence is not surprising either. Computation of "appropriate" responses to environmental inputs would be favored by thermodynamic-Darwinian evolution. Individual memory and learning would presumably utilize metastability and controlled mutability in a way much like that used by the hereditary mechanism, or by the immune system.

CONCLUSION

We believe that the foregoing arguments give a very general thermodynamic rationale for the evolution of prebiotic organization and the origin of life, and allows explanations for many of life's general characteristics. A continuous development from the molecular to the mental level can be envisaged. Darwinian evolution fits smoothly onto thermodynamic evolution, and the latter also explains the origin of complexity and individuality on a physicochemical basis. The need for a generalized thermodynamical supplementation of quantum mechanics as the foundation for theoretical biophysics is demonstrated on physical and methodological grounds.

Two philosophically exciting consequences of this approach are that the evolution of complexity, individuality, life and perhaps intelligence is a probable, if not inevitable, event under suitable conditions, and that the cosmos may contain many locations where exotic life forms have evolved. Von Neumann said in his Vanuxem Lectures at Princeton in 1952 (cited in Blum, 1961) that he could conceive of machines that could replicate themselves, but not one that could create itself. This difficulty does not arise in the development presented here, for the sun plays the role of an external intervenor, cooking up complex systems from an undifferentiated primeval broth. It is not necessary to assume any other acts of special creation for life.

The evolution of complex systems with individuality can be viewed as overwhelmingly probable, but detailed prediction of exactly what kinds of individuals will emerge, is not possible. Luck may be as important as fitness. Nor is it possible to predict the behavior of systems complex enough to do general computations. The generalized thermodynamics of well-informed heat engines admits the possibility of their having something akin to what one can call free will (Rothstein, 1964, 1979, 1982).

REFERENCES

Abu-Mostafa, ed., 1986: "Complexity in Information Theory," Springer, New York, Berlin.

Ballentine, L. E., 1970: "The Statistical Interpretation of Quantum Mechanics," Rev. Med. Phys., 42, 358-381.

Benioff, P. A., 1980: "The Computer as a Physical System," Jour. Stat. Physics, 22, 563.

Benioff, P. A., 1982: "Quantum Mechanical Hamiltonian Models of Discrete Processes that Erase their Own Histories," Int. Jour. of Theor. Physics, 21, 177.

Bennett, C. H., 1982: "The Thermodynamics of Computation - A Review," Internat. Jour. Theor. Physics, 21, 905.

Blum, H. F., 1955: Time's Arrow and Evolution, second edition, Princeton University Press, Princeton.

Blum, H. F., 1961: "On the Origin and Evolution of Living Machines," American Scientists, 49, 474-501. Surveys much of the early work and combines chemical and thermodynamic viewpoints.

Bohr, N., 1958: Atomic Physics and Human Knowledge, Wiley, New York.

Bohr, N., Essays 1958-1962: "On Atomic Physics and Human Knowledge," Interscience, New York.

Bonner, J. T., ed., 1982: Evolution and Development, Springer-Verlag, Berlin.

Bonner, J. T., 1988: The Evolution of Complexity, Princeton University Press, Princeton, New Jersey.

Buss, L. W., 1987: The Evolution of Individuality, Princeton University Press, Princeton, New Jersey.

Carter, F. L., ed., 1982: Molecular Electronic Devices, Marcel Dekker, New York.

Carter, F. L., ed., 1987: Molecular Electronic Devices II, Marcel Dekker, New York.

Chaitin, G. J., 1975: "A Theory of Program Size Formally Identical to Information Theory," JACM, 22, 329-340; 1977: "Algorithmic Information Theory," IBM Jour. of Research and Devel., 21, 350-359.

Champion, D. and Rothstein, J., 1987: "Immediate Parallel Solution of the Longest Common Subsequence Problem," Proc. 1987 Int. Conf. on Par Proc., Penn State University Press, University Park, PA 16802, 70-77.

Conrad, M., 1986: "The Lure of Molecular Computing" IEEE Spectrum, October 1986, 55-60.

de Baere, W., 1986: "Einstein-Podolsky-Rosen Paradoxon and Bell's Inequalities," Advances in Electronics and Electron Physics, vol. 68, Academic Press, New York, 245-336.

Deutsch, D., 1985: "Quantum Theory, The Church-Turing Principle and the Universal Quantum Computer," Proc. Roy. Soc. Lond., A400, 97.

Einstein, A., 1947: "Methods of the Sciences," Chapter 8, 2nd Edition, University of Chicago Press.

Elsasser, W. M., 1937: Phys. Rev. 56, 987.

Elsasser, W. M., 1958: The Physical Foundation of Biology, Pergamon Press, London.

Elsasser, W. M., 1961: Proc. Nat. Acad. Sci. 54, 1431.

Elsasser, W. M., 1961: "Quanta and the Concept of Organismic Law," Jour. Theor. Biol. 1, 27-58.

Feynman, R. P., 1985: "Quantum Mechanical Computers, Optics News, 11, 11.

Fox, S. W., 1960: "How did Life begin?" in Science, 132, 200-208.

Heisenberg, W., 1958: Daedalus 87.3, 95, 99.

Gierer, A., 1988: "Biological Pattern Formation and Physico-Chemical Laws," Synergetics and Dynamic Instabilities in C. Caglioti, H. Haken and L. Lugato, eds., Proceedings of the International School of Physics "Enrico Fermi," Course 99, North-Holland, Amsterdam, 151-174.

Jackson, J. B. C., Buss, L. W. and Cook, R. E., eds., 1985: Population Biology and Evolution of Clonal Organisms, Yale University Press, New Haven, Connecticut.

Jordan, P., 1941: "Physics and the Mystery of Organic Life" (in German), Braunschweig.

Katchalsky, A., Lifson, S., Michaeli, I. and Zurick, M., 1960: "Elementary Mechanochemical Processes," Contractile Polymers, Pergamon Press, New York.

Kirkpatrick, S., Gelatt, C. D. Jr. and Vecchi, M. P., 1983: "Optimization by Simulated Annealing," Science, 220, 671-680.

Kolmogoroff, A. N., 1965: "Three Approaches to the Definition of the Concept Quantity of Information," Probl. Pederachi Inform., 1, 3-11.

Kolmogoroff, A. N., 1968: "Logical Basis for Information Theory and Probability Theory," IEEE Trans. Info. Theory IT-14, 662-664.

Langton, C. G., ed., 1989: Artificial Life, Vol. VI, Sante Fe Institute, Studies in the Sciences of Complexity, Addison-Wesley, Reading, Massachusetts.

Lovelock, J. E., 1979: Gaia: A New Look at Life on Earth, Oxford University Press, Oxford.

Martin-Lof, P., 1966: "The Definition of Random Sequences," Information and Control, 6, 602-619.

Mehra, J., 1973: "Quantum Mechanics and the Explanation of Life," American Scientist, 61, 722-728.

Miller, S. L., 1953: "A Production of Amino Acids under Possible Primitive Conditions," Science, 117, 528-529; Miller, S. L. and Urey, H. C., 1959; Ibid., 130, 245-251.

Moshell, J. M. and Rothstein, J., 1979: "Bus Automata and Immediate Languages," Information and Control, 40, 88-121.

Neumann, J. V., 1932: Mathematische Grundlagenn der Quantenmechanik, Springer, Berlin. English translation by R. T. Beyer, Mathematical Foundations of Quantum Mechanics, Princeton University Press, Princeton, 1955. Chapters V, "Measurement and Reversibility" and VI "The Measuring Process" are particularly relevant and have been reprinted in "Quantum Theory and Measurement," J. A. Wheeler and W. H. Zurek, eds. Princeton University Press, 1983.

Onsager, L., 1931: _Phys. Rev., 37_, 405, 38, 2265.

Oparin, A. I., 1953: _The Origin of Life_, reprint of 2nd edition by Dover Publications, New York.

Pattee, H. H., Edelsack, E. A., Fein, L., Callahan, A. B., eds., 1989: _Natural Automata and Useful Simulation_, Spartan Books and MacMillan, Washington (1966).

Peres, A., 1985: "Reversible Logic and Quantum Computers," _Phys. Rev. A., 32_, 3266.

Polanyi, M., 1968: "Life's Irreducible Structure," _Science_, _160_, 1308-12.

Prigogine, I, Nicolis, G. and Babloyantz, A.: "Thermodynamics of Evolution," _Physics Today_, November 1972, 23-28, December 1972, 38-44.

Raff, R. A. and Raff, E. C., eds.: _Development as an Evolutionary Process_, MBL Lectures in Biology, Alan R. Liss, New York.

Rothstein, J., 1951: "Information, Measurement and Quantum Mechanics," _Science_, _114_, 171-175.

Rothstein, J., 1952: "Organization and Entropy," _Jour. Applied Physics, 23_, 1281-2.

Rothstein, J., 1952: "Information and Thermodynamics," _Phys. Rev., 85_, 135.

Rothstein, J., 1956: "Information Theory and Quality Control," _IRE Convention Record_, vol. 4, part 4, 3-11.

Rothstein, J., 1956: "Information, Logic and Physics," _Philosophy of Science, 23_, 31-35.

Rothstein, J., 1957: "A Physicist's Thoughts on the Formal Structure and Psychological Motivation of Theory and Observation," _Revue Internationale de Philosphie_, . 40, Fasc. 2, 211-266.

Rothstein, J., 1957: "Nuclear Spin Echo Experiments and the Foundations of Statistical Mechanics," _Amer. Jour. of Physics, 25_, 510-518.

Rothstein, J., 1958: "Metastability and Metric Decomposability," _Bull. APS II, 3_, 201, (abstract).

Rothstein, J., 1959: "Heuristic Application of Solid State Concepts to Molecular Phenomena of Possible Biological Interest," Proceedings of the First National Biophysics Conference (Columbus, Ohio, March 1957), H. Quastler and H. J. Morowitz, eds. 77-85, Yale University press, New Haven.

Rothstein, J., 1962: "Information and Organization as the Language of the Operational Viewpoint," Philosophy of Science, 29, 406-411.

Rothstein, J., September 1962: "Wiggleworm Physics," Physics Today, 28-32, 34, 36, 38.

Rothstein, J., 1963: "On Fundamental Limitations of Chemical and Bionic Information Storage Systems" in IEEE Transactions on Military Electronics, vol. MIL-7, 205-8.

Rothstein, J., 1963: Two papers in Summary of Proceedings, Late Irradiation Effects Conference, P. S. Henshaw, ed. Held December 4-6, 1963, at AEC Headquarters Aud. Germantown, Maryland. Printed as report WASH-1059 in May 1965, Division of Biology and Medicine USAEC and made available by the Clearinghouse for Federal Scientific and Technical Information, N.B.S., U.S. Department of Commerce. The papers are (1) An "Inventory" Model of Aging, 201-234, and (2) Thermodynamics, Evolution and Aging, 275-278.

Rothstein, J., 1964: "Thermodynamics and Some Undecided Physical Questions," Philos. of Science, 31, 40-48.

Rothstein, J., 1969: "Probability of Existence of Self-Reproducing Units," Bull. APS II, 14, 504 (abstract).

Rothstein, J., 1969: "Toward a Biostatistical Thermodynamics," Biophysical Journal, 9, Society Abstracts, A-101; "Non-Equilibrium Phases and Cooperative Effects in Biophysics" (abstract). Presented at the Third International Biophysics Congress (August 29-September 3, 1969, Cambridge, Massachusetts).

Rothstein, J., 1971: "Informational Generalization of Entropy in Physics," in Quantum Theory and Beyond, T. Bastin, ed., Cambridge University Press, Cambridge, 391-305.

Rothstein, J., 1974: "Loschmidt's and Zermelo's Paradoxes Do Not Exist," Foundations of Physics, 4, 83-89.

Rothstein, J., 1976: "On the Ultimate Limitations of Parallel Processing," Proceedings of the 1976 International Conference on Parallel Processing, 206-212. IEEE Catalog No. 76CH1127-0C. Best Paper Award.

Rothstein, J., 1977: "Toward an Arithmetic for Parallel Processing," Proceedings of the 1977 International Conference on Parallel Processing, 224-223. IEEE Catalog No. 77CH1253-4C. Most Original Paper Award.

Rothstein, J., 1977: "Transitive Closure, Parallelism, and the Modeling of Skill Acquisition," Proc. 1977 Int. Conf. on Cybernetics and Society, IEEE Cat. No. 77CH1259-1AMC, 232-236.

Rothstein, J., 1978: "Topological Pattern Recognition in Parallel and Neural Models on Bus Automata," Proc. 1978 Int. Conf. on Par. Proc., IEEE Cat. No. 78CH1321-9C, 92-107.

Rothstein, J., 1978: "Bus Automata for Intelligent Robots and Computer Vision," Proc. SPIE's 1978 Symp. on Advances in Intelligent Robotics Systems, D. P. Casasent, E. L. Hall, eds., Proc. SPIE 848, p. 299, 1988, Sixth in a Series.

Rothstein, J., 1979: "Generalized Life," Cosmic Search, 1, 35-38, 44-46, also relevant are refs. [10], [11] and [54].

Rothstein, J. and Davis, A., 1979: "Parallel Recognition of Parabolic and Conic Patterns by Bus Automata," Proc. 1979 Int. Conf. on Par. Proc., IEEE Cat No. 79CH1433-2C, 288-297.

Rothstein, J., 1979: "Generalized Entropy, Boundary Conditions, and Biology," The Maximum Entropy Formalism, R. D. Levine and M. Tribus, eds., The MIT Press, Cambridge, Mass., 423-468.

Rothstein, J., 1982: "Physics of Selective Systems: Computation and Biology, Internat. Jour. of Theoretical Physics 21, 327-50.

Rothstein, J., 1982: "Toward Pattern-Recognizing Visual Prostheses," Proc. IFAC Symp., "Control Aspects of Prosthetic and Orthotics," Pergamon Press, New York, 87-95.

Rothstein, J., 1985: "On the Scientific Validity and Utility of the Living Earth Concept and Its Further Generalization." Presented at the symposium "Is the Earth a Living Organism?" August 1-6, 1985, at the University of Massachusetts, Amherst, Massachusetts. Paper No. 57 in the Symposium Proceedings, published by the National Audobon Society Expedition Institute, Northeast Audobon Center, Sharon, Connecticut 06069. Also available as a technical report OSU-CISRC-TR-86-4.

Rothstein, J., 1986: "Bus Automata, Brains and Mental Models," Proc. 1986 IEEE Int. Conf. Systems, Man and Cybernetics, Systems, Man and Cybernetics, 18, 522 (1988).

Rothstein, J., 1986: "Non-Dichotomous, Relative, and Hierarchical Aspects of Life and Death." Presented at an international colloquium, "Molecular Theories of Cell Life and Death," Rutgers University, May 1-2. Proceedings in press.

Rothstein, J., 1988: "On Ultimate Thermodynamic Limitations in Communication and Computation," Performance Limits in Communication, Theory and Practice, in J. K. Skwirzynski, ed., NATO ASI Series, Series E, Applied Sciences, vol. 142, Kluwer Academic Publishers, Dordrecht, Boston, London, 43-58.

Rothstein, J., 1987: "Complexity and Individuality in Physics and Biology," Bull APS, 31, 290 (abstract).

Rothstein, J., 1987: "Bus Versus Cellular Automata, and Ultimate Limitations of Parallel Processing," Proceedings of the 6th International Conference on Mathematical Modeling, St. Louis. Published in Mathl. Comput. Modelling 11, 357-362 (1988).

Schroedinger, E., 1944: What is Life?, Cambridge University Press, Cambridge.

Schilpp, P. A. (ed.) 1957: A. Einstein: Philosopher-Scientists, Dover Publications, New York, 33, 53, 57.

Shannon, C. E., 1948: "Mathematical Theory of Communication," Bell Syst. Tech. Jour., 27, 379-423, 623-656.

Szilard, L., 1929: Zeit. f. Physik, 53, 840-856. English translation in Behavioral Science 9, 301-310 (1964), "On the Decrease of entropy in a Thermodynamic System by the Intervention of Intelligent Beings," reprinted in Wheeler and Zurek, ref. [60].

Wheeler, J. A. and Zurek, W. H., eds., 1983: Quantum Theory and Measurement, Princeton University Press, Princeton.

Wiener, N., 1950: Cybernetics, John Wiley, New York.

Wigner, E. P., 1961: "The Probability of Existence of a Self-Reproducing Unit," The Logic of Personal Knowledge (essays presented to M. Polyanyi), Routledge and Kegan Paul Ltd., London.

Wigner, E. P., 1965: "Remarks on the Mind-Body Question," The Scientist Speculates, I. J. Good, ed., Capricorn Books, New York, 284-302.

Wigner, E. P., 1969: "Are We Machines," Proc. Amer. Philos. Soc., 113, 95-101.

Wolkenstein, M. V., 1962: "The Theory of Cooperative Processes as the Foundation of Molecular Biophysics," Biophys. Jour., 2, 189-201.

Zotin, A. I. and Zotina, R. S., 1967: J. Theor. Biol., 17, 57.

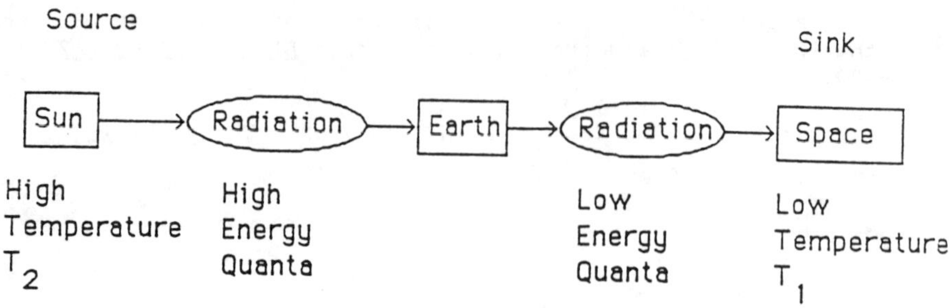

Fiqure 1. Net Energy Flux: Sun to Earth to Space

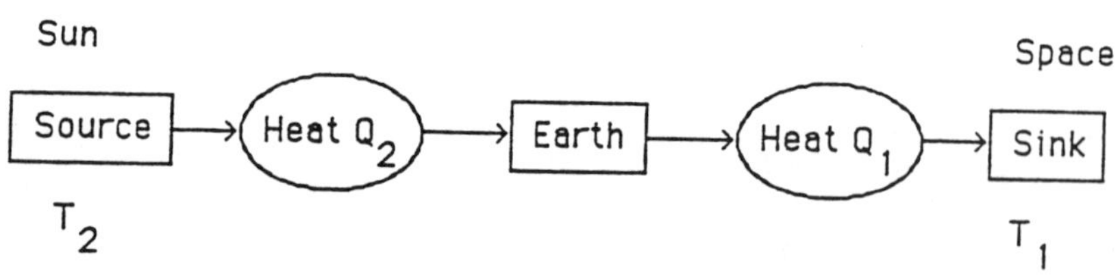

Fiqure 2. Net Heat Flux: Sun to Earth to Space

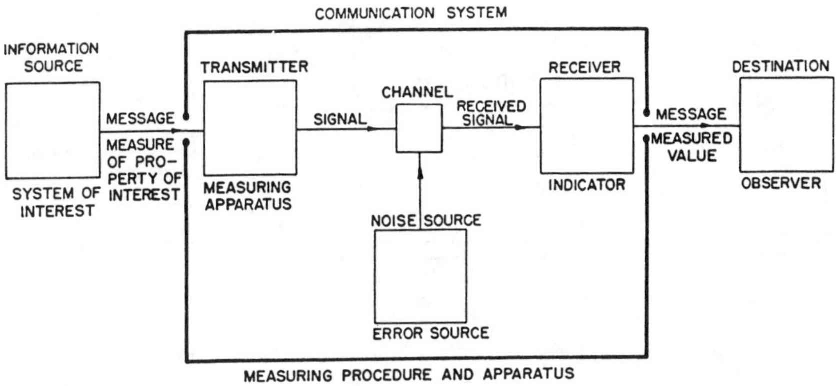

Figure 3. Analogy Between Communication and Measurement

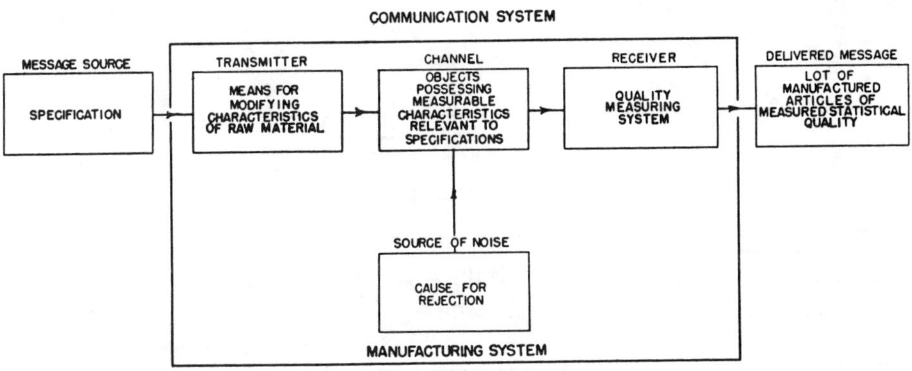

Figure 4. Analogy Between Communication and Manufacturing

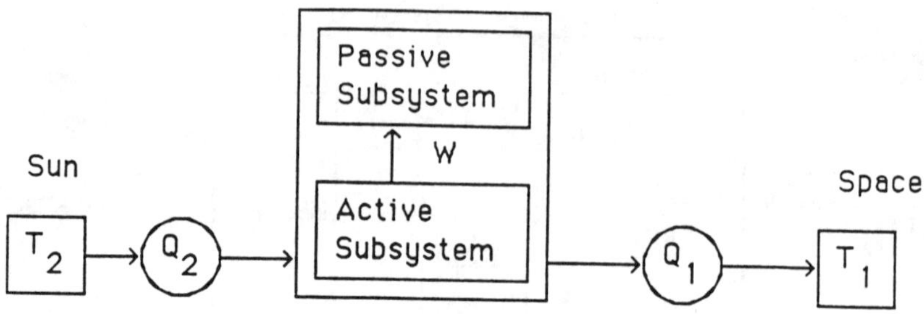

Figure 5. Formation of Terrestrial Heat Engines

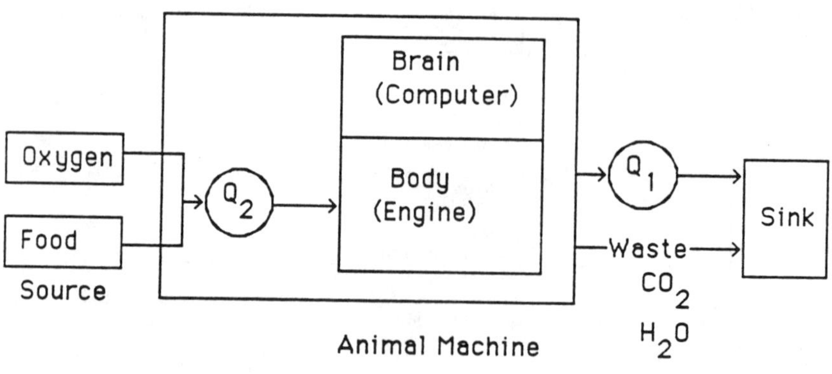

Figure 6. Animal Heat Engines

MECHANISMS OF ENTROPY DISPOSAL: THEIR ROLE IN SHAPING PREBIOLOGICAL AND BIOLOGICAL ORGANIZATION

Roger E. Clapp
The MITRE Corporation
Bedford, Massachusetts 01730

ABSTRACT

For each solar photon absorbed at the earth's surface, twenty photons are eventually emitted into the earth's infrared blackbody radiation field. The average entropy per photon is the same for both blackbody spectra. The extra nineteen photons, multiplied many times over, provide the entropy sink that is fundamental to biological organization. Accordingly, primitive cells have sizes to optimize infrared emission. But direct radiation is relatively slow; there is a much faster mechanism for immediate entropy transfer, utilizing as a temporary entropy buffer one or the other of two vibrational resonances in gaseous carbon dioxide, at 4.3 micrometers and at 15 micrometers in the infrared. Many primitive cells are sized to be half-wave resonators at one or the other of these wavelengths. The cell membranes are lipid bilayers, suggesting that life began in an amphipathic lipid monolayer on a water surface which would have been bombarded by solar ultraviolet quanta and by gaseous carbon dioxide. The ultraviolet rays could activate small molecules, entrained in the monolayer, which could come together into highly excited groupings. These would be disordered and likely to fall apart, but each impinging CO_2 molecule vibrating during recoil could increase its entropy and leave the groupings in a state of reduced entropy, a state of increased organization. The most typical lipids contain palmitic acid, whose chain length indicates the probable presence of electronic excitation modes matching the two infrared vibrational resonances in gaseous carbon dioxide. A scenario for the origin of life, based on the requirements for entropy disposal, is proposed.

INTRODUCTION

Life builds low-entropy structures out of high-entropy ingredients. Any full account of the origin and growth of living structures must make provision for the thermodynamic mechanisms that lead to this reduction of entropy. Important discussions of the

ISBN 0-937194-18-2

pertinent thermodynamics have been presented by Schrödinger (1944), Blum (1951), Katchalsky and Curran (1965), Gorski (1966), Prigogine (1967) and Morowitz (1968, 1970, 1978). Of these, only Blum (1951) directs our attention to the quantized character of entropy disposal through the emission of photons into the earth's black-body radiation field.

Entropy is a quantitative measure of disorder. In any closed system the second law of thermodynamics says that the entropy can only increase or at best remain constant. Clearly, life can only function in an open system where entropy may decrease by increasing elsewhere. Let us treat the earth as a whole, with the biosphere occupying a layer at the earth's surface. Life has existed here for several billion years. There is no place in this layer or in the earth beneath it where entropy can have been building up over this time period to compensate for the repeated steps of entropy reduction within the living structures themselves. We must depend on external flow into and out of the terrestrial system, where the terrestrial system includes both the earth's interior and the biosphere at its surface.

RADIATIVE DISPOSAL OF ENTROPY

The only significant flow into and out of the system of a planet is the flow of electromagnetic radiation. Does this permit a net increase of entropy?

The solar photons that reach the earth's surface (where the biosphere is located) are well described by a black-body distribution for an absolute temperature of roughly 6000° K. The earth itself emits a thermal black-body radiation spectrum that corresponds roughly to an absolute temperature of 300° K. The temperature ratio of 20:1 can be translated into a ratio of numbers of photons. As illustrated in Figure 1, for each solar photon absorbed at the earth's surface, some twenty terrestrial thermal photons will eventually be emitted, each carrying about one-twentieth the energy of a solar photon. The net outflow of nineteen photons "carries entropy away", and this permits "living" structures to reduce their entropy and become more ordered. Without this redistribution of entropy, life would be impossible (Blum, 1951; Clapp, 1981).

How much is the entropy increased extraterrestially by this radiative transfer? We can look at the black-body spectrum in Figure 2. This shows the spectrum for the terrestrial temperature of 300° K, but a similar curve would describe the spectrum for the solar temperature of 6000° K, if each number on the (abscissa) wavelength scale were reduced through multiplication by a factor of 1/20.

The curve in Figure 2 is a normalized plot of the Planck function, J_λ, (Allen, 1963), defined by

$$J_\lambda = \frac{2\pi c^2 h/\lambda^5}{\exp(ch/\lambda kT) - 1} \quad (1)$$

where c is the velocity of light, h is Planck's constant, and k is Boltzmann's constant. The heat Q carried by this spectrum can be obtained from

$$Q = \int_0^\infty J_\lambda \, d\lambda = \frac{2\pi k^4 T^4}{c^2 h^3} \Gamma(4) \zeta(4) \quad (2)$$

while the number of photons involved is given by

$$N = \int_0^\infty (J_\lambda / h\nu) d\lambda = \frac{2\pi k^3 T^3}{c^2 h^3} \Gamma(3) \zeta(3) \quad (3)$$

The mathematical functions that appear in (2) and (3) are the gamma-function and the Riemann zeta-function (Abramowitz and Stegun, 1970), defined for integer arguments by

$$\Gamma(n) = (n-1)! \quad (4)$$

$$\zeta(n) = 1 + 1/2^n + 1/3^n + \cdots . \quad (5)$$

To compute the redistribution of entropy by radiative heat transfer, we are on the safest ground if we consider a process that is reversible, or nearly reversible. In particular, let us consider the radiative heat exchange between two heat reservoirs whose temperatures differ by ΔT. The heat transferred is then given by ΔQ and the number of photons involved is ΔN, where these increments are expressed in terms of Q and N (above) through

$$\Delta Q - (4/T)Q\Delta T \quad (6)$$

$$\Delta N - (3/T)N\Delta T. \quad (7)$$

The entropy transferred is then $\Delta S - \Delta Q/T$, and the entropy per photon is

$$\Delta S/\Delta N = \Delta Q/T\Delta N = (4/3)Q/TN = 3.6016k. \quad (8)$$

We can see from Equation (8) that the entropy per photon, averaged over a black-body spectrum, does not depend on the temperature T for that spectrum. Accordingly, on the average the twenty terrestrial photons in Figure 1 are <u>each</u> processing about as much entropy as the single solar photon in Figure 1. The earth's black-body emission, in the infrared part of the spectrum (Figure 2), is therefore shifting about twenty times as much entropy as increased on earth with the absorption of solar black-body radiation.

The prebiological solar spectrum included ultraviolet radiation, and this was probably important to the origin of life. However, today the ultraviolet is largely screened out by ozone. The biologically important part of the solar spectrum is today mainly the emission in the visible part of the electromagnetic radiation spectrum, the part of the solar emission that penetrates easily through our atmosphere and reaches such molecules as chlorophyll that have adapted to receive and utilize this solar energy.

Life has adapted to receiving quanta on the level of solar radiation, but emits quanta in the infrared part of the spectrum in order to avoid build-up of the entropy that accompanied the absorbed solar radiation. Accordingly, we should expect to see adaptive mechanisms that optimize emission in the infrared part of the spectrum. In particular, the size of a primitive cell should correspond to the dimension that will best shift the infrared emission into the part of the spectrum where the infrared radiates into the outer sink.

Figure 3 is a diagram of a primitive bacterium, <u>Halobacterium halobium</u>, from Henderson (1977). The length of this bacterium (about 9 micrometers) is comparable to a half-wavelength of the infrared radiation that carries away entropy and permits life to exist. Figure 4 is a diagram of a chloroplast, from Wolfe (1972). This is the organelle that utilizes solar photons to build sugar molecules that living organisms then use for their energy storage. Once again, the dimension of this structure (about 8 micrometers) is comparable to a half-wavelength at the infrared part of the electromagnetic spectrum with which living structures transfer entropy.

The diagram of a chloroplast in Figure 4 has been simplified to show only the aspects that are thought to be essential. Stacks of disks forming the grana are shown, and a connection between separate grana is also shown. The resulting simplified diagram evokes a resemblance to the diagram in Figure 5, illustrating one of the antenna designs used by Heinrich Hertz in his pioneering investigations of electromagnetic radiation. What we can infer is that the structural aspects of the living cells may have evolved to optimize the emission of infrared electromagnetic radiation.

CONVECTIVE TRANSFER OF ENTROPY

Direct radiative emission of infrared quanta can be relatively slow, when compared with the faster steps in biological processes. Some of the initial steps in the processing of an absorbed solar photon take place in picoseconds or faster, whereas radiative emission by a biological structure may take many nanoseconds, even as long as microseconds. To ensure irreversibility, these steps must be dissipative and should be in the form of heat quanta that are sent out in times as short as a picosecond.

There is a convective mechanism that can be this fast or faster. Carbon dioxide gas in the atmosphere has two vibrational resonances, at 4.3 and 15 micrometers in the infrared, that are appropriately placed on the energy scale to transfer entropy into sinks. Figure 6 shows the infrared absorption of one meter of air. These two CO_2 resonances appear prominently. There are also water vapor absorption lines, most importantly the lines that are spread loosely over the range from 5 to 8 micrometers.

A gaseous molecule, impinging on a biological membrane, can couple at short range to the electromagnetic fields associated with quantum excitations in the infrared. The molecule can recoil vibrating, having picked up one or more quanta from the electromagnetic field through near-zone interaction. This short-range energy transfer is nonradiative, and can be much faster than a radiative process.

At today's concentration of carbon dioxide in the atmosphere, a cell-sized membrane surface will be struck by a CO_2 molecule some thirty times a picosecond. The prebiological atmosphere probably had a substantially larger concentration of CO_2. Of course there were no cells, but the organic compounds formed by such processes as lightning strokes would have included hydrophobic and amphipathic molecules. As a result of wave action, these molecules would have tended to collect in films at the borders of the primordial ocean. It is here that molecular condensation processes leading to biological polymers could have been frequent, and the attendant reduction of entropy could have proceeded by convective steps including the recoil of vibrationally excited CO_2 and H_2O molecules impinging from the air above.

The vibrational excitations could then have passed from one gaseous molecule to another, upward through the atmosphere. At any stage a resonant vibrational excitation could have been converted by a molecular collision into kinetic energy of translation and rotation. Eventually the heat added to the atmosphere would be radiated into space. What counted for prebiological organization was the rapid convective transfer of entropy to gaseous molecules in the air above the early ocean and, most importantly, above the surface of the quiet backwaters where the organic compounds may have collected.

ADAPTATION FOR ENTROPY DISPOSAL

Water vapor with its high heat capacity could have served as a significant entropy sink but, as evident from Figure 6, the water vapor absorption lines between 5 and 8 micrometers are so spread that it would be hard today to find evidence of <u>selective</u> adaptation. With carbon dioxide, on the other hand, the two resonance absorptions at 4.3 and 15 micrometers are narrow and deep, so that selective adaptation could have occurred.

For example, there is evidence of selection in primitive cells for entropy absorption and/or emission at 4.3 micrometers. Archean microfossils from the Swaziland System of South Africa, in rocks substantially more than three billion years old, have a size range of 1 to 4 micrometers, with the diameter distribution peaking sharply at 2.3 micrometers (Knoll and Barghoorn, 1977; Schopf, 1983). This is close to a half wavelength at 4.3 micrometers, suggesting that these primitive cells were selected to be half-wave resonators at an infrared wavelength where carbon dioxide molecules possessed a strong vibrational resonance. This adaptation would optimize the disposal of entropy into that vibrational mode of gaseous CO_2. The bacteria and chloroplasts illustrated in Figures 3 and 4 are somewhat larger, sized to be resonators for wavelengths closer to the 15 micrometer resonance in CO_2.

The membranes that enclose biological structures are phospholipid <u>bilayers</u>. This suggests that they originated as single layers of amphipathic lipids that were folded over to form bilayers. The single layers could date back to the above-mentioned prebiological films on the primordial ocean surface. The lipids themselves could then have been selected for entropy disposal.

Consider palmitic acid, the most prominent among the lipids in the biological membranes, with a length of 23 Angstroms. Let us compare this with the de Broglie wavelength of a fast moving electronic mass whose kinetic energy matches a CO_2 vibrational resonance. For the resonance at 4.3 micrometers, the de Broglie wavelength is 23 Angstroms; for the resonance at 15 micrometers, the de Broglie wavelength is 43 Angstroms. The following shows that this may be significant.

If a palmitate chain becomes positively charged, as for example under illumination by ultraviolet light, the ejected electron will leave a delocalized electron "hole." For the lowest electronic state, the quantum-mechanical wave function for this hole will have a constant phase. There will also be electronically excited states in which there is migration of this hole back and forth along the chain. There will be a first excited state in which the round-trip phase passes through 360°, and a second excited state in which the round-trip phase passes through 720°.

The migrating positive charge moves as a backward wave, with one electron at a time stepping into the hole and opening up a hole one step farther along. As a result only one electron is in motion at any one moment. The round-trip path length is 46 Angstroms, which will equal the de Broglie wavelength for the first excited state, or will equal twice the de Broglie wavelength for the second excited state. If the mass of this moving electron is roughly equal to the free-electron mass, then we get a close match between the kinetic energy of the moving electron in the first excited state and the CO_2 resonance at 15 micrometers (de Broglie wavelength is 43 Angstroms) and a <u>very</u> close match between the kinetic energy of the moving electron in the second excited state and the CO_2 resonance at 4.3 micrometers (de Broglie wavelength is 23 Angstroms).

It is therefore suggested that the need to dispose of entropy quickly and efficiently may have led to the selection of palmitic acid as a prominent constituent of biological lipid layers. The electronic excitations on the sixteen-carbon chain would then serve as intermediaries in absorbing quanta from growing biological polymers and in passing these quanta to molecules of atmospheric carbon dioxide.

Other possible instances of selective adaptation can be sought in later stages of evolution. Figure 7 shows the Z-diagram for green plant photosynthesis, adapted from Wraight (1982). The two long upward arrows represent the high-energy electrons produced by the absorption of solar photons. Other steps in the diagram are electron transfers. The larger downward steps indicate productive utilization, while some of the smaller steps are just dissipative. Increments along the ordinate scale in volts (at the left) translate into energy increments in electron volts for single-electron transfers. Some transfers are by pairs of electrons, and the vertical increments in volts are then doubled to give energy increments in electron volts. Except for the two long upward arrows, each of the steps on Figure 7 points downward. No steps are directly lateral. The smallest single-electron step is from PC (plastocyanin) to P700 (the reaction center for Photosystem I). This energy step is 0.08 electron volt, just equalling the energy in the 15 micrometer vibrational resonance in gaseous carbon dioxide.

We can see evidence for dissipation in these downward steps, the generation of loss in the form of heat quanta. The creation of a heat quantum raises the entropy of the universe. This is an irreversible process that makes the sequence of steps move only in the forward direction. The smallest package of generated heat evident on the Z-diagram of Fig. 7 just matches the sink provided by the 15 micrometer resonance in CO_2.

ENTROPY PACKAGING

As a result of equipartition, a molecule entering a gas phase at a temperature T will have an average heat energy of nkT/2, where n is the number of degrees of freedom that are actively involved in thermal motion. Division of this heat increment by T gives nk/2, a quantity with the units of entropy. We can suggest that such a molecule, leaving a surface, entering the atmosphere, and eventually radiating into space, may be carrying entropy of nk/2.

For the linear molecule, CO_2, n=5, with 3 degrees of freedom for translational motion and 2 for rotations. This results in a heat capacity of c_v=5k/2 per molecule. When the temperature rises to a value at which a vibrational mode is fully excited thermally, the heat capacity increases to c_v=7k/2 per molecule. From a classical point of view, these two vibrational contributions are highly correlated, so that on a time average only one additional degree of freedom is activated; that would imply an entropy of 6k/2=3k.

We should note that the CO_2 resonance at 15 micrometers is a transverse or "scissors" vibration. While vibrating in this mode, the molecule is no longer strictly linear. For the resonance at 4.3 micrometers, on the other hand, the vibration is longitudinal so that the molecule remains strictly linear.

The entropy acquired by a vibrating CO_2 molecule is not the same as the entropy of that molecule expressed as a state function. Bray and White (1966) give the entropy of gaseous carbon dioxide as 51.1 calories per mole per degree, which translates into 25.7k per molecule. This includes the entropy input needed to bring the substance up to room temperature from absolute zero. Wall (1974) gives the detailed functional dependence of the entropy of a gas, including the Sackur-Tetrode entropy of translation as well as the entropies of rotation and vibration.

When a non-vibrating CO_2 molecule impinges on a film at an air-water interface and recoils vibrating, much of its entropy (before and after) remains unchanged. The parameters that may differ are those that describe the translational and rotational velocity components and the vibrational mode. Counting up the degrees of freedom involved, we obtain 3k or 3.5k as the approximate amount by which the entropy of the molecule can be increased.

To estimate the amount of entropy to be disposed of in a single step of biological synthesis, we can examine the structural entropy involved in a biological polymer such as a protein chain. Bovine rhodopsin, for example, has been completely sequenced (Ovchinnikov, 1982; Hargrave et al., 1983; Nathans and Hogness, 1983). The protein chain contains 348 residues. The distribution among the twenty amino acids is: Ala(29), Arg(7), Asn(15), Asp(5), Cys(9), Gln(12), Glu(17), Gly(23), His(6), Ile(22), Leu(28), Lys(11), Met(16), Phe(31), Pro(20), Ser(16), Thr(27), Trp(5), Tyr(18), Val(31). The probability W_a that a chance selection of 348 residues would lead to this particular distribution is given by the reciprocal of the number of ways to arrange a row containing 348 items and 19 dividers:

$$W_a = 19!348!/(19+348)! \tag{9}$$

Once the distribution has been selected, the probability that a random arrangement of these 348 residues will happen to give bovine rhodopsin is given by:

$$W_b = 29!7!15!5!9!12!17!23!6!22!28!11!16!31!20!16!27!5!18!31!/348! \tag{10}$$

We can now use Boltzmann's expression for entropy (Mandl, 1971) to obtain

$$S = k \log_e(W_a W_b) = -1028.4\ k. \tag{11}$$

The quantity in (11) is negative. It represents an entropy reduction associated with the structural information stored in the protein chain by virtue of its linear arrangement of amino acid residues. Not included are all the steps involved in synthesizing the individual amino acids. Only the structural entropy is included. Dividing by the number of residues gives

$$-1028.4k/348 = -2.96k \tag{12}$$

as the entropy reduction per residue. This is the average informational entropy to be disposed of at each step of adding one more residue to the protein chain as it is assembled.

The entropy decrease given in (12) is comparable to that determined above for black-body photons and for atmospheric carbon dioxide molecules. Note that if in a protein each of the twenty amino acids were equally probable the entropy per residue would be

$$k \log_e(1/20) = -2.996\ k. \tag{13}$$

The genetic code actually has room for 63 different amino acids, but then the informational entropy per residue would be on the order of $k \log_e(1/63) = -4.143k$. This would exceed the available single-step entropy change and require two steps per residue for entropy disposal. Clearly this would be less effective as an evolutionary strategy.

SCENARIO

While lightning in a reducing atmosphere has been shown to produce small organic molecules, including amino acids, it is not likely that lightning would aid in lengthening a growing protein chain. More probably, lightning strokes would cause a growing polymer to break apart.

A different mechanism is needed for the prebiological assembly of biological polymers. This assembly will often involve the storage of energy; the peptide bond linking amino acids in a protein chain is a particular example of a linkage requiring energy for its formation. In addition, the assembly into ordered structures will require a loss of entropy.

It is suggested here that the prebiological assembly of organic polymers took place in a film or scum on the primordial ocean. Exposure to solar ultraviolet radiation would provide the high-energy quanta needed for peptide bonds or other linkages. Atmospheric molecules such as carbon dioxide would increase in entropy and stabilize the linkages. Lipids in the film would aid in the transfer of heat quanta from compound molecular structures to the gaseous CO_2 molecules. The probability of encounters between monomers and a growing polymer would be much greater in the two-dimensional film than in a three-dimensional "soup."

Figure 8 shows the elements of the postulated polymerization process. At an air-water interface there has been an accumulation of prebiotic organic molecules, including lipids and amino acids. Solar ultraviolet rays cause electrical charging and electronic excitations in the amino acids. Some of the excited molecules may then draw together into compound molecular structures that are internally disordered. These compound structures may become ordered by passing thermal quanta to the lipids and to CO_2 molecules which, now vibrating, carry heat quanta away as vibrational excitations. Eventually, they are radiated from the earth's upper atmosphere out into space.

The CO_2 resonance at 4.3 micrometers is the most intense, and we can expect it to play a dominant role in the _early_ evolution of biological polymers. On the other hand, the CO_2 resonance at 15 micrometers is more efficient as an entropy sink; we should expect it to enter strongly into the _later_ evolutionary processes, as seen in the photosynthetic machinery in Figure 7.

Wave action which folds the amphipathic lipid monolayers into bilayers, which now contain embedded polypeptides and other polymers, will form vesicles of various sizes. Those of appropriate dimensions to be infrared resonators tuned to a CO_2 absorption, at 4.3 or 15 micrometers, will be most effective at furthering the transfer of entropy to atmospheric carbon dioxide.

This scenario might bridge the gap between (a) the lightning-generated hydrocarbons and other monomeric organic molecules and (e) the evolving primitive cells. It examines the role of entropy loss by (b) the formation of polymeric structures in an organic film at an air-water interface, followed by (c) the folding of the film into a bilayer and (d) the folding of the bilayer to form vesicles. Evidence for this scenario is to be sought particularly in the matching of biological entropy reduction requirements to the entropy increase following the excitation of vibrational resonances of gaseous carbon dioxide at 4.3 and 15 micrometers in the infrared.

ACKNOWLEDGMENT

This work was funded by a grant to The MITRE Corporation from The Boston Foundation.

REFERENCES

Abramowitz, M. and Stegun, I. A., eds., 1970: Handbook of Mathematical Functions, 9th printing, National Bureau of Standards, Washington, D.C.

Allen, C. W., 1963: Astrophysical Quantities, 2nd ed., Athlone Press, London.

Alpert, N. L., Keiser, W. E. and Szymanski, H. A., 1970: IR: Theory and Practice of Infrared Spectroscopy, 2nd ed., Plenum Press, New York.

Blum, H. F., 1951: Time's Arrow and Evolution, Princeton University Press, Princeton, New Jersey.

Bray, H. G. and White, K., 1966: Kinetics and Thermodynamics in Biochemistry, 2nd ed., Academic Press, New York.

Clapp, R. E., 1981: J. Theor. Biol., 92, 15.

Górski, F., 1966: Plant Growth and Entropy Production, Zakład Fizjologii Roślin P.A.N., Kraków, Poland.

Hargrave, P. A., McDowell, J. H., Curtis, D. R., Wang, J., Juszczak, E., Fong, S.-L., Rao, J. K. M. and Argus, P., 1983: Biophysics of Struct. & Mech., 9, 235.

Henderson, R., 1977: Ann. Rev. Biophys. Bioeng., 6, 87.

Katchalsky, A. and Curran, P. F., 1965: Non-Equilibrium Thermodynamics in Biophysics, Harvard University Press, Cambridge, Massachusetts.

Knoll, A. H. and Barghoorn, E. S., 1977: Science, 198, 396.

Mandl, F., 1971: Statistical Physics, Wiley, New York.

Morowitz, H. J., 1968: Energy Flow in Biology, Academic Press, New York.

Morowitz, H. J., 1970: Entropy for Biologists, Academic Press, New York.

Morowitz, H. J., 1978: Foundations of Bioenergetics, Academic Press, New York.

Nathans, J. and Hogness, D. S., 1983: Cell, 34, 807.

Ovchinnikov, Yu. A., 1982: FEBS Letters, 148, 179.

Prigogine, I., 1967: Thermyodynamics of Irreversible Processes, 3rd ed., Wiley, New York.

Schopf, J. W., ed., 1983: Earth's Earliest Biosphere, Princeton University Press, Princeton, New Jersey.

Schrödinger, E., 1944: What is Life?, Cambridge University Press, Cambridge.

Wall, F. T., 1974: Chemical Thermodynamics, 3rd ed., W. H. Freeman, San Francisco.

Wolfe, S. L., 1972: Biology of the Cell, Wadsworth Publishing Company, Inc., Belmont, California.

Wraight, C. A., 1982: In Photosynthesis, edited by Govindjee, Academic Press, New York.

ADDENDUM

Clapp, R. E., 1989: The gibberellins: a proposed mechanism for their functioning as plant growth hormones. Biophys. J. 55, 570a.

Takahashi, N., 1986: Chemistry of Plant Hormones, CRC Press, Boca Raton, Florida.

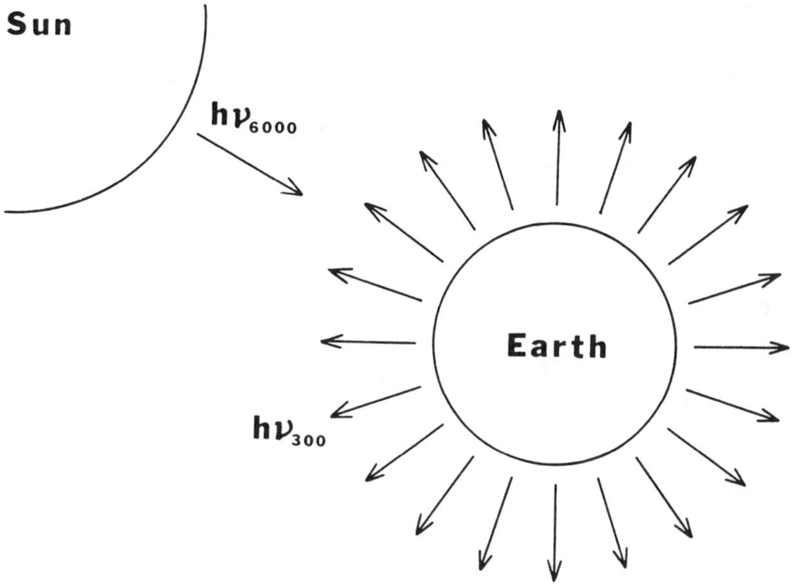

Figure 1. The energy balance for the sun-earth system, leading to an increase in the number of photons.

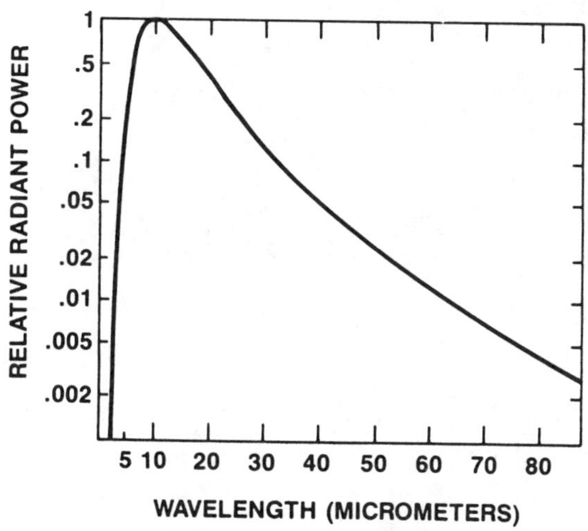

Figure 2. The black-body spectrum for a temperature of 300 K.

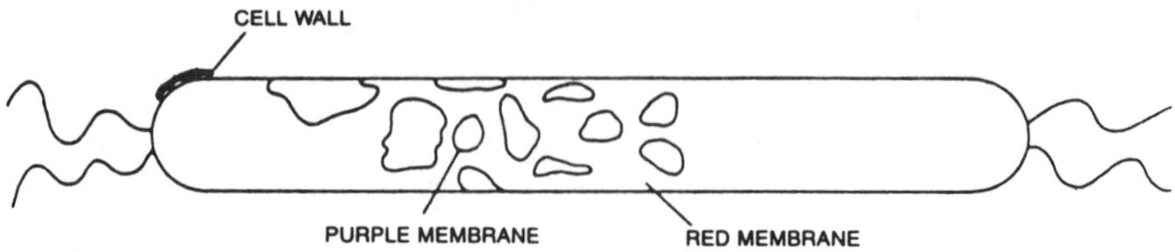

Figure 3. Scale drawing of one Halobacterium halobium cell. Adapted, with permission, from "The Purple Membrane from Halobacterium halobium," by Richard Henderson, in the Annual Review of Biophysics and Bioengineering, Volume 6, pp. 87-109. Copyright 1977 by Annual Reviews Inc.

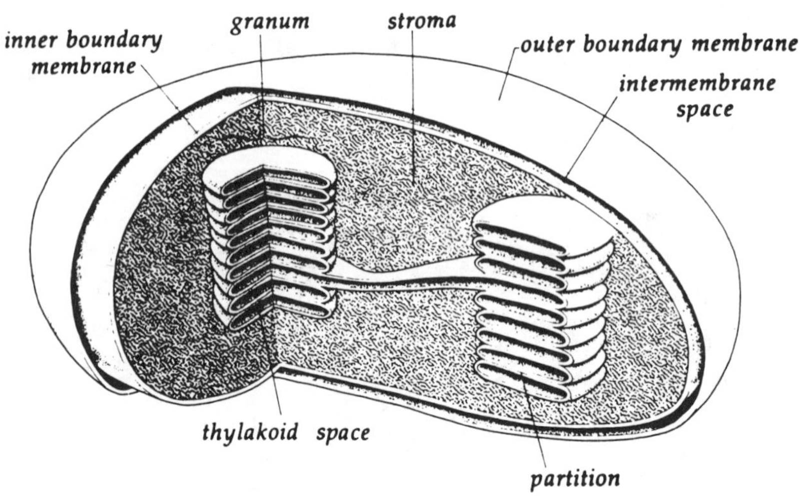

Figure 4. Diagram of a chloroplast. From Biology of the Cell by Stephen L. Wolfe, copyright 1972 by Wadsworth Publishing Company, Inc. Used with permission.

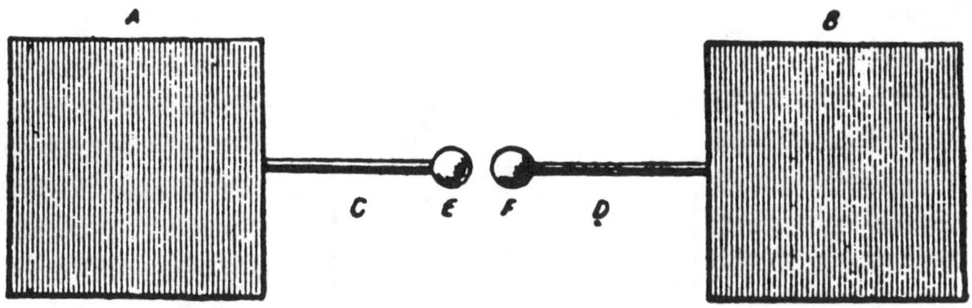

Figure 5. Scale diagram of a half-wave resonating antenna used by Heinrich Hertz (1857-1894) for his pioneering experiments on electromagnetic waves. From the Encyclopedia Britannica, 11th edition (1910-1911). Used with permission.

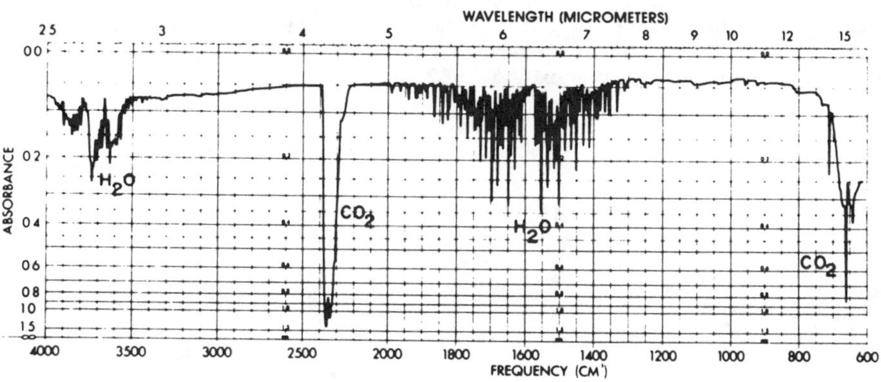

Figure 6. Infrared absorption in one meter of air. Reprinted with permission from N. L. Alpert, W. E. Keiser, and H. A. Szymanski, <u>IR: Theory and Practice of Infrared Spectroscopy</u>, second edition, copyright 1970 by Plenum Press, New York.

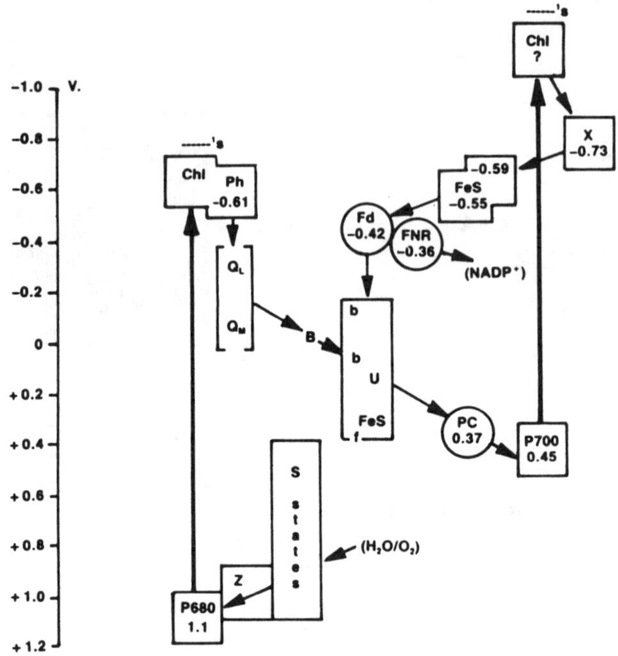

Figure 7. The Z-diagram Showing the Energy Steps in Photosynthesis." Adapted with permission from C. A. Wraight, "Current Attitudes in Photosynthesis Research," in Photosynthesis, Volume I, edited by Govindjee, copyright 1982 by Academic Press, New York.

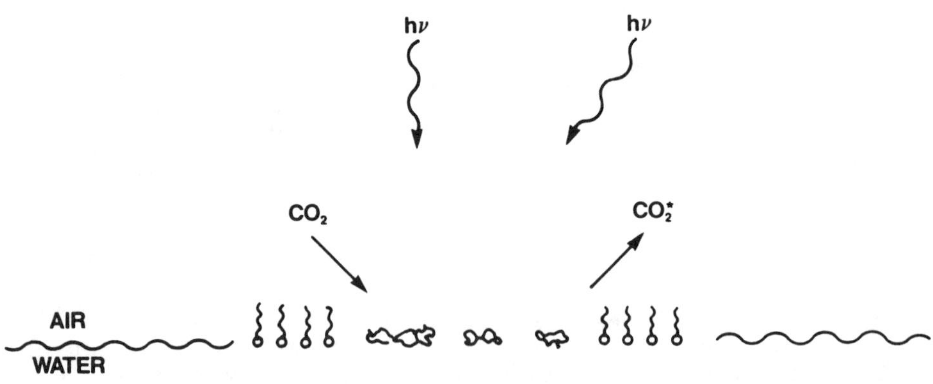

Figure 8. Prebiological polymerization at an air-water interface. A lipid monolayer contains entrained monomers such as amino acids, which are excited solar ultravioletrays and draw together into compound molecules. Impinging CO_2 molecules, recoiling as vibrationally excited CO_2* molecules, carry away entropy and stabilize the growing polymers.

MACROBIONT: THE POND AS PROTO-ORGANISM

Benton C. Clark
Planetary Sciences Laboratory (0560)
Martin Marietta Denver Aerospace
Denver, Colorado 80201

ABSTRACT

A highly heterogenous, shallow pond is hypothesized as a possible model for the origin of metabolic functions. The pond contains organic matter, some solubilized and some as condensed, adsorbed, or occluded phases in particulates. The composition of the matter is highly diverse, both chemically and physically. Depending upon particle size, density and hydrophobic interactions, some particles float and form a surface scum. Others sink to form a graded bed, and still others are colloidal in size and remain suspended for long periods of time. Solar heating and advective cooling provide thermal gradients which drive convection currents to maintain transport and suspension of the material, as well as some segregations on the pond floor. Wind action causes redistribution of matter, and provides a supply of exogenous mineral grains. Ultraviolet radiation-induced polymerization and degradation continuously processes near-surface organic material. Chemical weathering of minerals at the pond bottom, pond shoreline, and of eolian sedimented matter brings fresh ions into solution and provides new catalytic mineral forms and particle surfaces. Meteorological trends and events produce dehydration-rehydration and freeze-thaw chemical separations and physical migrations.

It is supposed that in the course of this stochastic sequence of events, semi-isolated chemical regimes form. Dynamic feedback chemical processes, such as in a natural reflux reactor, cause enhanced conversion of some compounds. Autocatalytic situations occur. For example, high molecular weight material selectively binds to stereo-specific sites on particulates situated on the pond bottom, thereby being withdrawn from the circulation. Other material, being unbound, is eventually exposed to the UV and perhaps oxidizing environment of the near surface.

The pond may be viewed as a macroorganism, with the various zones as poorly compartmentalized organelles. Catabolic and anabolic processes may be driven by the solar flux, by variations in thermodynamic equilibria in the ever changing thermal regions, and by

ISBN 0-937194-18-2

reactions with disequilibrium photochemical products from the atmosphere. Protoribosomal activity is exhibited by the myriad of microparticles. The pond exhibits the characteristics of energy capture and chemical conversion. It also exhibits other characteristics of a biotic entity. The pond itself is indeed bounded, as is a cell, but distinctly nonlocomotary. It can not reproduce. But as the natural processes which originated the segregated chemical pathways grinds to an inevitable conclusion of stagnation, the evolutionary pressure mounts for microminiaturization and juxtaposition of physiochemical processes. The macrobiont ultimately survives in an offspring of more efficient and prolific microforms. An example of how such a dynamic and complex pond may form will be given.

HYDROPHOBICITY AS AN ORGANIZING PRINCIPLE IN THE ORIGIN OF THE GENETIC CODE

James C. Lacey, Jr.
University of Alabama at Birmingham
Birmingham, Alabama 35294

ABSTRACT

The idea of hydrophobicity as an organizing principle is certainly not a new one. Its role in the formation of lipid bilayers (membranes) and the direction of the final secondary structure of proteins is well established. That it also played a role in the origin of the code and protein synthesis is suggested by statistically significant correlations between the hydrophobicities of amino acids and their anticodonic nucleotides. One cannot, however, reason directly from this observation to the mechanism by which the code and protein synthesis originated because these correlations are probably only a reflection of more fundamental realities which we must address more directly. Also, many of the amino acids, and CMP and UMP, are quite hydrophilic and, while the interaction of hydrophobic amino acids with hydrophobic nucleotides will be favored in aqueous solution, the interaction of hydrophilic amino acids and nucleotides in aqueous solution will not be favored. Because of the favored nature of hydrophobic interactions in aqueous solution, we have tended to concentrate on the hydrophobic side of the question in most of our recent research. A review of this work will reveal variations in binding constants and reaction rate constants between the hydrophobic nucleotide AMP and a series of amino acids and carboxylic acids of varying hydrophobicity. Further studies on molecular conformations show definite preferential intramolecular associations of the L-isomers of hydrophobic amino acids with adenine within aminoacyl adenylate anhydrides and esters. All of these observations relate to definite preferences in associations and reactions relevant to coding and protein synthesis and it has become increasingly obvious that hydrophobicity is one of the very important organizing principles in the origin of coding and protein synthesis. While the principles which govern the organization of hydrophilic moieties has not been adequately addressed, it is somehow implicit in the data that, while an aqueous milieu assists in the ordering of hydrophobic moieties, perhaps a hydrophobic milieu may be necessary for organizing hydrophilic moieties. It is suggested that an oil(lipid)-water interface may have been

PREBIOLOGICAL SELF ORGANIZATION OF MATTER
C. Ponnamperuma and F.R. Eirich (Eds.)

ISBN 0-937194-18-2

the ideal organizing locus for the origin of protein synthesis in that it is the one locus which could organize both hydrophobic and hydrophilic substances.

INTRODUCTION

Life, after all, is in the simplest terms the outcome of an organization of molecules. Our job in understanding how life originated, is to understand how the molecules organized themselves. This, I think was implicit in something I remember Sidney Fox saying and I paraphrase, "In the beginning life organized itself." He may have used the word assembled instead of organized, but the definite implication is that it was the molecules that organized themselves leading to life.

In our quest for understanding those origins we can observe that there are many different forces which are responsible for the organization of molecules. Among those are electrostatic effects, e.g. attraction of two opposite charges (or dipoles) or binding of two like charges by a divalent ion of opposite charge. Charge transfer complexes would certainly be included in this type. But perhaps the major organizer of biological molecules is <u>hydrophobicity</u>. The hydrophobic effect involves the exclusion of the hydrophobic moieties from the aqueous phase. Membranes can be formed more or less spontaneously under proper conditions to yield the well-known lipid bilayer. The tertiary and quaternary structures of many proteins are, at least partially dictated by the folding inward of hydrophobic amino acid side chains. Frequently active sites of enzymes are hydrophobic, especially when the substrate is also hydrophobic.

We became especially aware of the importance of hydrophobicity with respect to the origin of the genetic code when we noted that there is a direct correlation between the hydrophobicity of amino acids and the hydrophobicity of their anticodonic nucleotides (Weber <u>et al.</u>, 1978; Lacey <u>et al.</u>, 1983). Woese <u>et al.</u> (1966) had actually discussed the importance of hydrophobicity and the code origin much earlier. They made estimates of the polarities of the amino acids but did not present corresponding data for the nucleotides. Our basic finding was that, generally speaking, hydrophobic amino acids have hydrophobic anticodons (A-rich) and hydrophilic amino acids have hydrophilic anticodons (U-rich). The hydrophobicities of the nucleotides vary as A > G > C > U. The correlation of hydrophobicities itself was not responsible for the code origin, but the character of the amino acids and the nucleotides was responsible.

HYDROPHOBIC EFFECTS ON BINDING CONSTANTS BETWEEN AMINO ACIDS AND NUCLEOTIDES

Because of the observed correlations between amino acids and their anticodons (Figure 1), we have tended to think of the genetic anticode (Table I) rather than the genetic code. The logical nature of the Table, i.e. most of the hydrophobic amino acids have hydrophobic A as their middle and most important letter and most of the charged amino acids have hydrophilic U as their middle and most important letter, suggests what experiments should be tried. We should question the anticodon hypothesis by trying to determine whether there are selectivities in binding based on hydrophobicities. We have, because it is easier to study hydrophobes in aqueous solution, tended to concentrate on the A amino acids, those having A as the middle letter of their anticodons. After we had reported on the hydrophobicity correlation, Reuben and Polk (1980) studied the binding of sixteen amino acids to 5'-AMP. They observed an inverse correlation between the binding constants and our estimates of the hydrophilicity of the amino acids (Figure 2). We (Khaled <u>et al.</u>, 1983) similarly observed a direct correlation between the binding constants of the hydrophobic amino acids (Phe, Leu, Ile and Val) for polyadenylic acid and estimates of their hydrophobicity (Figure 3). We also found that, if one studies the binding of the four ribonucleotides to a singly hydrophobic amino acid Phe (Khaled <u>et al.</u>, 1984), we again see a correlation between the binding constants and the estimated hydrophobicities of the nucleotides (Figure 4). The general principle which Reuben and Polk (1980) elucidated and our data confirm is that, given either a hydrophobic nucleotide or a hydrophobic amino acid, the binding of a series of molecules to that hydrophobic moiety will vary directly as the hydrophobicity of the second kind of molecule. These observations would suggest that given a hydrophobic template (poly A), it would principally attract, associate with and organize hydrophobic amino acids. The ability to do would be Phe > Leu = Ile > Val. For the moment we will not carry it further as to how peptides could be synthesized. However, the kinds of results that have been obtained support the anticodon hypothesis (Lacey <u>et al.</u>, 1976).

EFFECTS ON MOLECULAR CONFORMATION

The intrinsic organizing ability of hydrophobicity has also been found to be important in establishing molecular conformations through intramolecular interactions. Every amino acid going into protein synthesis is first activated as the aminoacyl adenylate anhydride. We have found in a series of aminoacyl adenylate anhydrides of hydrophobic amino acids that even the aliphatic amino acid side chains tend to approach the face of the adenine ring, the closeness of approach being a function of the hydrophobicity of the side chain (Lacey et al., 1986). Furthermore, the L-isomer of Ac-L-Phe was found to associate with the adenine ring in these compounds more strongly than the D-isomer (Lacey et al., 1985). Because these compounds are the universal activated intermediates in protein synthesis, it is important that organizing influences are seen. Figure 5 shows a suggested conformation of phenylalanine adenylate. Given this structure, it is not difficult to imagine how stacked arrays could form (Figure 6) and these in turn could yield peptides more efficiently than if a hydrophilic amino acid were attached to AMP (Figure 6). Such self-organization would be possible with any combination of hydrophobic amino acid and 5'-AMP, but becoming more efficient as the hydrophobicity of the amino acid increases. NMR studies (Lacey et al., 1985) show that with AC-Phe-AMP the intramolecular associations are concentration independent but the intermolecular associations become noticeable only at >0.006 M adenylate. Such amino acid-anticodonic nucleotide organization at high concentration (> 0.006M) is imagined to yield peptides more efficiently than other combinations. The addition of a poly U template in this particular example should even enhance the association and allow efficient peptide formation at lower concentrations of adenylate. We have also proposed (Lacey et al., 1977) that simultaneous polymerization of the nucleotide portion and the amino acid portion would give rise to a peptide and its nucleic acid information equivalent.

Further work on the important adenylate anhydrides was carried out with a series of straight chain carboxylic acid adenylate anhydrides (Lacey et al., 1987). The purpose of this series (acetyl-, propionyl-, butyryl-, phenylpropionyl-) was to find out, using nuclear magnetic resonance, how long the aliphatic chain had to be before it associated with the hydrophobic adenine. The results showed that the acetyl- will not associate with the adenine ring but the propionyl- and all the rest of the series did. Consequently, we can say that the propionyl-chain (CH_3-CH_2-) is long enough and hydrophobic enough to associate with adenine.

The results above very definitely show that adenine can be viewed as an organizer of hydrophobic moieties, including hydrophobic amino acids. We propose that this fact is responsible for A being the central and most important anticodonic letter for Phe, Leu, Val, Ile and Met, all of which are hydrophobic amino acids. Two additional hydrophobic amino acids, Trp and Tyr, do not have A as their central anticodonic letter although they do have A at the 3', non-wobble position. Other work of ours (Staves et al., 1986, 1987) suggests that Trp and Tyr were not included in the proteinaceous amino acids until after the process of protein synthesis had been established, consequently their assignments were not necessarily arrived at through physicochemical relationships.

The compounds referred to above were all adenylate anhydrides(mixed anhydrides between carboxyl- and phosphoric acid group). These compounds are the first activated forms of all amino acids going into protein bio-synthesis. From that form, the amino acids pass to the 3′ terminus of its cognate tRNA where they become an ester (2′-3′) of the terminal AMP residue.

We have studied the aminoacyl adenylate esters as models of the 3′ end of aminoacylated tRNA. Again, we have found an extremely important organizing influence. Other workers have found that although amino acids esterified to the end of tRNA migrate back and forth between the 2′ and 3′ position of the ribose of AMP (Taiji et al., 1983), the amino acid is only incorporated into protein when it is in the 3′ position (Hecht, 1977; Sprinzl et al., 1979; Wagner et al., 1982). We have found (Lacey et al., 1988) that glycine and all L-amino acids we tried distribute to that position about 67% (i.e. k(3′/2′)=2), but that D-amino acids do not distribute uniformly. They distribute to that important position inversely as a function of the hydrophobicity of the amino acid side chain (Figure 7). That is to say that D-Trp and D-Phe distribute to the 3′ position much less well than do the corresponding L isomers. Consequently, we have proposed that the discrimination against the D-isomer being placed the 3′ position is responsible for the origin of our protein synthesizing system which exclusively uses this 3′ position for formation of the peptide bond. We also believe the consistency of the L-amino acids being in the 3′ position is one reason, but probably not the exclusive reason, for the preferential use of L-amino acids in protein synthesis. The point here is that again this discrimination is based on hydrophobic interactions between the adenine ring and the amino acid side chain. Preliminary NMR results suggest that the reason the hydrophobic D-amino acids do not favor the 3′ position is that, when D-amino acids are in the 2′ position, they can interact more strongly with the adenine ring than when they are in the 3′ position. Consequently the hydrophobic D-amino acids are drawn toward the 2′ position depending on the hydrophobicity of the side chain. Because the L-amino acids distribute uniformly (≈67%) to the 3′ position regardless of the side chain, there is then some intrinsic property of the L-configuration itself which is responsible for that distribution. It is certainly quite possible that because D-alanine also distributes 67% to the 3′ position, the same intrinsic property exists in the D-amino acids to associate with the 3′ position. However, when a very hydrophobic side chain is present, it associates with the adenine much more strongly when it is in the 2′ position, thus tending to hold the amino acid in the 2′ position. Because A is the middle anticodonic letter for most hydrophobic amino acids and A can discriminate against hydrophobic D-amino acids being in the 3′ position, there is an element of coding present. We have found that D- and L-amino acids when esterified to 5′-GMP have 2′-3′ distributions very nearly the same as 5′-AMP. However, 5′-UMP appears not to discriminate against the hydrophobic amino acids nearly so well as 5'-AMP and 5'-GMP. The similar effects of 5'-AMP and 5'-GMP are perhaps not surprising. It can be imagined that the amino acid side chain overlapping of the purine ring would "see" mainly the five membered ring with either 5'-AMP or 5'-GMP, i.e. the amino and keto groups would be distal from the point of association. The pyrimidines would have these more polar aspects of the bases nearer to any overlapping hydrophobic side chains.

EFFECTS ON REACTION RATES

We have also observed effects of hydrophobic interactions on the rates of chemical reactions between a hydrophobic amino acid, acetyl phenylalanine, and 5'-AMP compared with 5'-UMP (Lacey et al., 1985b). The esterification of 5'-AMP proceeds much faster than that of 5'-UMP (Figure 8). The difference in the rates of these reactions which we found by carrying out the reaction at three different temperatures, is due to differences in the entropy of activation for the reaction, i.e. due to orienting effects in the transition state. These effects, we propose, are principally due to differences in the binding constant of phenylalanine for 5'-AMP and for 5'-UMP (compare in ref. (Khaled et al., 1984), 6.6 M^{-1} with 1.1 M^{-1}). As pointed out earlier, the preference of Phe for 5'-AMP is based on hydrophobicity.

We have also found selectivities in reaction rates of the A amino acids for 5'-AMP (Mullins et al., 1986) and polyadenylic acid (Lacey et al., 1985a). The order of esterification is Ac-Phe > Ac-Leu > Ac-Val > Ac-Ile. This order is only partially one of decreasing hydrophobicity, but partially due to general reactivity.

Furthermore, the discrimination between hydrophobic Ac-Phe and hydrophilic Ac-Gly for esterification of the 2' OH group of polyadenylic acid (the most hydrophobic polyribonucleotide) could be enhanced in favor of Ac-Phe by carrying out the esterification reaction in the presence of an emulsion of hexane (Lacey et al., 1985a). In this experiment we believe the polyadenylic acid was at the interface. The adenine ring in the hexane and Ac-Phe could be partitioned into the hexane more so than Ac-Gly, thus increasing the rate of reaction of the Ac-Phe with the polyadenylic acid.

HYDROPHILIC AMINO ACIDS

All of these above results, as far as they go, are consistent with the anticodonic hypothesis for the origin of the genetic code (Lacey et al., 1976). But, of course, the genetic code contains representations of a number of hydrophilic amino acids and amino acids of intermediate hydrophobicity. We have addressed some of those amino acids only in experiments carried out with Sid Fox at Miami (Lacey et al., 1979). We prepared thermal proteinoids rich in Phe, Pro, Gly, or Lys, but also containing several other amino acids. Each of these preparations had at least 15 mole percent Lys to initiate binding to polyribonucleotides. We then tested the ability of each of these thermal proteinoids to coprecipitate with each of the four homopolyribonucleotides, poly A, poly G, poly C and poly U. The results were clearly anticodonic in preference for the Pro-rich, Gly-rich and Lys-rich proteinoids. With the Phe-rich proteinoids we again found a preference for poly U, the codonic nucleotide of Phe. Hypothesizing that the high Phe content of the proteinoid caused the hydrophobic residues to fold inward, thus hiding them from exposure to the solutions of polyribonucleotides, we ran that experiment again, but in the presence of 25% volume ethanol to unfold the proteinoid. Under those conditions, the Phe-rich proteinoid precipitated the most with the anticodonic nucleotide poly A. To ascertain the preference of Phe for A more clearly, we carried out the study of Phe binding to the four ribonucleotides and found a definite preference of Phe for 5'-AMP, its anticodonic nucleotide (Khaled et al., 1984). The results with the thermal proteinoids then do support the anticodonic hypothesis across the range of hydrophobicities. To our knowledge these were the first reports of anticodonic preferences in the interaction of amino acids and nucleotides (Lacey et al., 1979).

We have attempted to study the binding of Lys-O-Me to poly U and UpU and found very weak interactions at any salt concentration up to 0.25 M NaCl. The binding constants obtained were about 1 M^{-1} or less. This result raises a question about the general interaction of two hydrophilic molecules with each other in aqueous solution. 5'-UMP, UpU and poly U are all hydrophilic molecules. Lysine is a hydrophilic molecule. By definition these molecules like water. How can we expect Lys to associate with uracil in aqueous solution? My belief is that we cannot. Consider for a moment our studies with hydrophobic amino acids and the most hydrophobic nucleotide, 5'-AMP. The interaction of these molecules with each other is actually facilitated by the solvent, water. It would seem equally logical that the interaction of two hydrophilic molecules with each other would be facilitated by a hydrophobic solvent. This, however, generates the problem that we need a hydrophobic milieu for one system and a hydrophilic milieu for the other. The only obvious locus involving both is the oil(lipid)-water interface. With this thinking, is it not likely that primitive protein synthesis occurred at such interfaces? Our only results showing a lysine preference for its anticodonic nucleotide, U, was with thermal proteinoids (Lacey et al., 1976). These proteinoids, although Lys-rich, still contained about 30% of hydrophobic amino acids. Consequently, the proteinoid in solution could have had a hydrophobic core which was in close proximity to the lysine residues at the surface. Perhaps that experiment just accidentally furnished us with the right combination of hydrophobes and hydrophils juxtaposed next to each other.

It is, we think, very likely that the oil(lipid)-water interface was a very important locus for the early synthesis of proteins.

REFERENCES

Garel, J. P., Filliol, D. and Mandel, P., 1973: J. Chromat., 78, 381-388.

Hecht, S. M., 1977: Tetrahedron, 33, 1671-1696.

Khaled, M. A., Mullins, D. H., Jr., Swindle, M. and Lacey, J. C., Jr., 1983: Origin of Life, 13, 87.

Khaled, M. A., Mullins, D. W., Jr. and Lacey, J. C., Jr., 1984: J. Mol. Evol., 20, 66-70.

Lacey, J. C., Jr. and Weber, A. L., 1976: Protein Structure and Evolution, L. Fox, Z. Deyl, and A. Blaze, eds., Marcel Dekker, BAsel, 213-222.

Lacey, J. C., Jr., Weber, A. L. and Pruitt, K. M., 1977: In the Encyclopedia of Ignorance, R. Duncan and M. Weston-Smith, eds., Pergamon Press, New York, 219-223.

Lacey, J. C. Jr., Stephens, D. P. and Fox, S. W., 1979: Bio-syst., 11, 9-17.

Lacey, J. C., Jr. and Mullins, D. W., Jr., 1983: Orig. Life, 13, 3-42.

Lacey, J. C., Jr., Hall, L. M., Mullins, D. W., Jr. and Watkins, C. L., 1985a: Orig. Life, 16, 151-156.

Lacey, J. C., Jr. and Mullins, D. W., Jr., 1985b: J. Theo. Biol., 115, 595-601.

Lacey, J. C., Jr., Hall, L. M. and Mullins, D. W., Jr., 1985: Orig. Life, 16, 69-79.

Lacey, J. C., Jr., Mullins, D. W., Jr. and Watkins, C. L., 1986: J. Biomol. Struct. and Dynam., 3, 783-793.

Lacey, J. C., Jr., Mullins, D. W. Jr., Taylor, K., Hall, L. M. and Watkins, C. L., 1987a: In preparation.

Lacey, J. C., Jr., Hawkins, A. F., Thomas, R. D. and Watkins, C. L., 1988: Proc. Natl'l. Acad. Sci. USA, in press.

Mullins, D. W., Jr. and Lacey, J. C., Jr., 1986: J. Mol. Evol., 23, 76-79.

Reuben, J. and Polk, F. E., 1980: J. Mol. Evol., 15, 103-112.

Sprinzl, M. and Cramer, F., 1979: Prog. Nuc. Acid Res. Mol. Biol., 22, 1-69.

Staves, M. P., Bloch, D. P. and Lacey, J. C., Jr., 1986: Z. Naturforsch., 41C, 121-125.

Staves, M. P., Bloch, D. P. and Lacey, J. C., Jr., 1988: Origins of Life, 18, 97-105.

Taiji, M., Yokoyama, S. and Miyazawa, T., 1983: Biochem., 22, 3220-3225.

Wagner, T., Cramer, F. and Sprinzl, M., 1982: Biochem., 21, 1521-1529.

Weber, A. L. and Lacey, J. C., Jr., 1978: J. Mol. Evol., 11, 199-210.

Woese, C. R., Dugre, D. H., Saxinger, W. C., Dugre, S. A., 1966: Proc. Natl. Acad. Sci. USA, 55, 966.

ADDENDUM

Lacey, J. C., Jr., Hawkins, A., Thomas, R. D., and Watkins, C. L., 1988: Proc. Natl. Acad. Sci., 85, 4996-5000.

Sjostrom, M., and Wold, S., 1985: J. Mol. Evol., 22, 272-277.

Taylor, J. W., 1986: Adv. Space Res., 6, 19-21.

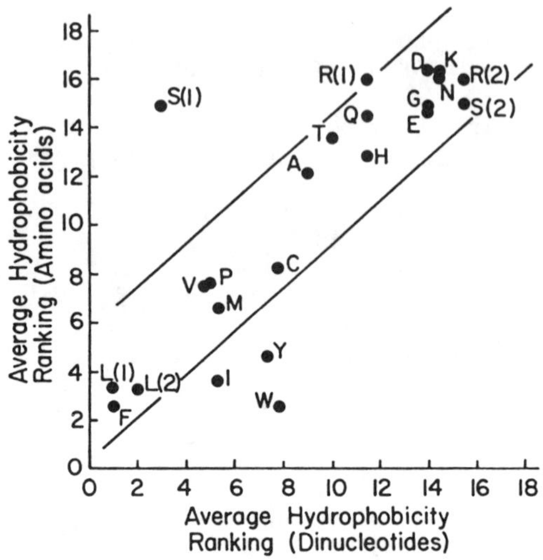

Figure 1. The average ranked hydrophobicity of the amino acids as a function of the average ranked hydrophobicity of their corresponding anticodonic dinucleoside monophosphates. The standard single letter abbreviation for the amino acids were used. Amino acids with accompaning numbers are those which possess two anticodonic assignments. The correlative coefficient for those points within the outlined band is 0.97 (Lacey and Mullins, 1983, for further details).

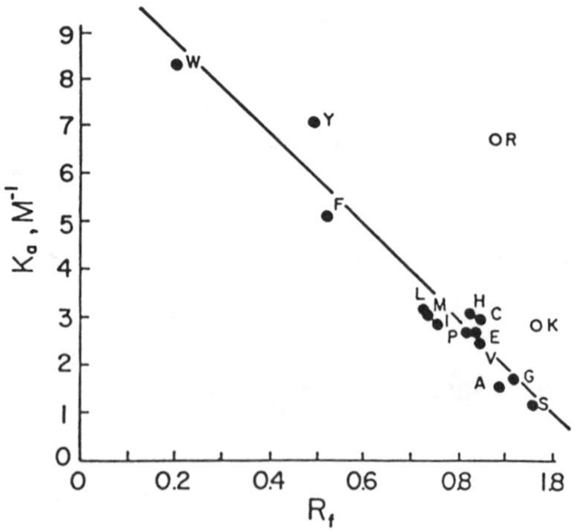

Figure 2. The binding constants (M^{-1}) of amino acids for 5'-AMP as determined by Reuben and Polk (1980) plotted vs. our estimated hydrophilicity of the amino acids (Weber and Lacey, 1978).

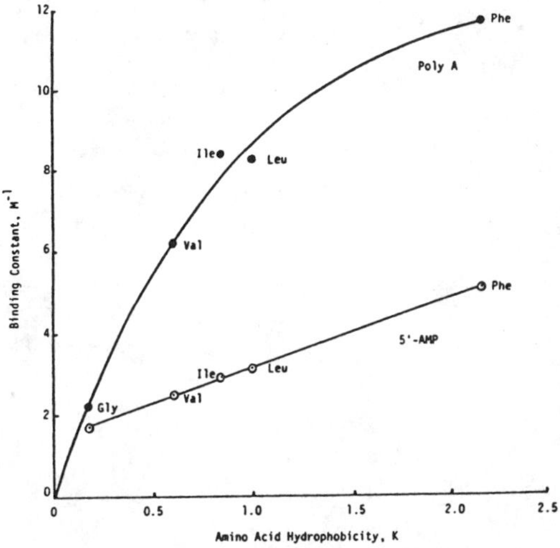

Figure 3. The binding constants (M^{-1}) for polyadenylic acid (Khaled et al., 1983) and for 5'-AMP (Reuben and Polk, 1980) of those amino acid (methyl esters) having A as middle letter anticodon vs. the estimated hydrophobicity of the amino acids (Garel et al., 1973).

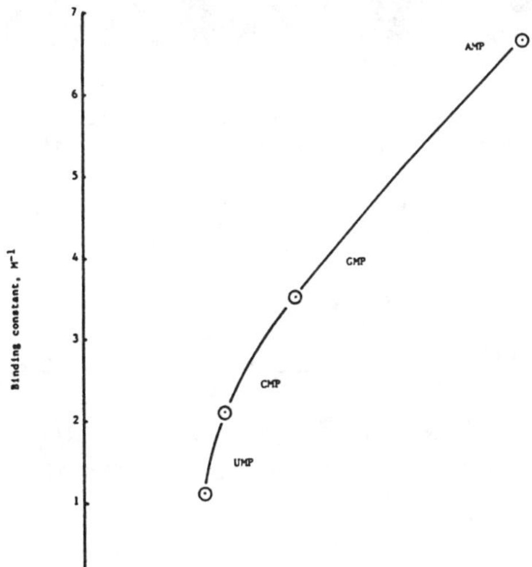

Figure 4. The binding constants (M^{-1}) for the mononucleotides of the methyl ester of phenylalanine (Khaled, et al., 1984) vs. the estimated hydrophobicity of the mononucleotides (Garel et al., 1973).

Figure 5. A possible structure of phenylalanine adenylate anhydride derived from NMR data (Lacey et al., 1985). The data also suggest that aliphatic amino acid side chains would associate with the face of the adenine ring in a similar manner. These associations are thought to be hydrophobic.

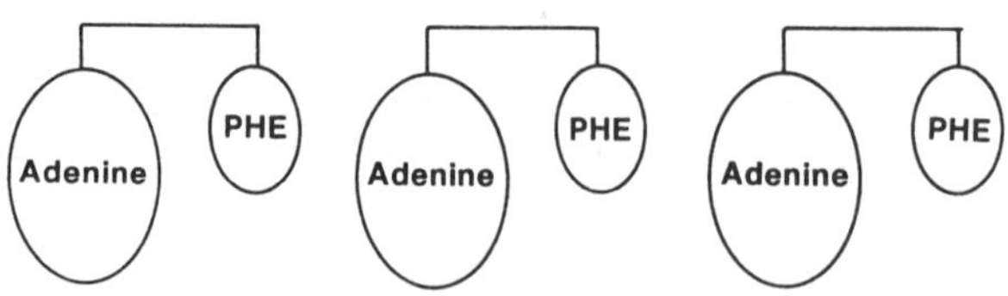

Figure 6. A conceptual self-organization (stacked arrays) of molecules of phenylalanine adenylate anhydride mediated through a cooperative effect of (Weber and Lacey, 1978) intramolecular ring-ring interactions of the phenyl ring with the adenine ring which then allows intermolecular interactions of the same type.

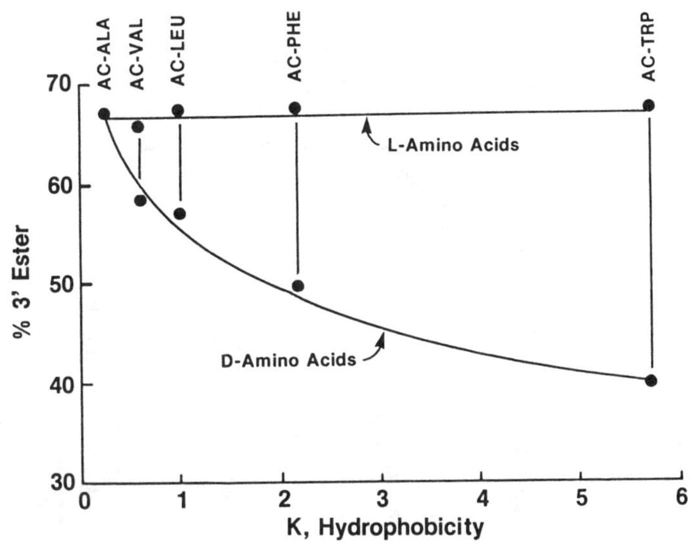

Figure 7. Data for the % 3' ester for Acetyl D- and L-amino acids esterified to the ribose of 5'-AMP (Lacey <u>et al.</u>, 1987) vs. the estimated hydrophobicity of the amino acids (Lacey <u>et al.</u>, 1977). NMR data suggest that when the more hydrophobic amino acids are in the 2' position they associate more strongly with the adenine ring and are consequently held more in that position.

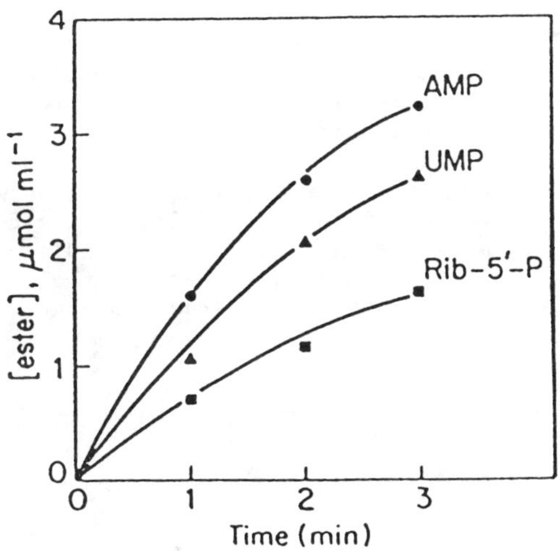

Figure 8. Rates of esterification of 5'-AMP and 5'-UMP and ribose-5'-phosphate in 13% DMF, 0.92 M acetylphenylalanine imidazole, 0.092 M nucleotide at 0 °C pH 7 (Mullins and Lacey, 1986).

Decreasing hydrophobicity →

5'-End	Middle position A	G	C	U	3'-End
A	Phe	Ser	Cys	Tyr	A
G	Phe	Ser	Cys	Tyr	
C	Leu	Ser	Trp	T	
U	Leu	Ser	T	T	
A	Leu	Pro	Arg	His	G
G	Leu	Pro	Arg	His	
C	Leu	Pro	Arg	Gln	
U	Leu	Pro	Arg	Gln	
A	Val	Ala	Gly	Asp	C
G	Val	Ala	Gly	Asp	
C	Val	Ala	Gly	Glu	
U	Val	Ala	Gly	Glu	
A	Ile	Thr	Ser	Asn	U
G	Ile	Thr	Ser	Asn	
C	Met	Thr	Arg	Lys	
U	Ile	Thr	Arg	Lys	

Decreasing hydrophobicity ↓ (3'-End)

Table I. The genetic anticode organized according to decreasing hydrophobicity of the mononucleotides. The assignments listed here are the anticodons predicted from Watson-Crick base pairing to the established codons. In reality, the anticodons found in tRNAs are seldom those shown. The variation is primarily in the wobble position (5'-end), where inosine is frequently found, and where A-U, G-C base pairing is not rigidly maintained. T signifies the anticodonic equivalent of the terminator codons. Variations from the universal code are not shown.

POLYMERIZATION OF NUCLEOTIDES IN AN AQUEOUS SOLUTION

H. Yanagawa, E. Kanaya, and H. Furuta
Mitsubishi-Kasei Institute of Life Sciences
11 Minamiooya, Machida, Tokyo 194, Japan

ABSTRACT

Cyanogen bromide (BrCN) condenses oligoadenylates [oligo(A)] on a poly(uridylic acid) [poly(U)] template in an aqueous solution. Imidazole and divalent metal ions such as Mn^{2+}, Co^{2+}, Ni^{2+}, Cu^{2+}, Zn^{2+}, Mg^{2+} and Fe^{2+} were required for the condensation. The chain length of oligo(A) and the reaction temperature affected the coupling yield. Hexaadenylate [$(pA)_6$] was converted to $(pA)_{12}$, $(pA)_{18}$, $(pA)_{24}$, $(pA)_{36}$, $(pA)_{42}$ and $(pA)_{48}$ in a up to 68 percent overall yield for 20 h at 25°C. The coupling yield increased with the poly(U) concentration. Five to sevenfold molar excess of uridylyl residues of poly(U) to adenylyl residues of oligo(A) gave the best yield (68 percent). Metal ions affected the formation of linkage isomers of the phosphate bonds: The 2′,5′- and 3′,5′-phosphodiester bonds were predominant in the presence of Co^{2+}, Zn^{2+} and Ni^{2+}, and the 5′,5′-pyrophosphate bond was predominant in the presence of Mn^{2+}. In particular, Ni^{2+} gave the highest ratio of the 3′,5′-phosphodiester bond (30 percent). N-Cyanoimidazole (1), N,N′-iminodiimidazole (2) and N-carboxamidoimidazole (3) were formed in a reaction of imidazole with BrCN in aqueous solution. 1 and 2 had much the same condensing activity for the polymerization of adenylates as BrCN. A reaction pathway is being proposed in which 1 and 2 are not only intermediates for the production of 3, but also the true condensing agent in the coupling reaction of oligo(A). Phosphorimidazolide derivatives were detected in a reaction of 5′-AMP with either 1 or 2. The condensation would proceed by way of N-cyanoimidazole-phosphate adduct, the phosphorimidazolide derivative, or both. A possible role of BrCN in the prebiotic synthesis of oligonucleotides will be discussed.

INTRODUCTION

Hydrogen cyanide (HCN) is present in the interstellar space and in the atmospheres of Titan and Jupiter. It is readily synthesized by ultraviolet light, spark discharge, and shock waves from gas mixtures that have been suggested as models for the atmosphere of the primitive earth. We were interested in the role of HCN and

PREBIOLOGICAL SELF ORGANIZATION OF MATTER
C. Ponnamperuma and F.R. Eirich (Eds.)

ISBN 0-937194-18-2

its derivatives in chemical evolution; in particular, in the role of diiminosuccinonitrile (DISN) as a prebiotic condensing agent. DISN has been readily synthesized by the oxidation of diaminomaleonitrile, a tetramer of HCN (Ferris et al., 1979; 1982). Therefore, we initiated a program of nucleotide condensation using DISN. Ferris et al. (1984) have recently shown that DISN and cyanogen bromide (BrCN) caused the conversion from 3′-AMP or 2′-AMP to 2′,3′-cyclic AMP in the presence of divalent metal ions and imidazole. Quite recently, we have found that BrCN and DISN brought about the coupling reaction of oligo(A) in aqueous solution (Kanaya and Yanagawa, 1985; Kanaya and Yanagawa, 1986). To gain further information, we have examined, in detail, the intermolecular condensation of oligo(A) on a poly(U) template using BrCN as a condensing agent. In this paper, we present a detailed account of the experiments and discuss a possible role of BrCN in the prebiotic synthesis of oligonucleotides.

MATERIALS AND METHODS

PREPARATION OF OLIGO(A)S

Oligo(A)s with chain lengths from 1 to 10 were prepared by nuclease SW (Seikagaku Kogyo Co., Ltd.) digestion of poly(A). Poly(A) (10 mg) was incubated with 14 units of nuclease SW (Mukai, 1965) in an aqueous solution (2 mL) containing 50 mM sodium carbonate buffer, pH 10.3, 100 mM NaCl and 2 mM magnesium acetate at 37°C. An adequate amount of the reaction mixture was taken out and dissolved in 100 mM NaCl, 10 mM Tris-HCl buffer, pH 7.0, every 10 min, and the increase in UV absorbance at 260 nm monitored. When the increase reached 30 percent, the reaction was stopped by neutralization with Tris-HCl buffer, pH 7.0, followed by extraction with 90 percent phenol. The aqueous layer was dialyzed against 1 mM Tris-HCl, pH 7.0 to remove phenol, followed by HPLC using MCI-GEL CQA-30S (0.76 x 60 cm; Mitsubishi Chemical Industries Ltd.). A linear gradient of 0.3-1.0 M ammonium formate was applied over 1 h at a flow rate of 1 mL/min. The progress of the effluent was followed at 270 nm. Peaks containing oligo(A) were collected, desalted by dialysis against 1 mM Tris-HCl, pH 7.0, lyophilized and stored at -20°C.

TEMPLATE-DIRECTED POLYMERIZATION OF OLIGO(A)S USING BrCN

The reaction was carried out in a 0.28 mL Eppendorf tube in a total volume of 20 μL at 4, 20, and 40°C. Details of the composition of the reaction mixture and the incubation time are given in the figure legends. The reaction was terminated by addition of enough Na_2EDTA to make complexes with all divalent metal ions. The reaction mixture was then lyophilized to remove the residual BrCN. After redissolution, the pH was adjusted to 8.0 with 1 M Tris-HCl buffer, pH 8.5. RNase A [Sigma, 0.25 mg per μmole of poly(U)] was added to digest excess poly(U), and then the solution was incubated at 37°C for 8 h. HPLC of the polymerized products of oligo(A) was performed on a RPC-5 column (0.4 x 25 cm, Nishio Kogyo Co. Ltd.) by using a linear gradient of 2-200 mM $NaClO_4$ in 10 mM Tris-acetate buffer, pH 7.5, for 40 min at a flow rate of 1 mL/min. The eluate was monitored at 260 nm with an absorbance detector connected to a recorder equipped with a data processor to determine each peak area quantitatively, corrected by using molar extinction coefficients of nucleic acids (Brahms et al., 1966). The efficiency of the reaction was estimated by comparing each peak area of the products in the chromatogram with the peak area of a corresponding starting material.

HYDROLYSIS OF DIMERIZED PRODUCTS OF TETRAADENYLATE WITH RNase T_2

For identification of the polymerized products of tetraadenylate [$(pA)_4$] on a poly(U) template using BrCN, the reaction mixture was applied to a CQA-30S column (0.4 x 25 cm) equilibrated with 0.3 M ammonium formate. The elution was carried out by a linear gradient of ammonium formate (0.3-1.0 M) over 40 min at 70°C. The flow rate was 1 mL/min. The compounds eluted at 39, 44, and 45.3 min were collected separately, dialyzed against 1 mM Tris-HCl, pH 7.0, and lyophilized. They were then redissolved in 10 μL of 50 mM ammonium acetate, pH 4.5, containing 20 percent (v/v) methanol and treated with RNase T_2 (0.02 units, Seikagaku Kogyo Co., Ltd.; Uchida, 1966). After incubation for 30 min at 37°C, the amount of Ap, A, A(2')p(5')Ap and pA(5')pp(5')Ap resulting from the hydrolysis was determined by HPLC performed on a C_{18} column (MCI-GEL ODS 1 MU, 0.46 x 15 cm) by using a mobile phase of 50 mM triethylammonium acetate containing 5 percent CH_3CN. Nuclease P_1 (Seikagaku Kogyo Co., Ltd.) digestion was carried out in an aqueous solution (10 μL) containing 0.1 mM oligo(A), 20 mM sodium acetate, pH 5.5 and 230 μg of nuclease P_1 at 37°C for 40 min. The digested products, pA and A(5')pp(5')A were analyzed in the same way as described above. A(2)p(5')Ap as an authentic sample was a gift from Dr. Sawai (University of Tokyo), and pA(5')pp(5')Ap was obtained by RNase T_2 digestion of p<A(5')pp(5')A>p purchased from P-L Biochemicals. A(5')pp(5')A was obtained from Sigma Chemical Co. A molar absorption coefficient of pA(5')pp(5')Ap was taken from published data (ϵ 27,700; Christie et al., 1953).

REACTION OF IMIDAZOLE WITH BrCN

A solution of BrCN (320 mg, 3 mmol) in 30 mL of 0.2 M imidazole-HNO_3, pH 6.2 was stirred for 24 h at 20°C. HPLC of the reaction products was performed on a C_{18} reverse-phase column (Develosil, Nomura Chemical, 1.0 x 25 cm) with solvent systems A (3 percent $AcCN/H_2O$) and B (50 percent $AcCN/H_2O$). The elution conditions were 0 percent B for 30 min followed by a linear gradient to 100 percent B, in 40 min. The flow rate was 2 mL/min, and the column effluent was monitored at 230 nm. Analysis by HPLC revealed four major peaks at 5, 13, 22, and 56 min.

N-Cyanoimidazole (1) was eluted as a peak with a retention time of 13 min. The eluate was freeze-dried, dissolved in acetonitrile, and flash chromatographed on silica gel. (1) was eluted with acetonitrile and the eluate was crystallized from benzene to give colorless needles: mp 59-60°C; ^{1}H NMR ($CDCl_3$) δ 7.16 (1H, d, H-4), 7.29 (1H, t, H-5), 7.94 (1H, br s, H-2); ^{13}C NMR ($CDCl_3$) δ 104.65 (cyano carbon), 119.98 (C-4), 130.70 (C-5), 138.92 (C-2); IR (KBr) 3160, 3130, 2270, 1725, 1690, 1640, 1530, 1470, 1285, 1225, 1060, 1000, 850, 750, 630, 590, 485 cm^{-1}; UV_{max} (AcCN) 218 nm (ϵ 4.4 x 10^3); mass spectrum, m/z 93 (M+), 68, 66, 41, 40. Anal. Calcd for $C_4H_3N_3$: C, 51.61; H, 3.25; N, 45.14. Found: C, 51.75; H, 3.23; N, 44.71. These spectral data were identical with those of an authentic sample that was synthesized from imidazole and BrCN in benzene at 60°C (Giesemann, 1955; Hagan, 1985).

N,N′-Iminodiimidazole (2) was eluted from the C_{18} column as a peak with a retention time of 22 min; mp 97-99°C; ^{1}H NMR ($CDCl_3$) δ 7.22 (1H, br s, H-4), 7.35 (1H, br s, H-5), 7.97 (1H, br s, H-2), 8.71 (2H, br s, =NH); ^{1}H NMR ($CDCl_3$ + 1 drop of D_2O) δ 7.20 (1H, t, H-4), 7.26 (1H, t, H-4′), 7.29 (1H, t, H-5), 7.39 (1H, t, H-5′), 7.91 (1H, t, H-2), 8.02 (1H, t, H-2′); ^{13}C NMR ($CDCl_3$) δ 118.14 (imino carbon), 121.82 (C-4), 131.21 (C-5), 134.95 (C-2); IR (KBr) 3160, 3110, 1685, 1655, 1475, 1410, 1330, 1230, 1215, 1110, 1030, 960, 905, 830, 765, 650 cm^{-1}; UV_{max} (AcCN) 201 nm (ϵ 1.14 x 10^4), 235 nm (shoulder); mass spectrum, m/z 161 (M+), 147, 146, 120, 94, 93, 68, 41, 40. Anal. Calcd for $C_7H_7N_5$: C, 52.16; H, 4.38; N, 43.46. Found: C, 52.01; H, 4.48: N, 43.43.

N-Carboxamidoimidazole (3) was eluted from the C_{18} column as a peak with retention time of 56 min: mp 91-93°C; ^{1}H NMR (CD_3OD) δ 7.10 (1H, t, H-4), 7.67 (1H, br s, H-2), 7.74 (1H, t, H-5); IR (KBr) 3320, 3120, 1690, 1655, 1640, 1475, 1370, 1320, 1245, 1210, 1160, 1110, 1000, 875, 740, 685, 650 cm^{-1}; UV_{max} (H_2O) 205 nm (ϵ 3.9 x 10^3), 241 (2.4 x 10^3); mass spectrum, m/z 111 (M+), 69, 68, 41, 40. Anal. Calcd. for $C_4H_5N_3O$: C, 43.24; N, 37.83: Found: C, 43.10; H, 4.51; N, 37.66.

RESULTS AND DISCUSSION

EFFECT OF CHAIN LENGTH OF OLIGO(A)S ON THEIR COUPLING REACTIONS

Figure 1 shows the effect of the chain length of the starting oligo(A)s on their coupling reactions at 4, 20, and 40°C at an oligo(A):poly(U) ratio of 1:5. With chain lengths greater than three, the tetramer, pentamer, and hexamer gave optimal coupling at 20°C; the pentamer in particular gave the best yield. However, chain lengths greater than six gave lower yields. On the other hand, at 40°C the yield increased markedly for the pentamer and reached a maximum (about 60 percent) for the hexamer. No decrease in yield was observed at chain length greater than six. At 4°C, the coupling yield was very low and the trimer gave the best yield (20 percent). A little decrease in the yield was observed at chain lengths larger than three.

Sawai (1982) has reported that the T_m of $A(pA)_2 \cdot poly(U)$ complex in 0.1 M NaCl solution was 25.7°C and the T_m of $A(pA)_4 \cdot poly(U)$ complex was 40.2°C. A marked increase in coupling yields in the trimer at 20°C and in the pentamer at 40°C in Figure 1 indicates that the tightness of binding between oligo(A) and poly(U), in other words, the T_m value, relates to the coupling yield of oligo(A) in the presence of poly(U). Thus, it appears likely that the formation of a duplex between oligo(A) and poly(U) is necessary to achieve a satisfactory yield in the coupling reaction. The reason why the yield decreased with oligomers longer than $(pA)_7$ at 20°C might be due to the reduction of the ease of moving of oligo(A) on a poly(U) template. We mainly used both, $(pA)_4$ and $(pA)_6$, in the following experiments, because $(pA)_4$, $(pA)_5$ and $(pA)_6$ gave sufficient coupling yields at 20°C.

EFFECT OF THE CONCENTRATION OF POLY(U) ON THE COUPLING REACTION OF $(pA)_6$

The coupling yields of $(pA)_6$ were examined at different poly(U) concentrations varying from 0.0375 to 1.05 mM at 4, 25, and 40°C (Figure 2). The yield was dependent on the poly(U) concentration at 25 and 40°C. The yield increased with the increase in the poly(U) concentration. When 5-7 times molar excess of poly(U) to oligo(A) was used, the maximum yield was about 60-70 percent. However, there was no effect of the poly(U) concentration on the yield at 4°C.

When poly(A) and poly(U) are mixed in a ratio of 1:2, a poly(U)•poly(A)•poly(U) triple helix is formed because the adenine heterocyclic is simultaneously able to engage in Watson-Crick and Hoogsteen base pairs (Blake et al., 1967). The structure of the ternary complex consists of an antiparallel poly(A) poly(U) double helix with Watson-Crick base pairs and the third poly(U) strand running parallel to the poly(A) strand (Arnott & Selsing, 1974). In the present experiment, a Hoogsteen-type base pair might be formed in some places, when the poly(U) concentration was raised. At a $(pA)_6$ to poly(U) ratio of 1:5, $(pA)_6$ is supposed to hybridize partially to poly(U) as shown in Fig. 3. The high concentration of poly(U) would present many reaction fields in which oligo(A)s are able to couple with one another.

Next, we should inquire into the concept of "sliding." When the poly(U) concentration was 5-7 times that of $(pA)_6$, the best coupling yield was obtained. How could the hexaadenylate molecules, which might be thinly distributed on the poly(U) template, come close to each other and combine with one another with the aid of BrCN? Pörschke (1974) has reported that "slidomers" play an important role in the formation and dissociation of double-helical complexes of short complementary homopolymers. Chen et al. (1985) have also introduced 'sliding' to explain the formation of oligomers one unit longer than a template. They considered that newly formed oligo(G) should slide on the template of oligo(dC). We suppose that oligo(A) molecules should slide on a poly(U) template; when they come close to each other, the condensation with BrCN could occur. In the case of an A-U base pairing, the sliding seems to be easier than in the case of G-C pairing, because an A-U pair is formed by two hydrogen bonds, whereas a G-C pair is formed by three hydrogen bonds.

EFFECTS OF DIVALENT METALIONS ON THE COUPLING RESEARCH OF $(pA)_6$

Divalent metal ions such as Mn^{2+}, Co^{2+}, Ni^{2+}, Cu^{2+}, Zn^{2+}, Mg^{2+} and Fe^{2+} had an appreciable effect on the coupling reaction of $(pA)_6$ (Table I). The effectiveness of divalent metal ions on the overall yield of the coupling reaction was in the order $Mn^{2+} > Co^{2+} > Ni^{2+} > Fe^{2+} > Zn^{2+} > Mg^{2+} > Cu^{2+}$. Among the metal ions, Mn^{2+} gave the best yield of the coupling reaction. When metal ions were removed from the reaction mixture, the yield was only 0.5 percent, and the product contained only dimers.

FORMATION OF LINKAGE ISOMERS IN THE DIMERIZED PRODUCTS OF $(pA)_4$

The dimerized products of $(pA)_4$ were found to be linked by three kinds of phosphodiester bonds, 5′,5′-, 2′,5′- and 3′,5′-linkages. As these dimers were separately eluted from the CQA-30S column at 70°C, the molar ratio of the three linkage isomers could be determined by comparing the corresponding peak areas. We assumed that there is no difference among the molar absorption coefficients of the three dimers of $(pA)_4$ at 70°C, because intra- and intermolecular interactions of oligo(A), such as base stacking or hydrogen bonding, should be disrupted at 70°C. The reason why the three linkage isomers could be separated by column chromatography is as follows: when a mixture of poly(U) and oligo(A) was applied to a CQA-30S column and the elution of oligo(A) was performed, the poly(U) was kept in the column. A difference in the affinities of three dimers for the poly(U) led to good separation. When the poly(U) was digested with RNase A prior to the application, three dimers of $(pA)_4$ were eluted from the column at a lower salt concentration, and no definite separation was observed. Table II summarizes the molar ratio of the 2′,5′- and 3′,5′-phosphodiester bonds and the 5′,5′-pyrophosphate bond formed with BrCN in the presence of Ni^{2+}, the Co^{2+}, Zn^{2+}, Mn^{2+} and Cu^{2+}.

In the presence of Mn^{2+}, the formation of a 5′,5′-pyrophosphate linkage was predominant (71.6 percent), whereas in the presence of Ni^{2+}, Co^{2+} and Zn^{2+}, most of the resulting linkage was 2′,5′- and 3′,5′-phosphodiester bonds. In particular, it is noteworthy that Ni^{2+} gave a 3′,5′-linkage of 30 percent and a 2′,5′-linkage of 61 percent. The effectiveness of metal ions on the formation of a 3′,5′-linkage was in the order $Ni^{2+} > Cu^{2+} > Co^{2+} > Mn^{2+}$. On the other hand, the effectiveness of metal ions on the formation of a 2′,5′-linkage was in the order $Ni^{2+} > Co^{2+} > Zn^{2+} > Cu^{2+} > Mn^{2+}$. This was just the reverse order as that seen in the formation of the 5′,5′-linkage. Thus, divalent metal ions affected the formation of the linkage isomers markedly.

It has been shown that metal ions influence the conformation of nucleic acids. Shin (1973) demonstrated that divalent metal ions such Ni^{2+}, Co^{2+} and Zn^{2+} affected the formation of the ordered structure of poly(U). Of the divalent metal ions, Ni^{2+} was most effective in producing the ordered structure of poly(U). The stimulatory effect of metal ions on the formation of the ordered structure of poly(U) increased in the order $Ni^{2+} > Co^{2+} > Zn^{2+} > Mn^{2+} > Cu^{2+}$. This order was in good agreement with the stimulatory effect of the metal ions other than Cu^{2+} on the formation of 2′,5′- and 3′,5′-phosphodiester bonds. In particular, Ni^{2+} was most effective in both cases. This fact probably indicates that the structure of poly(U) as a template affected the formation of the three linkage isomers; that is, when poly(U) formed ordered structures, like a helix, it promoted the formation of 2',5'- and 3',5'-phosphodiester bonds. This speculation is quite reasonable, because a 2',5'- or 3',5'-linkage must be formed between two oligo(A)s which form a line, one behind another, on a poly(U) strand. On the other hand, in the case of the formation of a 5',5'-pyrophosphate linkage, one of the two oligo(A)s is unable to hybridize to the same poly(U) strand with an ordered structure, because two 5'-terminal phosphate groups of oligo(A) must face each other. There is the possibility of the formation of a pyrophosphate linkage when poly(U) strands assume the state of random coils.

Metal ions such as Zn^{2+}, Ni^{2+}, Co^{2+} and Mn^{2+} possess a fairly strong affinity for both the base and the phosphate moieties of nucleic acids. The extent of the conformational change of nucleic acids caused by metal ions depends on the relative intensity of the binding of the ions to the base and phosphate moieties. In the case of Cu^{2+}, the binding to the base moiety predominates and prevents the stacking of bases and the formation of ordered structures of nucleic acids (Eichhorn, 1973). In our experiments, Cu^{2+} gave the lowest yield in the coupling reaction of $(pA)_4$ with BrCN (Table I), and many unidentified peaks were observed on the RPC-5 column chromatogram (data not shown). The appearance of these peaks may be due to the abundance of poly(U) random structures caused with Cu^{2+}.

Mn^{2+} gave the best yield of the coupling reaction of $(pA)_4$; however, 72 percent of the dimerized products had a 5',5'-pyrophosphate linkage (Table II). Shin (1973) has shown that Mn^{2+} had little stimulatory effect on the formation of an ordered structure of poly(U). It might indicate that moderate disruption of the poly(U) ordered structure with Mn^{2+} caused an increase in the formation of pyrophosphate linkages. Bridson and Orgel (1980) reported that Zn^{2+} was an efficient catalyst for the oligomerization of guanosine 5'-phosphorimidazolide on a poly(C) template, and the oligomeric products were predominantly 3',5'-linked. This was consistent with the fact that Zn^{2+} occurs at an active site of DNA and RNA polymerase (Mildvan, 1974). In our experiments using BrCN, the molar ratio of the 3',5'-linkage formed in the presence of Zn^{2+} was only 17 percent (Table II). This was inconsistent with the result obtained by Bridson and Orgel (1980). The discrepancy may be due to a difference between G-C and A-U pairings and to a difference in interaction between nucleotides and metal ions.

EFFECT OF pH ON THE COUPLING REACTION OF $(pA)_6$

The effect of pH on the coupling reaction of $(pA)_6$ was examined in the range of pH 3.5-9.0; 0.2 M sodium acetate buffer was used in the range of pH 4.1-5.9; 0.2 M imidazole nitrate in the range of pH 6.0-7.0; 0.2 M Tris-acetate in the range of pH 7.8-8.7. The optimal pH was 6.25. The yield was low below pH 5.9, or above pH 7.8.

FORMATION OF N-CYANOIMIDAZOLE, N,N'-IMINODIIMIDAZOLE AND N-CARBOXAMIDOIMIDAZOLE IN A REACTION OF IMIDAZOLE WITH BrCN, AND THEIR CONDENSING ACTIVITIES

To investigate a possible mechanism for the coupling reaction of oligo(A) on a poly(U) template, we tried to find products in a reaction of imidazole with BrCN. The reaction products were analyzed by HPLC performed on a C_{18} column. Four peaks appeared in the chromatogram at 2.0, 5.5, 9.0, and 33.2 min. The four compounds were isolated from the reaction mixture by preparative HPLC, and identified by their structure analyses with Mass, IR and NMR spectra: Imidazole nitrate from peak 1 (retention time, 2.0 min); N-cyanoimidazole (1) from peak 2 (retention time, 5.5 min); N,N'-iminodiimidazole (2) from peak 3; N-carboxamidoimidazole (3) from peak 4 (retention time, 33.3 min). (Scheme 1)

It was confirmed that 1 and 2 were unstable and gradually converted to 3 in an aqueous solution. The conversion was promoted by an acid catalyst such as silica gel. Figure 4, shows the time course of the formation of 1, 2 and 3. 1 and 2 were formed rapidly from imidazole and BrCN in an aqueous solution. (Figure 4) Their formation attained a maximum in 5-10 h and then gradually decreased. 3 appeared after 5 h and its formation then gradually increased. 1, 2 and 3 were obtained from imidazole and BrCN in 25 percent, 39 percent and 5 percent yields, respectively, after the reaction for 25 h. About 70 percent of BrCN was converted to 1, 2 and 3.

Table III shows the condensing activity of 1, 2 and 3 for the coupling reaction of $(pA)_4$ and $(pA)_6$. 1 and 2 had much the same activity for the coupling reaction as BrCN on the presence of imidazole. 1 and 2 were less effective in the absence of imidazole. BrCN showed no activity in the absence of imidazole. 3 had no activity even in the presence of imidazole. 1 and 2 contain a carbodiimide bond (-N=C=N-). The same bond was found in water-soluble carbodiimides or cyanoderivatives, such as cyanamide and dicyandiamide known as prebiotic condensing agents for nucleic acid and peptide synthesis (Steinman et al., 1964, 1965; Schimpl et al., 1965). An active central carbon atom of carbodiimide can be easily attacked with a nucleophilic oxygen atom of a phosphate group to form a chemical bond between an oxygen and a carbon atom. The phosphorus atom could then be activated by releasing electrons to the carbodiimide.

A POSSIBLE MECHANISM FOR THE COUPLING REACTION WITH BrCN

N-Cyanoimidazole (1) was first formed from imidazole and BrCN in an aqueous solution. An addition of one more imidazole molecule to 1 resulted in the formation of N,N-iminodiimidazole (2). 1 and 2 had a condensing activity for the coupling reaction of oligo(A). It is estimated that BrCN cannot condense the nucleic acids in the absence of imidazole, because the total conversion yield of 1 and 2 from BrCN was about 70 percent, and the lack of imidazole in the reaction mixture failed to induce polymerization (Table III).

Discussing the mechanisms for the activation of a phosphate group, two reaction mechanisms can be considered: One is a direct addition of the condensing reagent 1 or 2 to the phosphate group (route 1). The phosphorus atom can then be activated by releasing electrons toward an imidazole group through the newly formed P-O-C bond (Figure 5). The mechanism could be the formation of phospho-imidazolide (route 2). We examined this formation during a reaction of pA with BrCN in imidazole nitrate buffer, pH 6.0. A small amount of adenosine 5'-phosphorimidazolide was formed, but only when a large excess of BrCN was used as a condensing agent. Thus, mechanism 2 cannot be excluded. Metal ions might contribute to the stabilization of the structure of poly(U) templates by binding to phosphate groups. In addition, metal ions might interact with protons of a 2'- or 3'-hydroxy group of the ribose moiety through a hydrogen bond, and increase the nucleophilicity of the oxygen atom (Figure 5). On the other hand, imidazole is the starting material of both 1 and 2, and it may also play a role similar to that of the metal ions. This consideration stems from the result that the lack of imidazole in the coupling reaction of oligo(A) with 1 and 2 caused a decrease in yield. Since imidazole molecules can accept and donate protons from one to another, it may help to pull out a proton form a 2'- or 3'-hydroxy group, when the oxygen atom of a hydroxy attacks an activated phosphorus atom.

A POSSIBLE ROLE OF HALOGEN AND HYDROGEN CYANIDE IN THE PREBIOTIC SYNTHESIS OF OLIGONUCLEOTIDES

We estimated the composition of chlorine-hydrogen chloride and bromine-hydrogen bromide in the primitive atmosphere (Yanagawa

& Matsuo, 1987). A ratio of partial pressure of hydrogen chloride and chlorine was calculated using equilibrium constants. In a low oxidation-state magma, the ratio of P^2_{HCl}/P_{Cl2} was 2300 at 1000 K and 490 at 1200 K. In a high oxidation-state magma, the ratio was 2.3 at 1000 K, and 0.78 at 1200 K. Therefore, when the primitive atmosphere was in equilibrium with a high oxidation-state magma at 1000 K and over, chlorine was predominant. The molar ratio of H_2O to chlorine was 109, and bromine to chlorine was 8.5×10^{-3}. The concentration of chlorine in the ocean can be estimated to be 0.5 M and bromine 2 mM.

Finally, we would like to discuss the possible role of halogen, such as chlorine and bromine, as a starting material of condensing agents for the prebiotic synthesis of oligonucleotides. Halogens such as chlorine or bromine reacted with hydrogen cyanide in aqueous solution to give cyanogen halides, such as cyanogen chloride or cyanogen bromide. The cyanogen halides reacted with imidazole to form N-cyanoimidazole, which was further converted to iminodiimidazole. Mononucleotides then reacted with iminodiimidazole or N-cyanoimidazole to give active intermediates such as isourea-type compounds and phosphorimidazolide, which could be converted to oligonucleotides. Therefore, if halogens such as chlorine and bromine and hydrogen cyanide were present in the primitive ocean, formation of oligonucleotides would occur.

REFERENCES

Arnott, S. and Selsing, E., 1974: J. Mol. Biol., 88, 509-521.

Blake, R. D., Massoulie, J. and Fresco, J. R., 1967: J. Mol. Biol., 30, 291-308.

Brahms, J. Michelson, A. M. and Van Holde, K. E., 1966: J. Mol. Biol., 15, 467-488.

Bridson, P. K. and Orgel, L. E., 1980: J. Mol. Biol., 144, 567-577.

Chen, C. B., Inoue, T. and Orgel, L. E., 1985: J. Mol. Biol., 181, 271-279.

Christie, S. M. H., Elmore, D. T., Kenner, G. W., Todd, A. D. and Weymouth, F. J., 1953: J. Chem. Soc., 2947-2953.

Eichhorn, G. L., 1973: in Inorganic Biochemistry, Eichhorn, G. L., ed., Elsevier, Amsterdam, 1210-1243.

Ferris, J. P., Edelson, E. H., Mount, N. M. and Sullivan, A. E., 1979: J. Mol. Evol., 13, 317-330.

Ferris, J. P., Hagan, W. J., Jr., Alwis, K. W. and McCrea, J., 1982: J. Mol. Evol., 18, 304-309.

Ferris, J. P. and Yanagawa, H., 1984: J. Org. Chem., 49, 2121-2125.

Giesemann, H., 1955: J. Prakt. Chem., 1, 345-348.

Hagan, W. J., Jr., 1985: Ph.D. Thesis, Rensselaer Polytechnic Institute, New York.

Kanaya, E. and Yanagawa, H., 1985: Nucleic Acids Res. Symp. Ser., 16, 181-184.

Kanaya, E. and Yanagawa, H., 19864: Biochemistry, 25, 7423-7430.

Mildvan, A. S., 1974: Annu. Rev. Biochem., 43, 357-399.

Mukai, J. I., 1965: Biochem. Biophys. Res. Commun., 21, 562-567.

Pörschke, D., 1944: Biophys. Chem., 2, 83-96.

Sawai, H., 1982: Viva Origino, 10, 102-103.

Schimpl, A., Lemmon, R. M. and Calvin, M., 1965: Science, 147 (Washington, D.C.), 149-150.

Shin, Y. A., 1973: Biopolymers, 12, 2459-2475.

Steinman, G., Lemmon, R. M. and Calvin, M., 1964: Proc. Natl. Acad. Sci. U.S.A., 52, 27-30.

Uchida, T., 1966: J. Biochem., 60 (Tokyo), 115-132.

Yanagawa, H. and Matsuo, S., 1987: To be submitted to the Origin of Life.

TABLE I: Effect of Metal Ions on the Coupling Reaction of $(pA)_6$[a]

metal ion	yield of oligo(A) (%)				total yield (%)
	12 mer	18 mer	24 mer	30 mer	
Mn^{2+}	24.6	3.9	0.8		29.3
Co^{2+}	16.2	1.9	1.1	0.3	19.5
Ni^{2+}	11.5	4.0	1.9		17.4
Cu^{2+}	4.9	3.5	1.0		9.4
Zn^{2+}	7.9	2.6			10.5
Mg^{2+}	7.4	1.3	0.8		9.5
Fe^{2+}	5.8	4.5	1.2		11.5
metal free	0.5				0.5

[a]The reaction mixture (20 μL) contained 0.15 mM $(pA)_6$, 0.15 mM poly(U), 200 mM imidazole nitrate buffer, pH 6.0, 15 mM BrCN and 10 mM divalent metal ions, and was incubated at 25°C for 6 h.

TABLE II: Effect of Metal Ions on the Formation of Linkage Isomers of Octaadenylate[a]

metal	linkage isomers (%)			$\frac{2'\text{-}5' + 3'\text{-}5'}{5'\text{-}5'}$
	5'-5'	2'-5'	3'-5'	
Ni^{2+}	8.7	61.3	30.0	10.0
Co^{2+}	23.2	58.9	17.9	3.3
Zn^{2+}	39.5	44.0	16.5	1.5
Mn^{2+}	71.6	22.4	6.0	0.4
Cu^{2+}	48.5	30.5	21.1	1.1

[a]The reaction mixture (1.2 mL) contained 0.15 mM $(pA)_4$, 0.75 mM poly(U), 10 mM divalent metal ion, 200 mM imidazole nitrate buffer, pH 6.0, 15 mM BrCN, and was incubated at 20°C for 20 h.

TABLE III: The Coupling Reaction of $(pA)_4$ and $(pA)_6$ with BrCN, N-Cyanoimidazole and N,N'-Iminodiimidazole in the Presence or Absence of Imidazole.[a]

coupling agent	imidazole	yield (%)	
		from $(pA)_4$	from $(pA)_6$
BrCN	+	54.1	67.6
	-	0	0
N-cyanoimidazole	+	60.8	77.3
	-	44.0	47.5
N,N'-iminodiimidazole	+	64.5	63.5
	-	18.5	13.7
N-carboxamidoimidazole	+	0	0

[a]The reaction mixture (20 μL) contained 0.15 mM $(pA)_4$ or $(pA)_6$, 0.75 mM poly(U), 10 mM $MnCl_2$, 200 mM imidazole nitrate buffer, pH 6.0 or 200 mM sodium acetate buffer, pH 6.0, 10 mM BrCN, N-cyanoimidazole, N,N'-iminodiimidazole, or N-carboxamidoimidazole, and was incubated at 20°C for 20 h.

SCHEME 1.

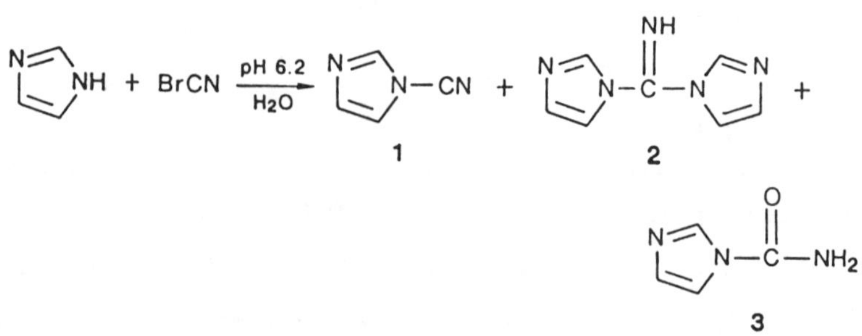

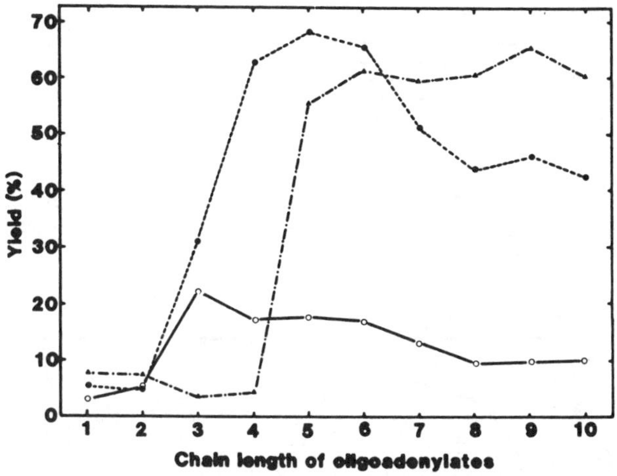

Figure 1. Effect of chain length of oligo(A)s on their coupling reactions. The reaction mixture (20 μL) contained 0.15 mM (pA)n (n=1-10), 0.75 mM poly(U), 50 mM $MnCl_2$, 200 mM imidazole nitrate buffer, pH 6.0, and 50 mM BrCN, and was incubated for 20 h at 4°C (o), 20°C (O), or 40°C (Δ).

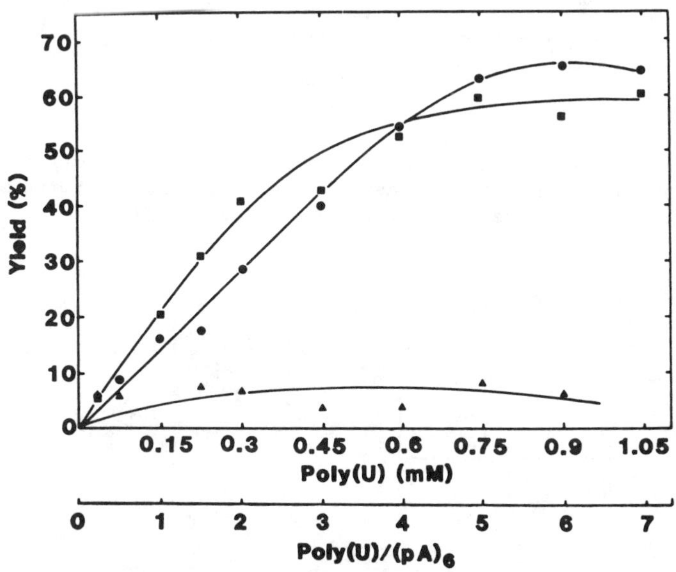

Figure 2. Effect of poly(U) as a template on the coupling reaction of $(pA)_6$. The reaction mixture (20 μL) contained 0.15 mM $(pA)_6$, 0.038-1.05 mM poly(U), 10mM $MnCl_2$, 200 mM imidazole nitrate buffer, pH 6.0, and 10 mM BrCN, and was incubated for 20 h at 4°C (Δ), 25°C (□), or 40°C (Δ).

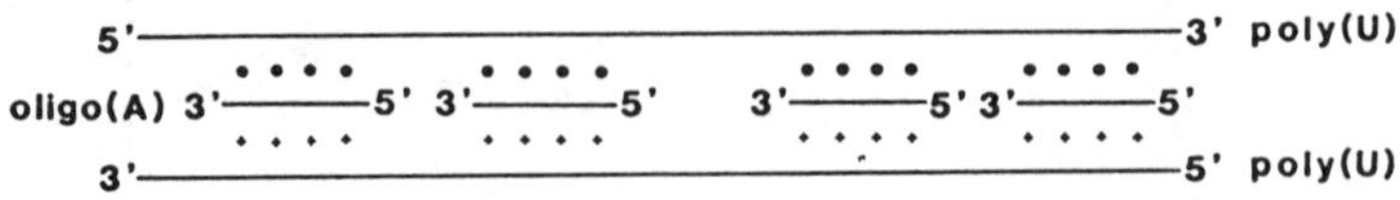

Figure 3. A possible interaction between oligo(A) and poly(U). ...., Watson-Crick base-pairs; ++++, Hoogsteen base-pairs.

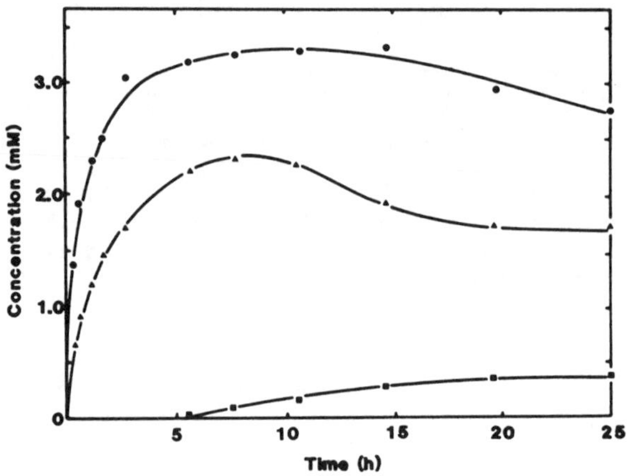

Figure 4. Time course of the formation of imidazole derivatives from imidazole and BrCN in an aqueous solution. N,N'-iminodiimidazole (O), N-cyanoimidazole (Δ), and N-carboxamidoimidazole (□).

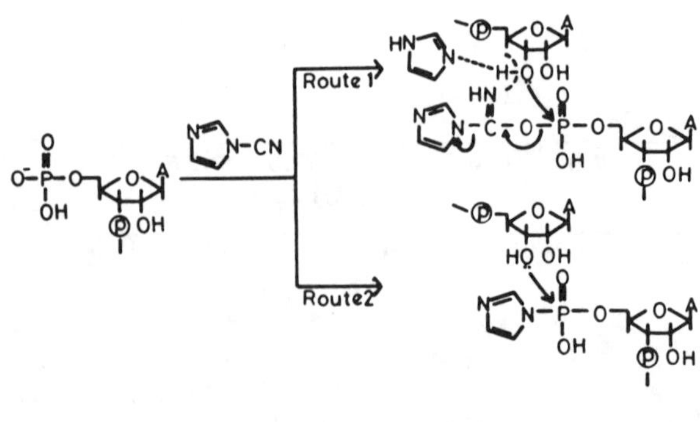

Figure 5. A possible reaction mechanism for the formation of the activated phosphate derivatives and the phosphodiester bond.

ARE RIBONUCLEOSIDES SPECIAL?

David Usher
Cornell University
Ithaca, New York 14853

ABSTRACT

Opinions differ as to whether nucleotides and nucleic acids would have formed easily on prebiotic Earth. If they are uniquely suitable for their job, then there would have been a strong selection for their synthesis, even though it may have been difficult. On the other hand, if other structures can be identified that do the job as well, then perhaps the nucleotides and nucleic acids were used because they were relatively abundant. The latter case would suggest that the correct choice of experimental conditions would lead to rediscovery of a ready synthesis of nucleotides and nucleic acids. Perhaps the nucleotides and nucleic acids were used because they were relatively abundant.

PREBIOLOGICAL SELF ORGANIZATION OF MATTER
C. Ponnamperuma and F.R. Eirich (Eds.)

ISBN 0-937194-18-2

DIRECT INTERACTIONS BETWEEN AMINO ACIDS AND NUCLEOTIDES AS A POSSIBLE EXPLANATION FOR THE ORIGIN OF THE GENETIC CODE

Nalinie Senaratne
Mitchell K. Hobish
Cyril Ponnamperuma
Laboratory of Chemical Evolution
University of Maryland
College Park, Maryland 20742

ABSTRACT

A study of the associations of amino acids and nucleotides was undertaken with a view to establish a possible physical basis for the origin of the codon assignment. The associations of homocodonic amino acids (L-Phe, L-Lys, L-Pro and Gly) and some hetercodonic amino acids (L-Ala, L-Arg, L-Cys, L-Glu, L-Met and L-Tyr) with various nucleotide sequences (nucleoside 2'-, nucleoside 3'-, or nucleoside 5'-monophosphates; dinucleoside (2'-5')-, or dinucleoside (3'-5')-monophosphates; and trinucleoside (3'-5')-diphosphates) were studied using 1H NMR spectroscopy. The strongest associations of all the amino acids investigated were observed with their respective anticodonic nucleotide sequences. The strength of association was seen to increase significantly with the increase in the chain length of the anticodonic nucleotide, from monomer, dimer through trimer. The conclusion reached as a result of these studies is that there is a likely anticodonic basis for the origin of the genetic code.

PREBIOLOGICAL SELF ORGANIZATION OF MATTER
C. Ponnamperuma and F.R. Eirich (Eds.)

ISBN 0-937194-18-2

ON A CO-EVOLUTION OF POLYPEPTIDES AND NUCLEIC ACIDS AND LIPOSOME ASSEMBLY

Frederick Eirich
Polytechnic University
333 Jay Street, Brooklyn, New York 11201

and

University of Maryland
College Park, Maryland 20742

ABSTRACT

This paper argues that neither the retropolation of the family trees of living organisms or of microfossils, nor the progress of experiments on chemical evolution, including the synthesis of most molecular building blocks of life under prebiotic conditions, promise a rapid progress in our understanding of what the first living forms may have been. Thus, in addition to these two approaches, one should begin a serious study of the development of life-mimicking chemical automatons. The self-synthesis of polypeptides and polynucleotides under plausible conditions appears to be within reach, and there exists an interesting proposal for a cyclic mechanism. Research on such reactions should help to establish the principles of the evolution of self-replicating macromolecular systems. Since spontaneous macromolecular association processes are beginning to be understood and the formation of vesicles and of their inclusion of macromolecules has been observed, work on the development of cellular automatons has become promising. The required steps are being discussed. Corresponding experiments would offer an operational avenue to solving the question whether the known physical and chemical principles suffice for a de novo formation of life, or at least of life mimicking systems.

INTRODUCTION

The rapidly increasing understanding of the world around us reveals greater complexity, but also more unifying principles than was ever imagined. This is particularly true for the riddle of the emergence of life. The foundations for an explanation were laid when single atoms were discovered to unite to molecules in interstellar space, and organic molecules were found in meteorites and comets as

ISBN 0-937194-18-2

well as in the laboratory under presumed prebiotic earth conditions, when it was shown that life is compatible with irreversible thermodynamics, and when common ground between genetics and information theory was discovered. Lastly, biochemistry, molecular biology, virology and genetic engineering proved that at least the elementary processes of life can be understood in chemical terms. However, the complexity of the chemistry and morphogenesis of living systems is so great that we can study life only by taking it apart, thereby destroying the very cooperative organization that constitutes the essence of life. Thus, the transition from inanimate to living systems is far from being experimentally accessible. We can merely speculate whether the occurrence of life was a chance event, or was inevitable as inherent in the blueprint of the periodic table, or both.

REPLICATING AUTOMATONS

Biological and paleontological extrapolation towards the earliest forms of life on the one hand, and prebiotic molecular evolution on the other, have therefore a long way to go before we may contemplate a synthetic approach to life as we find it today. It seems more expeditious to look for general laws of molecular organization and to try to construe simple analogs of self-maintaining and replicating chemical automatons. Preferably, though not necessarily, such automatons should be modeled along the lines of biochemistry, using the nucleotides which form the genetic alphabet for coding and replication, model polypeptides for catalytic action, simple phospholipids for membrane functions, and (sugar) phosphates as energy source and as markers. With their help one should attempt to design relatively simple cyclic processes which create molecular populations capable of some sort of evolution. Should we succeed in constructing a system that works like an analog, or fossil, of prebiotic reactions, we would have a better basis for further studies into the critical transition from non-living to living systems (Eirich, 1986).

CO-EVOLUTION

The intimate interaction between proteins and nucleic acids in our biology prompts one to believe that a chemical relation between them, or between their oligomeric building blocks, preceded the onset of life. Modelling a co-evolution of polypeptides and nucleic acids would be one way to explain the "chicken and egg" problem, i.e. that randomly evolving oligonucleotides happened to aid the synthesis of polypeptides which happened to catalyze the evolution of the nucleic acids; or, vice versa, how polypeptides catalyzed polynucleotides which accidentally carried information on the amino acid sequences. More likely would have been an evolution of polypeptides and nucleic acids during which the formation of one affected the order within the other. However, the chemistry of such a process, while thermodynamically and kinetically possible, has yet to be established.

Many workers (Westheimer, 1986; Gilbert, 1986; Zaug and Cech, 1986; North, 1987) are persuaded that the key to the chemistry on which life on earth is based were RNA's primitive informational molecules. This argument was reinforced by the finding that RNA also showed activity as ribozymes. This overlooks that, to be the key to life, RNA must have invented the relevant information, or received it from some template, leaving the old problem of the origin of the code. Moreover, the catalytic ability of ribozymes appears to be restricted to ester exchanges. Yet the key arguments against an "RNA-world" are that the current findings of ribozymic activity are not reactions under presumed prebiotic conditions, but the in vitro observations on biological material in contemporary biological processes, such as the excisions of introns in tetrahymena ribosomal RNA, these intron's catalytic effects on cytidylic acid oligomer exchanges, or the ribozymic release of t-RNA's from their parent RNase-P. Thus, despite the value of ribozymatic action for the progress of bioengineering, speculations on its effectiveness for protein syntheses must be considered unfounded. Last not least, the prebiotic synthesis in vitro of the nucleotides, and their simulated prebiotic oligo-, or polymerization (Orgel, 1986), involve some unresolved steps, Shapiro (1986), as does the development of optical uniformity. See Dyson (1985), for error-prone and-tolerant all-protein cell.

Amino acids on the other hand, were certainly prebiotically available and were likely to polymerize under drying conditions. The multifaceted structure and flexibility of polypeptide chains make it likely that among the possible isomers some will have developed folding which affected the chemistry of cosolutes catalytically. It is thus justifiable to assume, based on observed affinities (Senaratne, 1986) that activated anticodonic nucleotides (see below) became associated alongside polypeptides and that they were then polymerized (see also Yanagawa, this issue) in situ during drying or freezing cycles. Such polymerizates would be partly ordered as a result of the foregoing association, and be subsequently re-arranged by transesterification under the influence of attached oligo amino acids. Alternatively, during drying, the amino acids could have formed mixed carboxy-phophoanhydrides with e.g., glycerin nucleotides (Schwartz 1986). Thus activated, the amino acids could have reacted as proposed by the feedback mechanism of Figure 1, shown below.

In the search for the experimental self-generation and evolution of nucleic acids, Eigen and co-workers (1983) elaborated Spiegelman's

(1965) simulation in vitro of the evolution of a viral RNA with the help of its Qβ-replicase. Eigen's proposal for a replication machine (Kuepers, 1983) is based on an information transfer from the Qβ-replicase to the growing RNA, as in Spiegelman's study. However, whether the replicase was conjugated with very small amounts of nucleotides, had its information content previously received from a nucleic acid, or was organized otherwise, the information on the sequence of RNA's was transmitted by the replicase. It may then be not too far-fetched to expect that information flow of such nature may have been more widespread in the past, i.e. that amino acid sequences were able, through coupled or sequential reactions, to lay down information on their order in form of sequences of the nucleotides of nucleic acids. Such a finding would amount to a natural explanation for the genetic code.

A mechanism constituting a co-synthetic process has been postulated by Lacey et al. (1977), Figure 1, and been discussed further by Saxinger and Ponnamperuma (1981) and Lacey et al., (1985, 1986). Concerning the postulated steps, it was assumed that the tendency towards molecular association between amino acids and their anticodonic nucleotides (Seneratne, 1986) helped the formation of their respective acylates during dehydration cycles of their solutions (Honda, 1986), and that subsequently these acylates polymerized.

Lacey et al. (1977) did not indicate a mechanism by which the amino acid acylates might generate polypeptides and their polynucleotides. but one possibility consists of the fast, clay catalyzed, polymerization of amino acid adenylates and guanylates, Figure 2, discovered by Paecht, Berger, and Katchalsky (1970). Such polycondensation of activated amino acids, adsorbed or in solution, could have been random but more likely was responsive to natural inter-amino acid affinities, as proposed by Fox (1980) and others (Mutter, 1974; Hartmann et al., 1984). We explored the details of the clay induced mechanism together with M. Paecht-Horowitz (Eirich, 1981; Paecht, 1984; Paecht and Eirich, 1988), including the mode of copolymerization and the role of different clays. However, the ordered polymerization of the nucleotide moieties of the acylates of Figure 2, which would implement the postulates inherent in Figure 1, has been established only to a limited extent on preformed nucleotide templates (Lohrmann and Orgel, 1979; Bridson and Orgel, 1980; Orgel, L. E., 1986). One might conceive a less specific mechanism than ordered polymerization by which the nucleotide moieties of Figure 1 stayed sufficiently long adsorbed to yield nucleic acids of the right numerical composition. Given the self-replicating and self-splicing tendencies of RNA (Cech, 1981; Westheimer, 1986) and the ordering influence of polypeptides through their association with nucleotides (Seneratne, 1986), the evolution of today's anticodonic relationships would have been possible.

Tyagi et al. (1986) will report during this Colloquium on their failure to find enhanced polymerization. Their experiments indicate that the expected collection and alignment of amino acid nucleotides alongside their complementary nucleic acids may not have taken place. Although thus, only 2 of the 5 essential steps of Figure 1 have been made likely, Lacey's scheme remains so far the only one which allows a reciprocal exchange of information between peptides and nucleotides, as well as their cyclic replication. Also, the secondary peptides of Fig. 1 could play the role of primeval RNA polymerases which has been argued for by Lazcano (1986) and Lazcano et al. (1986), or together with nucleotides could act as primitive ribosomes.

Figure 1, allows one also to speculate, in as much as enzymes today are catalyzing by surrounding their substrates in chemically active pockets, that amino acid acylates could become adsorbed around a substrate before their polymerization. An example of a pocket in two dimensions was given by Follmann (1981) in which a polypeptide sequence of 15 amino acids forms a catalytic fold. Thus, co-adsorption of substrates and amino acid nucleotides and the latter's polymerization would point to a 2-dimensional origin of a linear code for developing enzymatic activity in 3-dimensions (Eirich, 1986, 1987).

POSTULATED CELL EVOLUTION

The chemistries embodied in Figures 1 and 2 point out possibilities how the major building blocks of our life, including its genetic information system, could have self-assembled on the basis of inherent chemical affinities. Together with the accumulated knowledge of macromolecular chemistry, biochemistry and molecular biology, this should enable us to account for the processes that brought the emerging macromolecules together and allowed them to engage in further catalysis. It is therefore now possible, see *a* to *f* below, to update past postulates (Haldane, 1954; Oparin, 1961; Bhargava, 1986) of the steps of molecular self-assembly, and also of the formation of cellular units (Deamer and Barchfeld, 1982) which were required for the appearance of individuality and the premises of Darwinian evolution.

a. Linear flexible macromolecules have the tendency to entangle, aggregate and thus to engage in concerted interaction even in the absence of specific forces.

b. Single, as well as entangled macromolecules assume a huge number of conformations, among them some prone to possessing pockets of specific steric or chemical functions (Challa, 1983). Permanent conformations of single chains, or of partly ordered coacervates or microgels may be the result of ion pairing, of hydrogen-, or polar-bonding, of hydrophobic clustering, or of the oxidative joining of 2 juxtaposed -SH groups.

c. Some of the coacervating-aggregating macromolecules are polypeptides and polynucleotides preformed, e.g., by amino acid nucleotide polymerization on interfaces, as in Figures 1 and 2.

d. Macromolecular chains, such as polypeptides form mixed micelles with surfactants including phospholipids. Micelles of the latter become liposomes at the right pH's, Fig. 3.

e. During their formation, or on drying and re-hydration, the liposomes enclose cosolutes, or may engorge macromolecular aggregates, Fig. 4.

f. The liposomic membranes are capable of active ion transport (Wilson et al., 1976) and catalytic and autocatalytic cycles by virtue of macromolecular polypeptide channels; thus, they become semipermeable for nutrient molecules and waste products. The minimum requirements for autopoiesis and for establishing an energy gradient have been discussed by Fleischaker and Margulis (1986) and by Tabushi (1986), and the preparation and properties of active synthetic cells by Okahata (1987).

REVIEW OF SOME RELEVANT EXPERIMENTS ON MACROMOLECULAR INTERACTIONS

The following is an account of the author's and his coworkers' results on some aspects of the processes a - d.

Macromolecules, when entangled or adsorbed at several points along their chains, take a long time to separate again, even in the absence of other than van der Waals forces. The slowness of reaching equilibrium after dilution, and the slowness of their desorption (Simha et al., 1953; Koral et al., 1958) are the result of the improbability of a simultaneous multiple dissociation. The attainment of equilibrium following macromolecular association and folding by specific attractive forces, may also take a very long time. Eventually, the positions of free energy minima of all components, including ions, other cosolutes and water, will be attained (Thomas, 1971) by place exchange, sliding, or inch-worming off, or on, the chains.

The relative position of the chains oscillating around a dynamic average, can be determined by marking the end-groups (Shiah and Morawetz, 1984; Chang and Morawetz, 1987). Aggregating polypeptides reach geometries in accordance with their longitudinal order (sequences), ranging from linear and sheetlike arraignments (Brack,

1981) to loops and helices. Adsorbed, they may be present as coils with long pending tails (Scheutjens et al., 1986). Thus, the formation of catalytically active pockets (see Figure 5) that act on substrates that happen to diffuse into the chain aggregates, must be a frequent occurrence. It follows also that macromolecules, entangled or adsorbed on interfaces or membranes, are more likely to be catalytically active than their single, freely dissolved, members of the species.

When charged macromolecules, i.e. polyelectrolytes, meet, pairing of the opposite charges occurs. Left-over charged groups on one molecule remain available for pairing with third macromolecules, or with small ions. Any groups that were blocked beforehand, say by bound ions, are unavailable, unless they produce ion pairs that lower the free energy of the whole system. Should the charge-paired aggregates be limited -, or in-soluble, which depends on their relative structures, on the ionic strength and on the presence of other solutes, coacervates or microgels are being formed. These processes were studied in detail with G. Thomas (1971) (see Figure 6) and S. Bekelnizky (1974). See also Koetz et al. (1986).

The pairing of polyampholytes occurs strictly between their net charges. Thus, when different kinds of electrically neutral ampholytes are mixed, they do not interact though each may carry dozens of positively and negatively charged groups; but net oppositely charged polyampholytes, e.g., one polypeptide at $pH<pK$ with another one at $pH>pK$, or with polynucleotides at $pH>6$, interact rapidly until optimal charge compensation is achieved, Figure 7. Katcher (1964) has also shown how two different gelatins will interpenetrate rapidly their molecular coils, or into the top layer of their gels. Again, the number of residual net charges, by modifying a given hydrophobic-hydrophilic balance, decides whether the compensating polyampholyte aggregates stay soluble, or form gels. When the flexibility of the polyelectrolytes (ampholytes) permits the hydrophobic groups to associate and to form islands that are surrounded by charged groups, rather tight spheroidal conformations of single macromolecules, or aggregates, may result, yet remain soluble. Tight conformations were also observed following the interaction in water of polar, but themselves electrically neutral, macromolecules with electrolytes, e.g. poly(vinylpyrrolidone) (PVP) with small negative ions such as I_3 (Barkin, 1955), or azodyes (Frank, 1957).

We investigated (Rosen et al., 1951; Fishman et al., 1975) PVP, or poly(acryl amide) as polypeptide models for their interactions with cosolutes. PVP adsorbs negative surfactants, e.g., sodium dodecyl sulfate, and turns into a polyelectrolyte. Surfactants may become activated by macromolecules so that the onset of their, usually mixed, micellization advances to lower concentrations. Depending on the relative amounts of the partners during the stages of micelle formation, a macromolecular chain may act as a nucleus, may transgress the micelles of a surfactant, or be folded and contained within the liposome and/or in the latter's bilayer membrane, see again Figure 3.

DISCUSSION

Equilibrium aggregates, coacervates, microemulsions, micelles, liquid crystals, etc., are individually organized condensed microphases. Normally they exist only in a specific range of temperature and concentration (activity) prior to the onset of crystallization in 3-dimensions, but may become stabilized, e.g., by charges or by adsorbates. Liposomes are bilayered closed spheroidal sheets derived from micelles at pH's of about the halfpoint of dissociation of the micelle-former. As microphases, liposomes are in dynamic equilibrium with their environments, capable of incorporating or losing membrane-soluble molecules (Adamson, 1967), and of contracting or expanding in size as a result of the interplay of osmotic pressure, molecular supply, and electrostatic and molecular free energy. Liposomes enclose the solution in which they are formed and thus contain also their cosolutes. The latter, however, in contrast to when they are free in solution, come under geometrical constraint by the bounding membrane (Fendler, 1982) and may form internal coacervates or ordered gels, especially when the liposomes shrink or dry.

Baeza, Deamer, Oro, and coworkers demonstrated (1980, 1982, 1986) that macromolecules may be included in liposomes during drying and rehydration cycles. This is illustrated in Figure 4a, which is Figure 2 of Deamer and Barchfeld (1982). One may assume that amino acids, nucleotides and bentonite-type clay particles were present during prebiotic freezing or drying cycles. Oligopeptides and their anticodonic oligonucleotides were produced independently, but later taken up by the liposomes, Figure 4b, within which they formed coacervates and catalyzed the formation of ATP, polypeptides and nucleotides. The result of such events would be life mimicking cellular automatons. Speculating on further developments, provided that their membrane permitted an ongoing inward diffusion of CO_2, H_2S, and of nutrients such as phosphorylated glucose (whose prebiotic synthesis was studied by Degani and Halman, 1972), and of nucleotides, amino acids, phosphates and ions, the liposomes themselves will grow anisotropically because of the linear growth of their contents. The membrane will become anisodiametrically extended and, as an ellipsoidal thin shell, will turn unstable and form a waist. Division of the cell follows. When the daughter cells undergo the same type of growth and division, while subject to molecular evolution of their macromolecular contents, a family of related cells would be generated. Molecular evolution, actually a misnomer (Margulis, 1972), is the result of a competition between rates of synthesis and decomposition of replicating molecules, which in open solution leads merely to the production of a more and more uniform collection of copies of equivalent molecules. True evolution needs the selective survival of suitable members of a population (Oparin, 1972; Oro, 1980; Ferris and Usher, 1983). Cellular development may well have been the result of a see-saw between a Darwinian evolution of cell populations and molecular evolution of groups of polypeptides and polynucleotides inside cells (Dyson, 1986).

Returning to v. Neumann's concept (1966) of automatons which consist of a factory, a set of instructions, a transcriber and a controller, as long as they are mechanical, they must have been prefabricated and primed into action. A series of chemical processes, however, can develop and prime themselves. Reactions between units of reagents go on unobserved all the time as the result of the collisions between molecules of a given activation energy and molecules of a sufficiently energized population. They remain inconsequential unless they initiate a chain of steady state reactions. Thus, as spontaneous dissipation processes that exist since the emergence of atoms in our cosmos, which must have prevailed in the thin broth of high-energy compounds on primitive Earth, unbounded chemical reactions may have played the role of delocalized factories, and their products have acquired the qualities of instructions, transcribers and controllers, but were too dispersed to engage in orderly, or feed-back functions. Enclosed, the same reactions and their products became sufficiently concentrated to support a subsequent chemistry whose results eventually reinforced or controlled the first reaction, creating autocatalytic cycles that lasted as long as reagents were supplied. Within the confined volume of a liposome, such cycles established individual cellular automatons. It may well be that the development of spatially enclosed chemical loops which triggered protein production and nucleic acid coding and replication, marked the transition from non-living to living matter.

ADDENDUM

Parenthetically, the problem of the self-organization of dissipative structures within disordered, chaotic systems has been explored by mathematical (bifurcation) theories and been applied to fluid dynamics, steady state chemical reactions, and problems of theoretical biology (P. H. Anderson, 1983; Holden, 1986; Nicolis and Prigogine, 1977). Ocean currents, weather cyclones, the Benard phenomenon of the development of convection cells in heated fluid layers, population dynamics, and the Zhabotinsky-Belousov reaction are examples of (according to their boundary conditions) deterministic, periodic or aperiodic structured fluxes; thus, spontaneously ordered molecular entities, functioning as "hardware", may have been able to facilitate steady state chemistries.

Ordering, in dynamic equilibrium with unstructured phases, occurs at constant temperature during the formation of condensed units, e.g., the formation of aggregates, micelles, and liquid and solid crystals. Since the loss of entropy equals the enthalpy losses, the free energy of the self-ordering mass stays unchanged. The net out-flow of (latent) heat from the system does not in general start chemical reactions. The development of ordered reactions may, however, be assisted when the newly ordered phase erects constraints on potential reagents. Examples are: epitactic crystallization, chemical reactions induced by a heterogeneous catalyst, polymerization within a gel which inhibits the diffusion of the growing chain ends, and encapsulated enzymes for which the membrane permits the inward flow of the substrate and the outflow of the reaction products. If the enclosure is the result of the phase change when micelle formers reach the CMC, the act of micellization may switch the spatially chaotic reaction of the previously dispersed enzyme into locally (i.e. within the micelle or liposome) constrained reactions that continue a dissipative flux of energy and mass. Order out of disorder thus appears whenever condensed phases emerge. Within their confines, the latter may then steer reactions into going into their steady states.

CONCLUSION

In conclusion, one may postulate that the self-synthesis, breakdown, and re-synthesis of organic compounds on primitive Earth led to the formation of activated amino acids, to oligo- and polypeptides and their clusters with some catalytic activity. We may further assume that these prebiotic reaction mixtures became enclosed in phophatidyl liposomes, enhancing molecular ordering and cyclic reactions as forerunners of early cell metabolism. Independently, during heating and freezing cycles in the presence of catalytic surfaces, amino acids were activated, e.g., by phosphoderivatives of anti-codonic nucleotide bases. Their polycondensation, and the partial replication of the resulting polypeptides and nucleotides amounted to the first primitive imprinting and information transfer. Liposomes that happened to enclose proto-enzymatic oligo amino acids as well as replicative components, could begin to evolve.

Thus, besides the ongoing work and efforts that go into studying a nucleic acid replication without proteins that may never have happened (one might suppose that otherwise viruses should be able to replicate without biogenic material), one should emphasize two other approaches. Our results point to the need of studying potential activities of polypeptides within liposomes, and likewise the coupled or sequential selfsynthesis of polypeptides and nucleotides on solid surfaces. Success in the creation of the resulting automatons and the study of their functions would offer a promising avenue to answering the question whether the known physical and chemical principles suffice to construct life mimicking systems.

REFERENCES

Adamson, A. W., 1967: "The Physical Chemistry of Surfaces," 2nd Ed., Interscience-Wiley, New York.

Anderson, P. H., 1983: "A Model for Prebiotic Evolution: The Use of Chaos," Natl. Acad. Sci. USA, 80, 3386.

Baezo, I., Ibanez, M., Santiago, J. C., Wong, L., Lazcano, A. and Oro, J., 1986: Adv. in Space., 6, No. 11, 39.

Barkin, S., Frank, H. P. and Eirich, F. R., 1955: Ricerce Scientifica, Suppl., 25, 844.

Bhargava, P. M., 1986: Adv. in Space., 6, No. 11, 7.

Bekelnitzky, S., 1974: Dissertation, Polytechnic Institute of New York, Brooklyn, New York.

Brack, A., 1981: in "Origins of Life," Wolman, Ed., D. Reidel Publ., Dordrecht and New York, 487.

Bridson, P. K. and Orgel, L. E., 1980: Journal Mol. Evol., 14, 243.

Cech, T. R., Zaug, A. J. and Grabowsky, P. L., 1981: Cell, 27, 487.

Challa, G., 1983: J. Molecular Catalysis, 21, 16.

Chang, L. P. and Morawetz, H., 1987: Macromolecules, 20, 428.

Degani, Ch. and Halmann, M., 1972: Nature (New Biology), 235, 171.

Deamer, D. W. and Barchfeld, G. L., 1982: J. Mol. Evol., 18, 203.

Deamer, D. W. and Oro, J., 1980: Biosystems, 12, 167.

Dyson, F., 1986: in "The Origins of Life," Cambridge University Press, Cambridge.

Eigen, M., 1983: in "Evolution from Molecules to Man," Bendall, D. S., ed., Cambridge University Press, Cambridge, 105.

Eirich, F. R., 1981: in "Origins of Life," Y. Wolman, ed., D. Reidel Publ., p. 296.

Eirich, F. R., 1986: Adv. Space., 6, 3.

Eirich, F. R., 1987: J. Polymer Sci., 75, 219.

Fendler, J. H., 1982: "Membrane Mimetic Chemistry," Wiley Interscience, New York, Chapter 6.

Ferris, J. P. and Usher, D. A., 1983: in "Biochemistry," G. Zuleay, ed., Addison-Wesley, Reading, Massachusetts, 1191.

Fishman, M. L. and Eirich, F. R., 1971, 1975: J. Phys. Chem., 75, 3135; 79, 2740.

Fleischaker, G. R. and Margulis, L., 1986: Adv. in Space., 6, No. 11, 53.

Follmann, H., 1981: "Die Chemie and Biologie der Evolution," Quelle and Meyer, Heidelberg, 115.

Fox, S. W., 1980: Naturwissensch., 67, 378.

Frank, H. P., Barkin, S. and Eirich, F. R., 1957: J. Phys. Chem., 61, 1375.

Gilbert, W., 1986: Nature, 319, 618.

Haldane, J. B. S., 1954: New Biology, 16, 12.

Hartmann, L., Nawroth, F. and Dose, K., 1984: in "Origins of Life," Dose, Schwartz, Thiemann, eds., D. Reidel Publ., 213.

Helmann, M., Sanchez, R. L., Orgel, L. F., 1969: J. Org. Chem., 34, 3702.

Holden, A. V., ed., 1986: "Chaos," Princeton University Press, New Jersey.

Honda, Y., 1986: Laboratory of Chemical Evolution, University of Maryland, private communication.

Katcher, J., 1964: Thesis, Polytechnic Institute, Brooklyn, New York.

Koetz, J., Linow,K. J., Burkart, Ph., Hu, L. P. and Vogl, O., 1986: Polymer, 27, 1574.

Koral, J., Ullman, R. and Eirich, F. R., 1958: J. Phys. Chem., 62, 541.

Kuepers, B-O., 1983: "The Molecular Theory of Evolution," Springer-Verlag, Berlin.

Lacey, J. C., Jr., Weber, A. L. and Pruitt, K. M., 1977: in "The Encyclopedia of Ignorance," Duncan, R., Webster-Smith, M., eds., Pergamon Press, New York.

Lacey, J. C., Jr., Hale, L. M., Mullins, D. W., Jr. and Watkins, C. L., 1985: Origins of Life, 16, 151.44

Lacey, J. C., Jr., Mullins, D. W., Jr. and Watkins, C. L., 1986: J. Biomol. Struct. and Dynam., 3, 783.

Lazcano, A., 1986: in "Origina de la Vida," Lazcano, Guerrero, Margulis, eds., Trabella Soc., Catalana Biologica, 39.

Lazcano, A., Valverde, V., Fastag, J., Gariglia, D., Ramirez, C. and Oro, J., 1986: This Volume.

Lohrmann, R. and Orgel, L. E., 1979: Journ. Mol. Evol., 14, 243.

Margulis, L., 1972: in "Exobiology," C. Ponnamperuma, ed., North Holland Publ. Co., Amsterdam, 342.

Mutter, M., 1974: Internat. Journal Peptide and Protein Res., 13, 274.

V. Neumann, J., 1966: "Self-Reproducing Automata," University of Illinois Press.

Nicolis, G. and Prigogine, I., 1977: "Self-organization in Non-Equilibrium Systems," Wiley Interscience, New York.

North, A., 1987: Nature 328, 18.

Okahata, Y., 1987: "Current Topics in Polymer Science," Ottenbrite, Utracki, Inoul, eds., Hanser Publ., Munich, Vienna, New York, p. 299.

Oparin, A. I., 1961: "Life: Its Nature, Origin, and Development," Oliver and Boyd, Edinburgh.

Oparin, A. I., 1972: in "Exobiology," C. Ponnamperuma, ed., North Holland Publ. Co., Amsterdam, 1.

Orgel, L. E., 1986: Journ. Theor. Biology, 123, 127.

Oro, J., 1980: "The Origin of Life and Evolution," H. O. Halvarsen, K. E. V. Hole, eds., A. Liss, Inc. Publ., New York, p. 97.

Paecht-Horowitz, M., 1983: Origins of Life, 14, 307.

Paecht-Horowitz, M., Berger, J. C. and Katchalsky, A., 1970: Nature, 228, 639.

Paecht-Horowitz, M. and Eirich, F. R., 1988: Origins of Life, 18, 359.

Rosen, B., Kamath, P., Eirich, F., 1951: Faraday Society Discussions, No. 11, 135.

Saxinger, C. and Ponnamperuma, C., 1981: J. Mol. Evol., 1, 63.

Scheutjens, J. M. H. and Fleer, G. J., 1985: J. Colloid Interf. Sci., 111, 446.

Schwartz, A. W., 1986: Adv. in Space Res., 6, No. 11, 23.

Seneratne, S. M. N., 1986: Dissertation, University of Maryland.

Shapiro, R., 1986: "Origins," Summit Books, New York.

Shiah, T. P. J. and Morawetz, H., 1984: Macromolecules, 17, 792.

Simha, R., Frisch, H., Eirich, F. R., 1953: J. Phys. Chem., 57, 541.

Spiegelman, S., Haruna, I., Holland, I. B., Beaudreau, G. and Mills, D., 1965: Procced. Nat. Acad. Sci., USA, 54, 919.

Tabushi, I., 1986: "Minimum Requirements of Single Cell Activity," in Advances in Space Research, Vol. 6, No. 11, 45.

Thomas, G. B., 1971: Dissertation, Polytechnic Institute of Brooklyn, New York.

Tyagi, S., this issue.

Westheimer, F. H., 1986: Nature, 319, 534.

Wilson, T. H. and Maloney, P. C., 1976: FASEB Proceed., 35, 2174.

Winfree, A. T., 1984: "Rotating Chemical Reactions," Scient. Amer., 23330, 82.

Zaug, A. J. and Cech, T. R., 1986: Science 231, 470.

Figure 1. Schematic representation of processes through which a system of amino acids and nucleotides could generate and replicate information do novo as a result of intermolecular affinities. Adopted from Lacey et al., 1977.

A. Amino acids are activated by reaction with specific nucleotides; e.g. ATP activates the hydrophobic amino acids, GTP activates hydrophilic, and UTP activates charged amino acids. The proposed activated forms are the aminoacyl nucleotidyl mono-or-diphosphate anhydrides in which the carboxyl group of the amino acid is attached to the terminal phosphate of the nucleotide, e.g. Phe-AMP.

B. After having been adsorbed on 3-layer clay the amino acids of the nucleotide acylates polymerize at neutral pH in the order of their adsorption, yielding an oliogopeptide; the nucleotides polymerize at acid pH in their order to generate a nucleic acid which is anticondonically related to the oligopeptide.

C. When those oligonucleotide chains collect complementary nucleotides and the latter become polymerized, codonic chains are produced.

D. When the latter collect the complementary acylates, and they polymerize, the original oligopeptides and anticondonic nucleic acids (of B) are reproduced.

Whenever the adsorptive order of the original acylates is influenced by co-adsorption around S, the subsequently produced nucleic acids would code for S-specific oligopeptides.

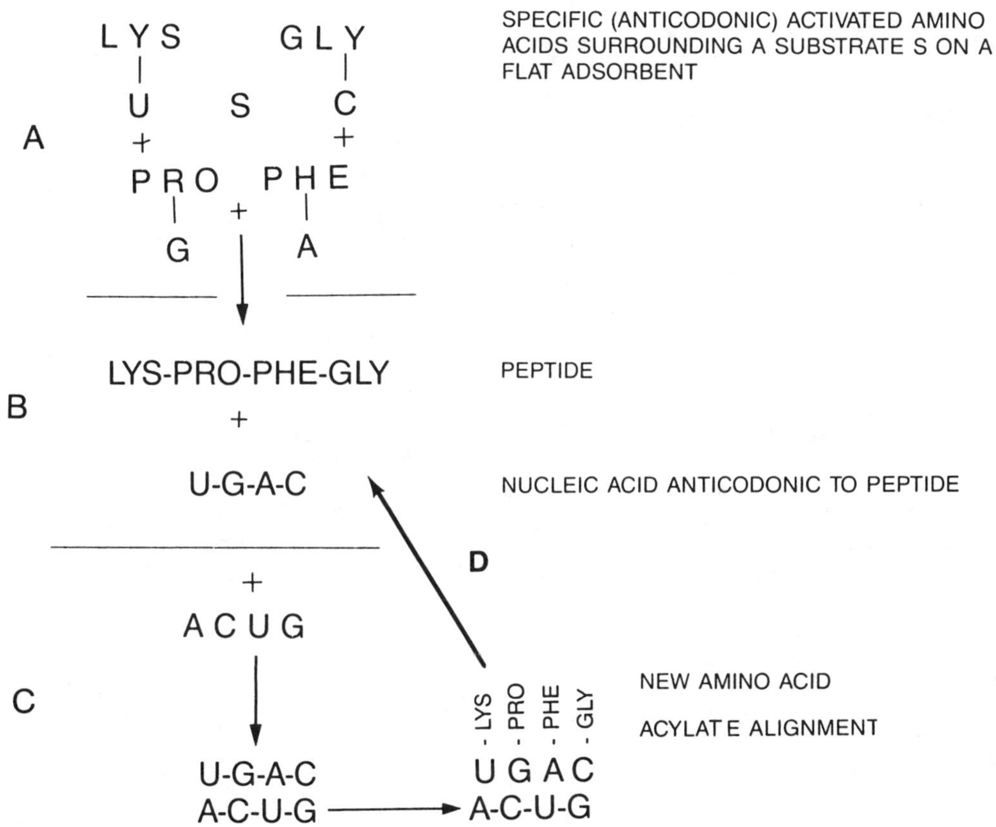
LYS
GLY
U
S
C
A
PRO
PHE
G
A
SPECIFIC (ANTICODONIC) ACTIVATED AMINO ACIDS SURROUNDING A SUBSTRATE S ON A FLAT ADSORBENT
LYS-PRO-PHE-GLY
B
PEPTIDE
U-G-A-C
NUCLEIC ACID ANTICODONIC TO PEPTIDE
D
ACUG
C
- LYS - PRO - PHE - GLY
NEW AMINO ACID
ACYLAT E ALIGNMENT
U-G-A-C
A-C-U-G
U G A C
A-C-U-G

FIGURE 1

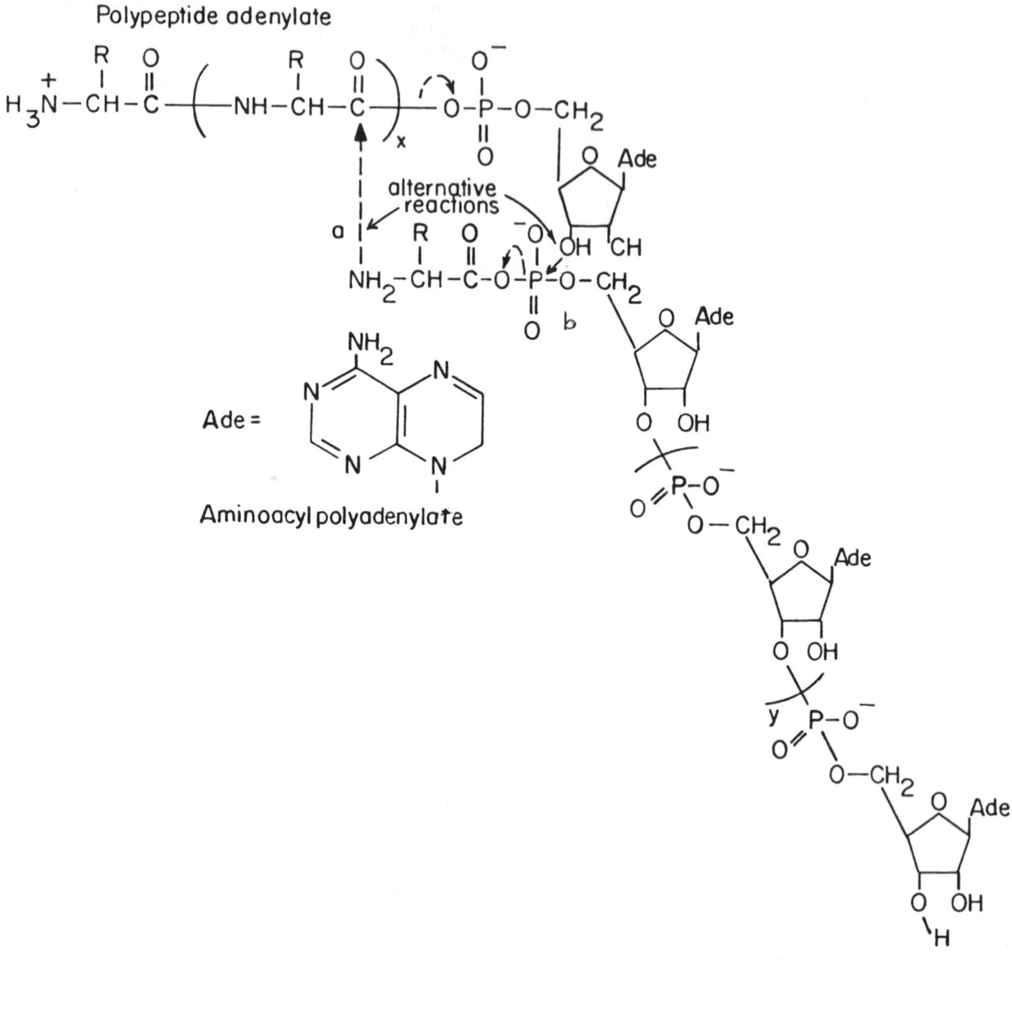

Figure 2.

A. Reaction path of amino acid adenylates adsorbed on clay by nucleophilic attach onto the carbonyl carbon atom of the anhydride bridge. Oligoamino acids are generated and AMP displaced.

B. The hypothetical alternative reaction path, when the 3′-O- become involved in nucleophilic attacks on the P-atoms, and oligonucleotides would form.

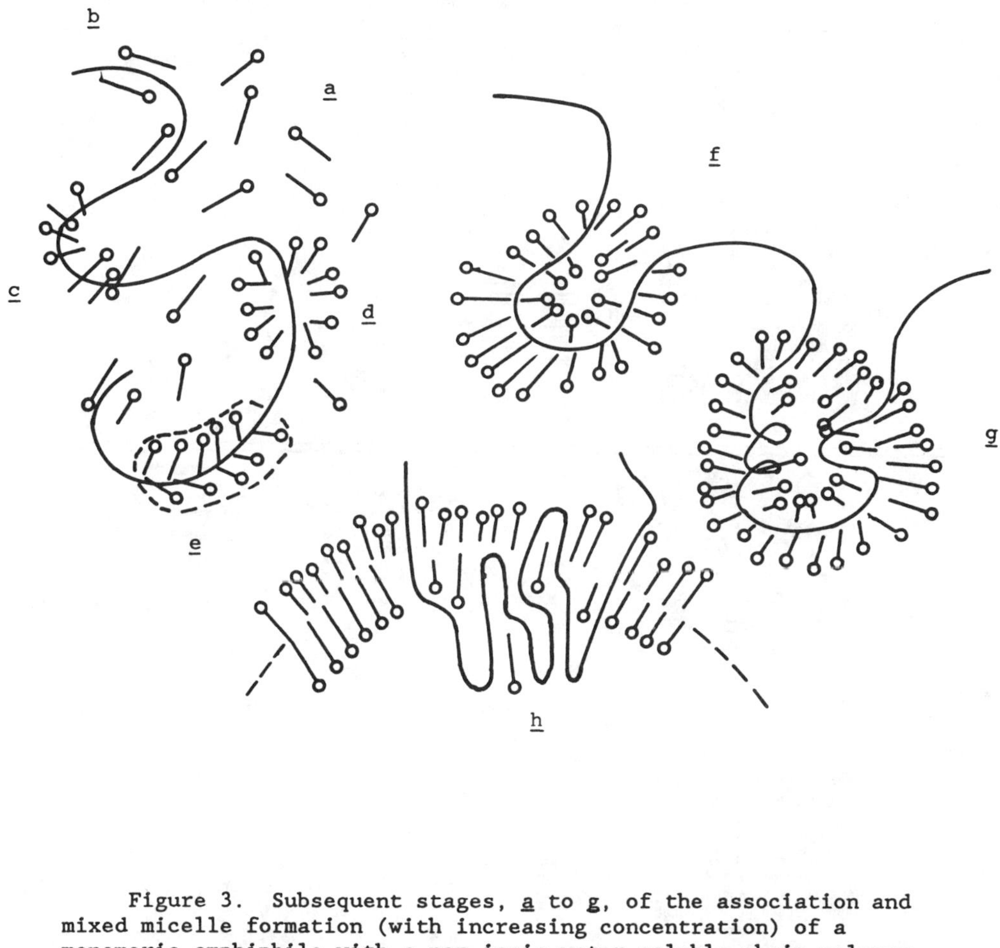

Figure 3. Subsequent stages, a to g, of the association and mixed micelle formation (with increasing concentration) of a monomeric amphiphile with a non-ionic water soluble chain polymer (e.g., PVP) leading to a liposomic vesicle, h, with polymeric chains embedded in the bilayer membrane.

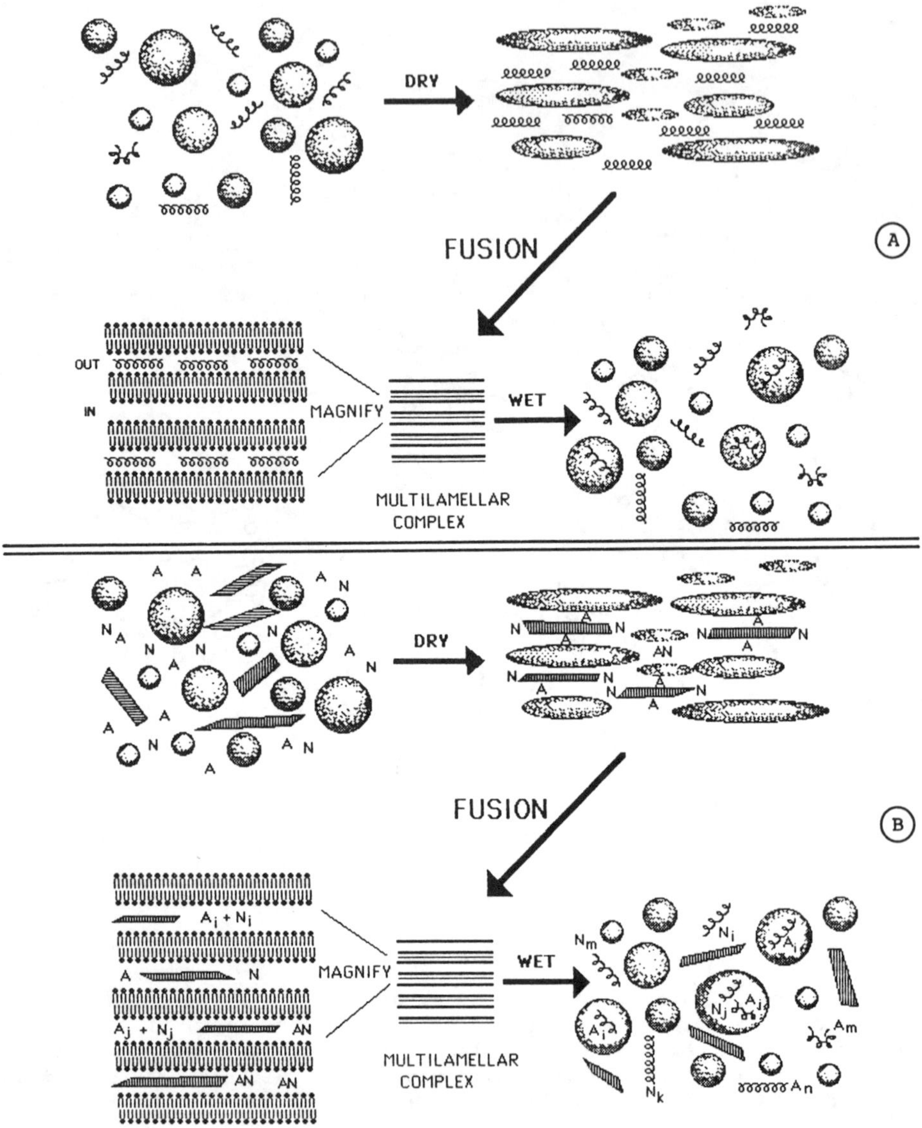

Figure 4.

A. Reproduction of figure 2 of the referenced paper by Deamer and Barchfeld (1982), which shows those author's concept of the uptake of macromolecules into rehydrating liposomes.

B. Additional diagram. proposing how during drying cycles in the presence of clay, amino acids and nucleotides could have produced the anticodonic oligopeptides and -nucleotides according to <u>our</u> Figure 1, which were then enclosed in liposomes according to Deamer and Barchfeld.

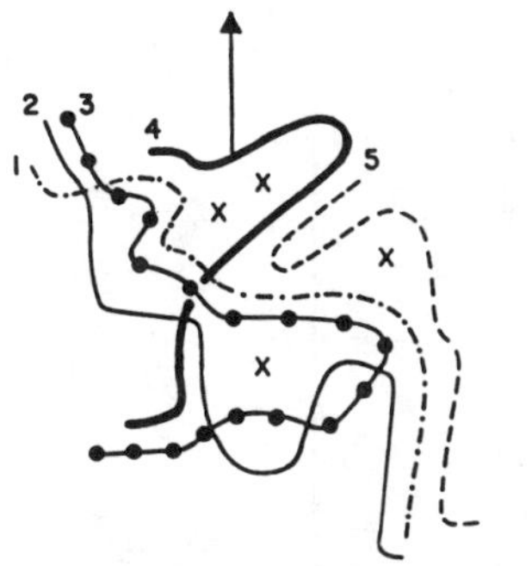

Figure 5. A schematic 2-dimensional representation of a cluster of five chain molecules. The parallel regions symbolize ordered ranges such as helices, or β-sheets. The loops symbolize coiled ranged which form pockets such as those shown, +. Every pocket provides a special environment that may act catalytically for incoming molecules.

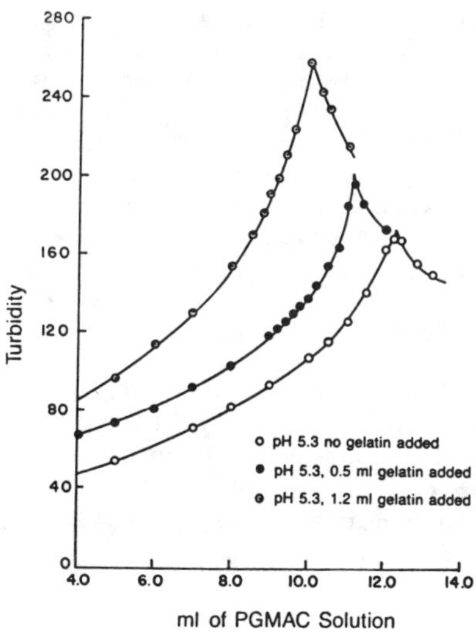

Figure 6. Course of Solution Turbidity during Titration of 15ml of $7.3.10^{-5}$ equ. poly (styrene sulfonate) sodium salt (PSSNa) with a solution of $8.1.10^{-5}$ equ. poly (trimethylammonium hydroxy propylmethacrylate (PGMac). 0. 0.6 and 1.2ml of P-1 gelatine of 4.10^{-5} equ. positive charge added show a that the gelatin effectively neutralizes the negative polyelectrolyte, and b that the gelatin joined the polyelectrolyte complex and reduced its solubility.

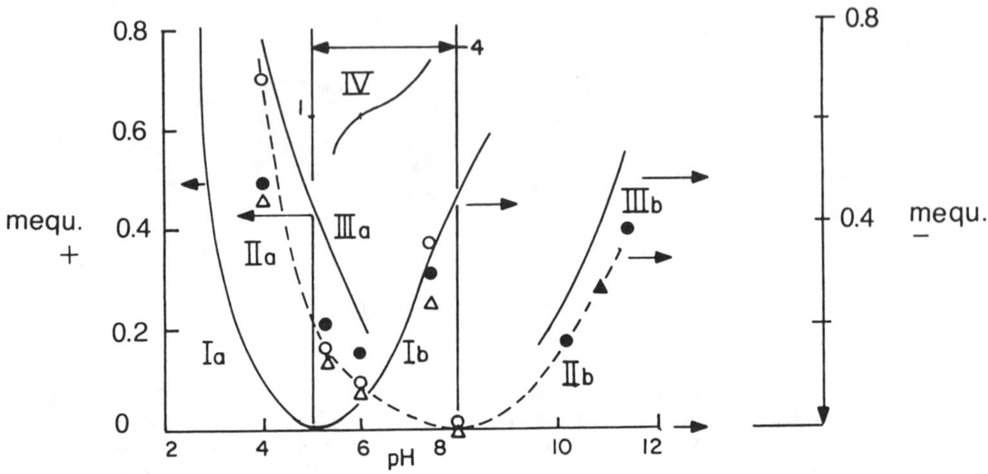

Figure 7. Positive and Negative net charges on C-1 and P-1 gelatins in millimoles per g gelatin*, at 25°C, as f(pH) determined by different methods: curves la and lb, the charges on C-1; curves IIa and IIb, on P-1. Open points (o), determined turbidimetrically; full points (•), potentiometrically; triangles (Δ) calculated from amino acid composition (these points are shown only for P-1). Curves IIIa and IIIb, were obtained from titrations with Na-poly(styrene-sulfonate) and poly(styrene/vinyl benzene ammonium chloride), respectively. Curve IV shows the relative amounts of P-1 to cause the turbidity maximum when added to C-1. The equivalent point in g, of the gelatins is shown as one.

*C-1 is an alkali-treated calfskin gelatin of M.W. = 30,000, and pI=5. P-1 is an acid treated pigskin gelatin of M.W. of about 300,000 and pI≈8. The measurements were carried out in 1% gelatin solutions by addition of 0.1 N NaOH or HCl. Reaction between the two gelatins occurs only between the respective isoelectrical points, i.e., between pH 5 and 8.

SIMULTANEOUS ORIGIN OF POLYPEPTIDES AND POLYNUCLEOTIDES: HYDROGEN CYANIDE POLYMERS AS PREBIOTIC CONDENSING AGENTS

Clifford N. Matthews
University of Illinois at Chicago
Chicago, Illinois 60680

ABSTRACT

Evidence from laboratory and extraterrestrial chemistry suggests that hydrogen cyanide polymerization is a truly universal process yielding conjugated ladder structures as well as polyamidines readily converted by water to polypeptides. This plausible route for the origin of proteins in the reducing environment of primitive Earth could also have led to the appearance of the first nucleic acids. In the absence of water -- on land -- polyamidines could have been the original condensing agents, directing the synthesis of nucleosides and nucleotides from available sugars, phosphates and nitrogen bases. Most significant would have been the parallel synthesis of polypeptides and polynucleotides arising from the dehydrating action of polyamidines on nucleotides:

A-A-A-A-A-A-A polyamidines		P-P-P-P-P-P-P polypeptides
+	---------->	+
N N N N N N N nucleotides		N-N-N-N-N-N-N polynucleotides

On our dynamic planet, this polypeptide-polynucleotide symbiosis mediated by polyamidines may have set the pattern for the evolution of protein-nucleic acid systems controlled by enzymes, the mode characteristic of life today.

ISBN 0-937194-18-2

INTRODUCTION

It is widely assumed that the prebiotic formation of primitive proteins occurred in two steps: α-amino acid synthesis brought about by the action of natural high energy sources on the components of a reducing atmosphere, followed somehow by polycondensation of the accumulated monomers in the oceans, or on land (Miller, 1984). The pioneering demonstration by Miller and Urey (Miller, 1953; Urey, 1952) that α-amino acids are readily obtained from methane, ammonia and water subjected to electric discharges, taken with subsequent synthesis of peptides from amino acids under many sets of conditions (Miller, 1984), appear to be in accord with this view. A more critical examination of the experimental evidence, however, raises many questions concerning selection, concentration, purification and interaction of such products. How plausible are these attempted simulations as models of prebiotic chemistry? On primitive Earth, the inherent thermodynamic barrier to spontaneous polymerization of α-amino acids might not have been so easily overcome.

An alternative pathway for the origin of proteins proposed by Matthews and coworkers (Matthews *et al.*, 1966, 1967) bypasses this problem by postulating the direct synthesis of polypeptides from hydrogen cyanide and water without the prior formation of α-amino acids. Polymerization of hydrogen cyanide is the key step. In this paper, I discuss the chemistry of cyanide polymers, their probable extraterrestrial existence, and their possible role in the events leading to the origin of life on Earth.

HYDROGEN CYANIDE POLYMERS: SYNTHESIS AND STRUCTURE

Liquid HCN polymerizes spontaneously to a brown-black solid at room temperature in the presence of a base such as an amine or ammonia (Matthews *et al.*, 1966, 1967, 1984). Polymerization also occurs readily in non-aqueous solvents, or in water (Matthews *et al.*, 1966, 1967, 1984; Völker, 1960). Two types of structural units appear to be present in these solid materials. Most stable are the ladder polymers shown in Figure 1, formally derived from the olefinic tetramer of HCN, diaminomaleonitrile (A), which is usually found among the products. Probably, however, polymerization to the substituted polymethylene (B) proceeds by way of an HCN dimer (Völker, 1960). Cyclization to (C) and (D) then leads to ladder structures possessing conjugated -C=N- bonds, as proposed by Völker (1960) on the basis of extensive physical and chemical investigations which have been recently supplemented by the work of Umemoto *et al.* (1986).

More controversial is the existence of the polyamidine structures shown in Figure 2 (Matthews et al., 1966, 1967; Ferris, 1979). Although polyaminomalononitrile (F) can be formally derived by addition polymerization of the reactive trimer aminomalononitrile (E), it seems again more likely that polymerization occurs through an HCN dimer (Matthews et al., 1966, 1967; Kliss et al., 1962). Further reactions of HCN with the highly activated nitrile groups of (F) yield the heteropolyamidines (G), which are readily converted by water to heteropolypeptides (Matthews et al., 1966, 1967). Overall, this series of reactions constitutes a route for the direct synthesis of heteropolypeptides without the intervening formation of α-amino acids.

Evidence for this cyanide model of the origin of proteins has been provided by several kinds of experiments (Matthews, 1984, 1975). In general, water-soluble, yellow-brown solids can be extracted from the products of each of the following types of reactions:

1. base-catalyzed polymerization of liquid HCN, alone, in water, or in non-aqueous solvents (Matthews et al., 1967);

2. electric discharge experiments producing HCN from methane-ammonia mixtures (Matthews et al., 1966);

3. alkaline hydrolysis of aminoacetonitrile (Moser et al., 1968), aminomalononitrile (E, HCN trimer) (Moser et al., 1968) and diaminomaleonitrile (A, HCN tetramer) (Moser et al., 1968), all of which are ready sources of HCN at high pH;

4. HCN modification of the reactive nitrile side chains of poly-α-cyanoglycine, a polyamide analog of polyaminomalononitrile (F) (Minard et al., 1975).

As predicted, acid hydrolysis of these yellow-brown polymers yields not just glycine, the major product, but other α-amino acids as well, such as alanine, aspartic acid, glutamic acid, serine, threonine and leucine, together with some α-amino acids not found in proteins. Also as predicted, the glycine is perdeuterated when D_2O/DCl is used for hydrolysis instead of H_2O/HCl (Matthews et al., 1977), an indication of the highly acidic nature of the methine group of (F), and of the great reactivity of its nitrile side chains.

Comparable polypeptides with molecular weights of 4000-20000 daltons, obtained by Draganic, Niketic et al. (1980, 1984) in their studies of ^{60}Co irradiation of aqueous cyanide solutions, yielded up to 20 percent amino acids after hydrolysis. The polymers gave positive biuret tests and could be hydrolyzed by proteolytic enzymes. Although radical mechanisms were proposed for this γ-radiation chemistry, it seems reasonable to suppose that polyaminomalononitrile

(F) would have been an important intermediate, as in the thermal polymerization reactions.

Infrared analysis of these polymeric HCN products shows the expected peaks for hydrocarbon (≈3μ), nitrile (4.5 μ) and amide (≈6μ) groups. Further non-destructive analysis became possible with the advent of cross-polarization magic-angle spinning solid-state NMR (^{13}C and ^{15}N) (Matthews _et al._, 1984; Schaefer _et al._, 1982; McKay _et al._, 1984). In particular, the unambiguous presence of secondary amide groups, as in peptides, has been established by double-cross-polarization studies on polymers synthesized from equimolar amounts of $H^{13}CN$ and $HC^{15}N$. Also significant is the NMR detection of a substantial peptide presence in the water-insoluble material, as well as in the aqueous extracts. This suggests that some peptides are attached to the other HCN polymers, probably through their amino groups. Taken together, these GCMS, IR and NMR studies point to the presence of peptide polymers, free or hybrid, in these HCN products.

HYDROGEN CYANIDE POLYMERS IN THE SOLAR SYSTEM

Most striking have been the direct images obtained by the Vega (Sagdeev _et al._, 1986) and Giotto (Reinhard, 1986) spacecraft showing that the nucleus of comet Halley is covered with a surface layer as black as any body in the solar system. This non-volatile insulating crust of dark material may be largely composed of hydrogen cyanide polymers and related compounds (Matthews _et al._, 1986, 1987), since the expected predominance on cometary nuclei of frozen volatiles -- methane, ammonia, and water -- subjected to high energy sources makes them ideal sites for the formation and condensed-phase polymerization of HCN. Dust emanating from the nucleus, contributing to the coma and tail, would also arise partly from the polymer. Results of the recent Halley missions support this as follows:

1. For the first time, the nucleus of a comet was detected unambiguously. Following Vega observations (Sagdeev _et al._, 1986) of a surface largely masked by clouds, the Giotto multicolor camera showed the presence of a potato-shaped core (≈15 x 10 km) with a very low albedo (2-4%) (Reinhard, 1986; Matthews _et al._, 1986). Most of the cometary activity appeared to originate from a few discrete sources on the nucleus, notably two major, bright dust jets.

2. Giottos' neutral mass spectrometer confirmed that Halley's exhalations consisted of water vapor (at least 80% by volume) together with methane, ammonia, and carbon dioxide (Krankowsky, 1986). The infrared observations of Vega 1 (Combes _et al._, 1986)

showed the surface temperature to be above 300°K, suggesting that the frozen volatiles were covered with a layer of debris perhaps 1 cm thick.

3. Among the most prominent emissions in the near ultraviolet and visible spectrum of Halley were those of the CN radical (Moreels et al., 1986; Levasseur-Regourd et al., 1986). Ground-based studies of Halley showed the presence also of HCN (Schloerb et al., 1986). From their investigations of CN jets in the inner coma, A'Hearn et al. (1986) conclude that, while some of the CN radicals of Halley might be split off from HCN, other CN sources must exist, tied to very small grains.

4. Mass spectra from Halley's coma sent back by the Vega (Kissel et al., 1986) and Giotto (Kissel et al., 1986) missions showed most of the cometary dust grains to be rich in H, C, O, and N, indicating the presence of polymeric organic material. The close approach of Giotto to the dust-covered nucleus enabled particle types to be distinguished. In addition to silicate grains and particles of mixed composition, there appear to be populations consisting only of H, C, O, N, and others only of H, C, N (Clark et al., 1986). Both kinds could be closely related to HCN polymers.

The possibility of an extensive cyanide chemistry on Halley and a myriad other comets in the Oort cloud strongly reinforces the view (Matthews, 1984, 1985) that HCN polymerization is a truly universal process, a preferred pathway within the solar system and beyond. HCN polymers appear red in the photovisual spectral region (0.3 - 1.0 μm) (Cruikshank, 1986) and may account for the dark red colour of D-type asteroids and other outer solar system bodies. VJHK photometry by Cruikshank et al. (1985) indicates that the bare nucleus of Halley and other comets resembles such asteroids (Hartmann et al., 1982). Carbonaceous chondrites from the asteroid belt might therefore, also possess HCN products. Studies of the Murchison meteorite have shown that free α-amino acids are present (Kvenvolden et al., 1973; Ponnamperuma, 1983) together with acid-labile amino acid precursors of undefined structure (Cronin, 1976). Deuterolysis of these water-soluble yellow-brown solid extracts yields perdeuterated glycine as would be expected from peptide segments derived from HCN polymers (Matthews et al., 1980). Hydrogen cyanide polymerization could account, too, for much of the yellow-brown-orange coloration of Jupiter and Saturn. Indeed, HCN has been found in Jupiter's reducing atmosphere and in the atmosphere of Titan, the largest moon of Saturn. Most intriguing is an orange haze high in Titan's stratosphere that may consist of organic polymers (Owen, 1982). These aerosols could be polycyanides formed directly from HCN (Matthews, 1982). While life would not be expected to arise in such environments, the visible cyanide chemistry on these bodies is a continuing reminder that hydrides of the elements oxygen, carbon and nitrogen are a ready source of prebiotic molecules, as Urey had originally suggested. In particular, the ubiquitous presence of HCN polymers invites the reexamination and possible reinterpretation of almost all studies concerned with the origin of proteins. Which came first (Figure 3), amino acids or their polymers?

HYDROGEN CYANIDE POLYMERS AND THE ORIGIN OF LIFE

In the Miller-Urey experiment, for example, it seems from our reinvestigations that the primary products were not α-amino acids, as claimed, but rather HCN polymers, the HCN being formed from methane and ammonia by electric discharge reactions and by elimination from intermediates such as aminoacetonitriles and HCN oligomers. The polymers then became hydrolyzed to amino acids either during reflux in the reaction flask, or later during the working up procedure. The same conclusion applies to virtually all simulations of the chemistry of primitive atmospheres, as well as to studies of aqueous cyanide chemistry by Oro (1980), Ferris (1984) and others. In our view, these investigations ostensibly yielding α-amino acids actually supply evidence for the abundant prebiotic existence of polymeric protein ancestors -- heteropolypeptides synthesized directly from hydrogen cyanide and water.

The ready conversion by water of polyamidines to polypeptides demonstrated by these investigations suggests that the polyamidines I and II might have played a further essential role in chemical evolution (Matthews, 1986). In the absence of water -- on land -- they could have been the original condensing agents of prebiotic chemistry. Their reactive amidine groups, liable to become amides, would have brought about the stepwise formation of nucleosides, nucleotides and polynucleotides from available sugars, phosphates and nitrogen bases. Most significant would have been the parallel synthesis of polypeptides and polynucleotides arising from the dehydrating action of polyamidines on nucleotides:

```
A-A-A-A-A-A-A        P-P-P-P-P-P-P
polyamidines         polypeptides

     +      ------->     +

N N N N N N N        N-N-N-N-N-N-N
nucleotides          polynucleotides
```

Optimum conditions might well have existed on a gradually warming Earth where prebiotic synthesis of organic molecules proceeded in the atmosphere in three overlapping stages defined by the relative volatility of methane, ammonia and water (Kliss *et al.*, 1962). First, hydrocarbons were formed following the photolysis of methane to acetylene and to the tautomeric C_3H_4 molecules methylacetylene and allene. Isoprene oligomers and polymers -- the foundation for today's Isoprene Rule -- were among the important products.

During the second stage, with ammonia becoming more involved in atmospheric photochemistry, hydrogen cyanide and cyanoacetylene were major reactants. Polymeric peptide precursors (F) and (G) were formed in the stratosphere together with nitrogen heterocycles (Oro et al., 1980; Ferris et al., 1984) possessing the basic skeletons of purines, pyrmidines, pyridines and porphyrins. Hydrogen sulphide, also present, took part in the modification of side-chains of the polymers. When most of the ammonia had been used up -- its action as an ultraviolet screen thereby considerably diminished -- photolysis of water vapor that had been confined to lower levels became possible, leading to the third stage, when formaldehyde and sugars were synthesized as well as phosphates from phosphine. Meanwhile, as the macromolecules and heterocycles of the second stage gradually settled, contact with water introduced oxygen functionality mainly by hydrolysis of azomethines to carbonyl groups, and of nitriles to amides and carboxylic acids.

As the Earth's surface became covered with this organic shower, potential membrane material -- carboxylic acids, carbohydrates, polypeptides -- accumulated in lakes and oceans, while on land the simultaneous synthesis of polypeptides and polynucleotides was promoted by polyamidines (F,G) -- cyanide polymers -- perhaps assisted by clays. On our planet this polypeptide-polynucleotide symbiosis mediated by polyamidines set the pattern for the evolution of protein-nucleic acid systems controlled by enzymes, the characteristic mode of life today.

REFERENCES

A'Hearn, M. F., Hoban, S., Birch, P. V., Bowers, C., Martin, R. and Klinglesmith, D. A., 1986: Nature, 324, 649.

Clark, B. C., Mason, L. W. and Kissel, J., 1986: Proc. 20th ESLAB Symposium on the Exploration of Halley's Comet, ESA SP-250, p. 353.

Combes, M. Moroz, V. I., Crifo, J. F., Lamarre, J. M., Charra, J., Sanko, N. F., Soufflot, A., Bibring, J. P., Cazes, S., Coron, N., Crovisier, J., Emerich, C., Encrenaz, T., Gispert, R., Grigoryev, A. V., Guyot, C., Krasnopolsky, V. A., Nikolsky, Yu. V., Rocard, F., 1986: Nature, 321, 266.

Cronin, J. R., 1976: Origins of Life, 7, 337, 343.

Cruikshank, D. P., 1986: Advances in Space Research, in press.

Cruikshank, D. P., Hartmann, W. K. and Tholen, D. J., 1985: Nature, 315, 122.

Draganic, Z., Niketic, V., Jovanic, S. and Draganic, I. G., 1980: J. Mol. Evoln., 15, 239.

Ferris, J. P., 1979: Science, 203, 1135.

Ferris, J. P. and Hagan, W. J., 1984: Tetrahedron, 40, 1093.

Hartmann, W. K., Cruikshank, D. P. and Degewij, J., 1982: Icarus, 52, 377.

Keller, H. U., Arpigny, C., Barbieri, C., Bonnet, R. M., Cazes, S., Coradini, M., Cosmovici, C. V., Delamere, W. A., Huebner, W. F., Hughes, D. W., Jamar, C., Malaise, D., Reitsema, H. J., Schmidt, W. K. H., Seige, P., Whipple, F. L., Wilhelm, K., 1986: Nature, 321, 320.

Kissel, J., Sagdeev, R. Z., Bertaux, J. L., Angarov, V. N., Audouze, J., Blamont, J. E., Büchler, K., Evlanov, E. N., Fechtig, H., Fomenkova, M. N., von Hoerner, H., Inogamov, N. A., Khromov, V. N., Knabe, W., Krueger, F. R., Langevin, Y., Leonas, V. B., Levasseur-Regourd, A. C., Managadze, G. G., Podkolzin, S. N., Shapiro, V. D., Tabaldyev, S. R. and Zubkov, B. V., 1986: Nature, 321, 280.

Kissel, J., Brownlee, D. E., Büchler, K., Clark, B. C., Fechtig, H., Grün, E., Hornung, K., Igenbergs, E. B., Jessberger, E. K., Krueger, F. R., Kuczera, H., McDonnell, J. A. M., Morfill, G. M., Rahe, J., Schwehm, G. H., Sekanina, Z., Utterback, N. G., Völk, H. J., Zook, H. A., 1986: Nature, 321, 336.

Kliss, R. M. and Matthews, C. N., 1962: Proc. Natl. Acad. Sci. U.S., 48, 1300.

Krankowsky, D., Lämmerzahl, P., Herrwerth, I., Woweries, J., Eberhardt, P., Dolder, U., Herrman, U., Schulte, W., Bertheier, J. J., Illiano, J. M., Hodges, R. R., Hoffman, J. H., 1986: Nature, 321, 326.

Kvenvolden, K. A., Lawless, J. G., and Folsome, C. E., 1973: Scientific American, 227, June 38.

Levasseur-Regourd, A. C., Bertaux, J. L., Dumont, R., Festov, M., Giese, R. H., Giovane, F., Lamy, P., LeBlanc, J. M., Llebaria, A., Weinberg, J. L., 1986: Nature, 321, 341.

Matthews, C. N. and Moser, R. E., 1966: Proc. Natl. Acad. Sci. U.S., 56, 1087.

Matthews, C. N. and Moser, R. E., 1967: Nature, 215, 1230.

Matthews, C. N., Ludicky, R. A., Schaefer, J., Stejskal, E. O., and McKay, R. A., 1984: Origins of Life, 14, 243.

Matthews, C. N., 1984: Proc. Roy. Inst. Gt. Britain, 55, 199.

Matthews, C. N., 1975: Origins of Life, 6, 155.

Matthews, C. N., Nelson, J., Varma, P. and Minard, R. D., 1977: Science, 198, 622.

Matthews, C. N. and Ludicky, R. A., 1986: Proc. 20th ESLAB - Symposium on the Exploration of Halley's Comet, ESA S.P. 250, p. 273.

Matthews, C. N. and Ludicky, R. A., 1987: Polymer Preprints, 28, No. 1, 104.

Matthews, C. N., Nelson J. E., and Minard, R. D., 1980: Abstracts, 6th Intl. Conf. Origins of Life, Jerusalem, 100.

Matthews, C. N., 1982: Origins of Life, 12, 281.

Matthews, C. N., 1985: In M. D. Papagiannis (ed.) The Search for Extraterrestrial Life: Recent Developments, Reidel, Dordrecht, Holland, 151.

Matthews, C. N., 1986: Origins of Life, 16, 500.

McKay, R. A., Schaefer, J., Stejskal, E. O., Ludicky, R. and Matthews, C. N., 1984: Macromolecules, 17, 1124.

Miller, S. L., 1984: In G. Nicolis (ed.), Aspects of Chemical Evolution, Wiley, New York, 85.

Miller, S. L., 1953: Science, 10, 528.

Minard, R. D., Yang, W., Varma, P., Nelson, J. and Matthews, C. N., 1975: Science, 190, 387.

Moreels, G., Gogoshev, M., Krasnopolsky, V. A., Clairemidi, J., Vincent, M., Parisot, J. P., Bertaux, J. L., Blamont, J. E., Festou, M. C., Gogosheva, Ts., Sargoichev, S., Palasov, K., Moroz, V. I., Krysko, A. A., Vanysek, V., 1986: Nature, 321, 271.

Moser, R. E. and Matthews, C. N., 1968: Experientia, 24, 658.

Moser, R. E., Claggett, A. R. and Matthews, C. N., 1968: Tetrahedron Lett. 1599, 1605.

Niketic, V., 1984: Origins of Life, 14, 251.

Owen, T., 1982: Scientific American, 246, February, 98.

Oro, J. and Lazcano-Araujo, A., 1980: In Vennesland, B., Conn, E. E., Knowles, C. J., Westley, J. and Wissing, F. (eds.), Cyanide in Biology, Academic Press, New York, 517.

Ponnamperuma, C. (ed.), 1983: Cosmochemistry and the Origin of Life, Reidel, Dordrecht, Holland, Ch. 1.

Reinhard, R., 1986: Nature, 321, 313.

Sagdeev, R. Z., Blamont, J., Galeev, A. A., Moroz, V. I., Shapiro, V. D., Shevchenko, V. I. and Szegö, K., 1986: Nature, 231, 259.

Schaefer, J., Stejskal, E. O., Jacob, G. S., and McKay, R. M., 1982: Appl. Spectr., 36, 179.

Schloerb, F. P., Kinzel, W. M., Swade, D. A., and Irvine, W. M., 1986: Proc. 20th ESLAB Symposium on the Exploration of Halley's Comet, ESA SP-250, p. 577.

Umemoto, K., Takahashi, M., Yokota, K. and Noto, N., 1986: Origins of Life, 16, 508.

Urey, H. C., 1952: Proc. Natl. Acad. Sci. U.S., 38, 351.

Völker, T., 1960: Angew. Chem., 72, 379.

ADDENDUM

Matthews, C. N., 1988: "Cosmic Metabolism: The Origin of Macromolecules," in G. Marx (ed.), Bioastronomy--The Next Steps, Kluwer, Dordrecht, Holland, 167.

Matthews, C. N. and Ludicky, R., 1988: "Hydrogen Cyanide Polymers: Solid State NMR(^{15}N, ^{13}C)Investigations," Polymer Preprints, 29, 84.

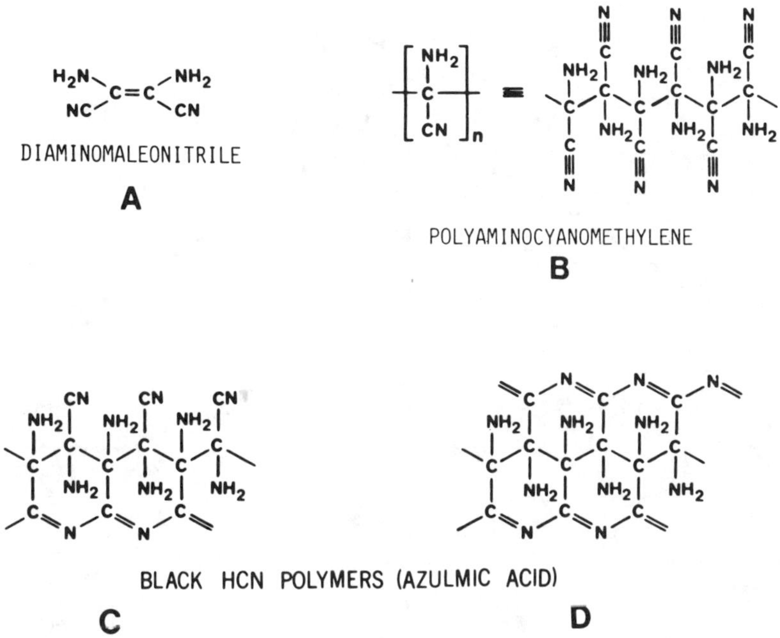

Figure 1. HCN polymers formally derived from HCN tetramer (diaminomaleonitrile).

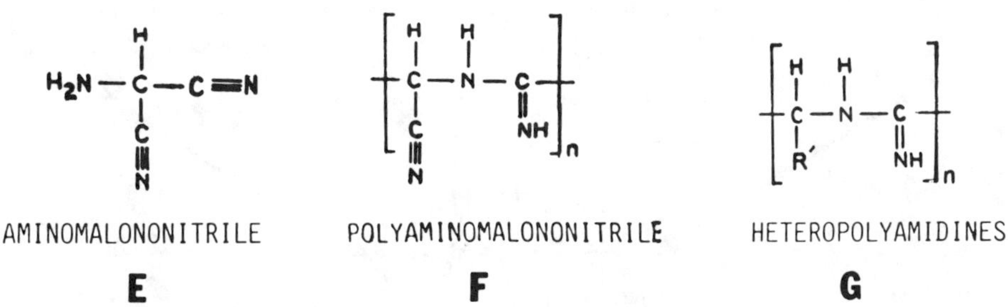

Figure 2. HCN polymers formally derived from HCN trimer (aminomalononitrile).

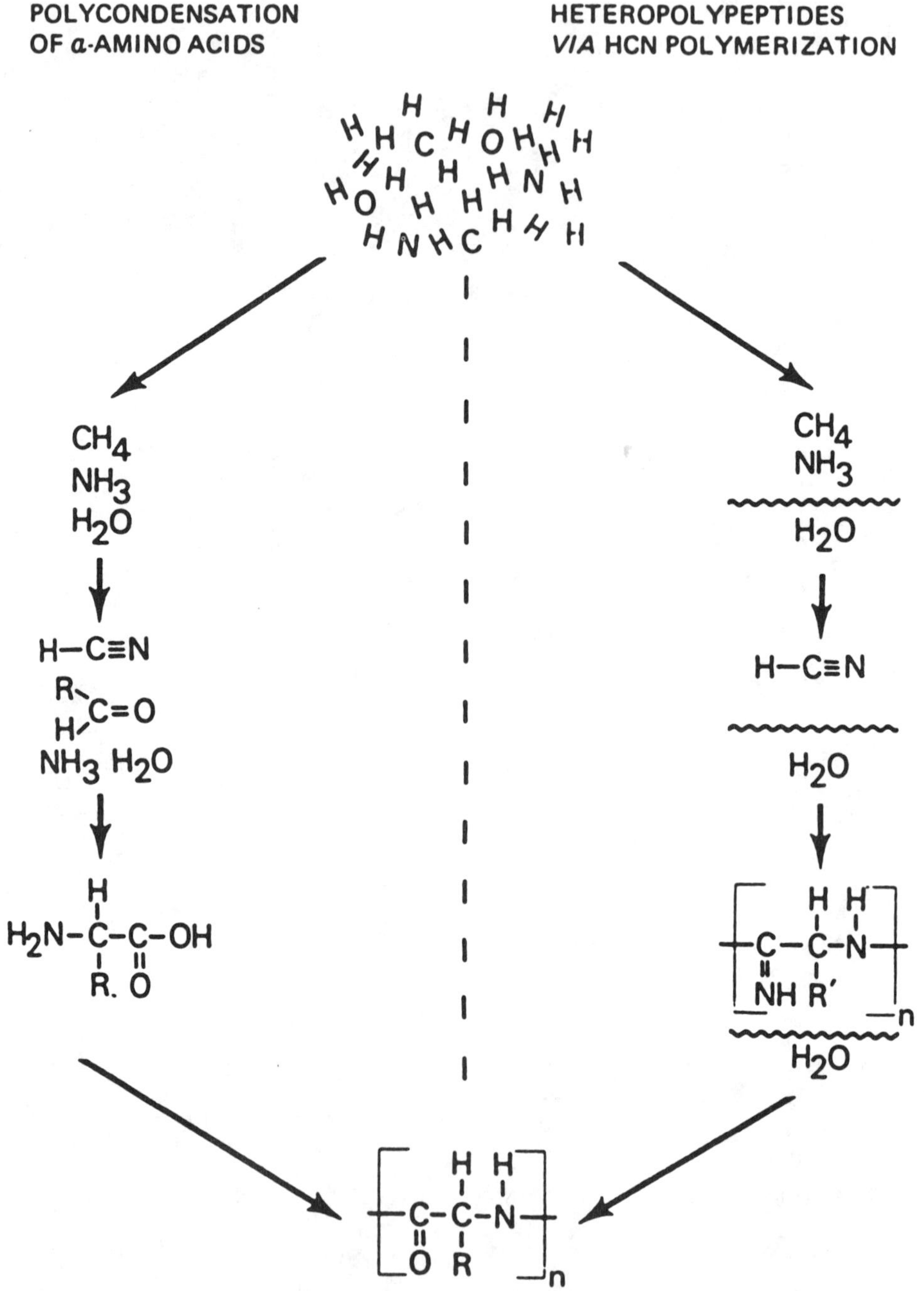

Figure 3. Two opposing models for the origin of proteins. Which came first, amino acids or their polymers?

OPARIN, THE FIRST CELLS, AND SELECTION PROCESSES

Sidney W. Fox
Institute for Molecular and Cellular Evolution
University of Miami
Miami, Flordia 33177

ABSTRACT

Two models for protocells that appear in a number of text books are the coacervate droplet and the proteinoid microsphere. The coacervate droplet composed of gelatin and gum arabic was recognized by Oparin as quite unstable. Oparin proposed that coacervate droplets would have acquired stability by natural selection operating in a prebiological version of Darwinian natural selection.

Proteinoid microspheres are the result of self-organization of thermal proteins, and can fill the role of protobiological cells. They have stability as soon as they are formed. The thermal proteins (proteinoids) result from the polycondensation of amino acids which yield highly non-random peptides. Relics of such primordial self-ordering of amino acids have recently been reported by Ivanov and Förtsch for modern proteins.

The self-ordering is also a kind of natural process; it has been designated as molecular selection. The origin of individuals and species, according to the experiments, involves molecular selection followed by Darwinian natural selection. This kind of sequence of processes to yield evolving cells is in accord with the experimental evolutionary flow-sheet, whereas coacervate droplets are anachronistic in their employment of already evolved polymers.

INTRODUCTION

Aleksandr I. Oparin properly receives credit (Fox, 1974) for stimulating interest in, and making contributions to, the general theory of the origin of life (Oparin and Gladilin, 1980). Statements of a more analytical nature acknowledge his emphasis on the role of organization of molecules and for his contribution to the views, novel in the 1930s-1940s, that cells are not simply bags of enzymes in solution.

ISBN 0-937194-18-2

Negative criticisms of his interpretations largely focus on inferences he drew from experiments with coacervate droplets. Lehninger (1975), for example, recognized that Oparin's coacervate droplets could not be valid models of protocells since they were made from modern biopolymers. Woese (1980) rejected "Oparinism" because "Basic biochemical reactions, by and large, are dehydrations; they cannot reasonably happen in an ocean," and for other reasons. Woese's criticism relates, as will be shown, to Lehninger's comment inasmuch as the primary need for our understanding this subject concerns the first biopolymers that preceded the first cells.

Woese objected in general that "Oparinism" tended to stultify advances on the problem. The role of coacervate droplets alongside that of proteinoid microspheres (Fox and Dose, 1977) is considered here in the context of that general thinking, partly inasmuch as the coacervate model and the proteinoid microsphere appear in tandem in numerous textbooks. Most of these appearances are without discriminating comments such as those of Lehninger.

PRODUCTION AND COOPERATIVE CHEMISTRY

As Lehninger stated, the coacervate droplets are typically made from gum arabic and gelatin, and sometimes enzymes have been included (Oparin, 1971). All of these materials, including the enzymes, are obtained from modern organisms. For this reason they cannot adequately represent the precells that they are often thought of as modelling; Oparin was correct in his later statements that coacervate droplets could not represent either precells or protocells.

The precursors for protocells, as modelled by proteinoid microspheres (Fox and Nakashima, 1980), are thermal copolymers of amino acids produced under conditions that are and almost certainly were geological (Fox and Dose, 1977). The polymers have been studied in many laboratories. They have been indexed collectively under proteins, thermal in Chemical Abstracts since 1972; Chemical Abstracts represents the single most complete source of information on both the polymers and the microspheres. The known properties are now numerous (see Fox, 1980a for a selective review). The availability of the amino acids on and off the Earth have been reviewed (Fox *et al.*, 1976; Fox, 1973).

The key finding for the polymerization of sets of amino acids in the laboratory, and presumably therefore on the Earth, was the thermal *co*polymerization of other amino acids with aspartic acid. While other amino acids were known mostly to decompose on heating (Katchalski, 1957), aspartic acid appeared to give what was perhaps (Katchalsky, 1957) a peptide. This product was subsequently shown to be a true polypeptide (Vegotsky _et al._, 1958) as are the copolymers (Fox and Harada, 1960).

The thermal protein and a large range of thermal copolyaspartic acids were, after the thermal protein was made, soon observed to yield cell-like structures. However, the cataloguing of the properties of the microstructures required more than another two decades. Their existence in the laboratory was nevertheless made possible by prior chemistry. In comparison with coacervate droplets, we shall be concerned here especially with stability, biofunctionality, original biological information, and the relationships to natural selection of the proteinoid microspheres.

STABILITY

Oparin (1957) acknowledged the lack of stability of coacervate droplets, as contrasted to true cells, for which the droplets were regarded as models of direct precursors. In his last review of the subject, with Gladilin (1980), before his death, Oparin restated his earlier visualization of how such instability would be overcome by natural selection: "The origin of life per se was bound to be associated with the natural selection of integral phase-related systems---." Oparin thus relied on the concept of natural selection as a process to introduce stability into the first cells. He seems to have regarded the assigned role of coacervate droplets as models of the first cell more seriously than their physical nature was subsequently shown to merit.

All those who have worked with proteinoid microspheres know them to be robust as soon as they are made. Dr. Laura Hsu prepared some aseptically since they are easily food for wild bacteria from the air; her preparation lasted six years (Przybylski and Fox, 1986) before she discarded them, having learned at that time that they could last almost indefinitely.

Some confusion has been created by one writer (Day, 1979) who did not work with proteinoid microspheres. Day said, "Coacervates are notoriously unstable and microspheres exist only in saturated solutions." The statement for coacervate droplets is correct, but the comment about microspheres is purely imaginative. It is _unsupported_ by experiments, and in the context of Day's several other unsupported statements, capable of being misleading. The stability is already clear from numerous experiments such as that of Hsu, quoted here.

Another assumption is that polypeptides in aqueous solution tend toward an equilibrium in which they are substantially broken down (Miller and Orgel, 1974). On the basis of purely thermodynamic data (Huffman, 1942), this is theoretically true. The theory is substantially incomplete since it fails to take into account the existence of stabilizing intramacromolecular interactions. Indeed, it is common experience among biochemists that aseptically prepared aqueous solutions of many proteins will last indefinitely.

More significantly, microspheres made from both acidic and lysine-rich proteinoids were shown by 1963 (Fox and Yuyama) to be stable over a wide pH range from about 4 to 8.5. Inclusion of both acidic and basic thermal proteins furthermore mimics modern cells more closely and more generally than does inclusion of the acidic type alone. Another mode of production that yields stable microspheres follows from making the proteinoids in the presence of seawater salts (Snyder and Fox, 1975). The microspheres assembled from such thermal proteins are very stable in the presence of either water, or seawater. Saturated proteinoid solution is not needed for stability in either case. Other stabilizing agents include lipids and calcium ions (Brooke and Fox, 1977). A wide range of proteinoid microspheres are, accordingly, stable over a very wide range of conditions, in contrast to the coacervate droplets that were for long used as a guide to the understanding of the first cells.

BIOFUNCTIONALITY

The microspheres have virtually all of the properties of the thermal proteins from which they are assembled. This array is "remarkable" (Lehninger, 1975). The catalytic activities have been especially studied (Rohlfing and Fox, 1969; Dose, 1984; Nakashima, 1987). The most significant in a forward evolutionary direction are the anabolic activities, in which suspensions of proteinoid microspheres catalyze the synthesis of peptide bonds in aqueous solutions containing energy-rich phosphates (Nakashima nd Fox, 1980) and in which the microspheres catalyze the synthesis also of the phosphodiester bonds of polynucleotides. Such reactions have been shown to occur in suspensions of microspheres containing lysine-rich proteinoids, or in solutions of the same lysine-rich proteinoid. The fact that the synthesis of both, junior proteins and junior polynucleotides, can occur in the same microlocale as the result of action of the same catalytic agent has suggested that such a site would have been ideal for the initial working of the genetic code (Fox, 1981a) at the stage of input of information.

Oparin (1971) has reported captivating experiments of enzyme action within coacervate droplets. These models, however, use enzymes from modern organisms and are therefore deficient as models in the same way as are the gum arabic and the gelatin used for structural purposes.

Since the thermal proteins have themselves a large array of enzymatic activities (Dose, 1984; Fox and Nakashima, 1984; Nakashima, 1987), even though at much lower levels of activity than modern enzymes, the microspheres provide valid models for this aspect of the first cells in which metabolism started. Other model protocellular activities have been catalogued extensively for the microspheres (Fox, 1980a; Fox and Nakashima, 1980; Fox and Dose, 1977).

THE FIRST BIOLOGICAL INFORMATION

Biofunctional proteins and biofunctional cells require specificity of amino acid residue sequence in the proteins. We know that in many instances activity does not become appreciable unless there exists a multiplicity of the same type of molecule. For both bequeathing the specific sequence to successive generations and for multiplicity of molecules within a single generation, a mechanism for stamping out the molecules is needed. As Eigen has pointed out, random arrays of functional molecules would not support reproducibility of function between molecules (Fox and Windsor, 1984). Non-random, or determinate factors were accordingly needed (Eden, 1967). In modern organisms, the reproducibility is attained by the DNA/RNA mechanism. To demonstrate or postulate an initial precursor for this, involving two or three different types of macromolecule simultaneously, would be extremely difficult. What has been learned instead as feasible is that comparable ordering of amino acids is self-instructed (Fox, 1981b). Other cellular qualities could have been added to such a beginning situation by stepwise evolution.

Oparin postulated in general terms that biofunctional molecules would have been the product of prolonged evolution: "All that we can expect from the relatively simple and thermodynamic laws which prevailed on the surface of the primeval Earth is that they should explain the formation of organic polymers -- having as yet no orderly arrangement of amino acid and nucleotide residues adapted to the performance of particular functions."

Undoubtedly the largest question for the relevance of a model that relies on the self-instruction of amino acids over three billion years ago has been whether data could overcome the uncertainties caused by the huge lapse in time involved. Indeed, Calvin (1969) suggested that the finding of evolutionary relics of such a primordial self-ordering would be helpful to the understanding. If such could be found, one could state with increased assurance that the source of bioevolutionary information was indeed a primordial self-ordering of amino acids.

This kind of possibility has been identified and reported with the complete protein data base available in 1984, and helped by advice of Gerhard Braunitzer. Ivanov and Förtsch (1986) found that after examining 2898 proteins and 591,717 amino acid residues, "Vast majority of cases amino acids show a preference in bonding (self-ordering) and periodicity." They have begun to limn an evolutionary tree for proteins, with the most self-ordered thermal protein as a likely candidate for the initial evolutionary ancestor.

THE PROTEINOID MICROSPHERES

The proteinoid microspheres are of highly regular appearance, probably reflecting the highly uniform, non-random nature of the polymers from which they assemble. The uniformity is evident in the scanning electron micrograph of Fig. 1. The infrastructure (Fig. 2) is less uniform in micrographs throughout each individual, although the infrastructure from microsphere to microsphere is also highly non-random. The double layer in the transmission micrograph is often remarked on, probably especially due to the fact that the microsphere contains no discrete lipid. Herein lies a principal issue on which the significance of an archetypal cell hinges. Like modern proteins, the thermal proteins contain substantial proportions of apolar sidechains in such amino acid residues as tryptophan, leucine, or phenylalanine. The microsphere is in fact lipoidal, even though the boundary of lipoidal proteinoid provides a less efficient barrier than a boundary of compositionally dominant lipid (Fox <u>et al.</u>, 1969).

THE FLOWSHEET OF THE EVOLUTIONARY SEQUENCE

The flowsheet (Fig. 3), updated from Lehninger (1975), constitutes a comprehensive sequence of events for the emergence of life and beyond, as explained in this article and elsewhere (Fox, 1986a). Rather remarkable was Lehninger's perception that the events of molecular evolution resulting in stable protocells would take the amino acid sequence into the phase of Darwinian evolution. Since 1975, a number of authors have recognized this (see for example, Ho and Saunders (1984), and Szent-Györgyi, 1972). Szent-Györgyi defined the problem: "We know that Nature can build complex organic molecules without the intervention of life . . . Another paradox still awaits solution. Mutation is a random process. How could a random process lead to such highly ordered structures as a multicellular living organism? The usual answer to this question is that there was plenty of time to try everything. I could never accept this answer. . . . A certain selection was already supposed by Darwin who made the struggle for life responsible for selection.

The selection, however, in my opinion, has to take place much earlier, not only with the final product."

The selection principle in evolution has been parsed into molecular selection and natural selection (Ho and Saunders, 1984; Fox, 1986b). Oparin's emphasis on natural selection drawn from coacervate droplets thus missed, and probably clouded, the vision (Woese, 1980) of subsequent awareness of both the initial microsphere stability and its origin from orderly polymers that had gotten their instructions from precursor amino acids. What is now perceivable is that evolution proceeded from ancient (Fox and Nakashima, 1980) protein to modern (Ivanov and Förtsch, 1986) protein and from ancient (Fox and Dose, 1977) cells to modern cells (Francis, et al., 1979) in which at each stage protein was able to organize into cells.

OPARIN AND DYSON

A recent proponent of Oparin's thinking is Dyson (1985) who, as a theorist, proposes experiments be done in the future to validate the sequence of cell-->enzyme-->gene. He properly chose to apologize "in advance" to six authors he named but had not read. Had he read most of these authors, he would have found that the state of the art in 1985 was already way beyond Oparin, or Dyson-modified Oparin, and that significant experiments that he proposed that experimentalists should do had already been done. These experiments led to the Dyson-modified Oparin sequence and more: enzyme (ordered protein) -->cell-->cellular enzyme-->cellular gene. The crucial difference in these two sequences is that in attempting to learn the sequence, Dyson, Oparin, Eigen, et al., back-extrapolate in different ways in a subject matter that requires forward processes from primordial events rather than analysis and back-extrapolation of modern events.

With the analytical approach, by contrast, we can see that the proteinoid model has come full circle in two major aspects.

1. The proteinoid model, and indeed a modern view of Darwinian evolution (Ho and Saunders, 1984) rest on the laboratory demonstration of the self-ordering of amino acids (Fox, 1981). The theory states that the evolutionary precursor of the complex genetic coding mechanism was the process of self-ordering (self-sequencing) of amino acids. As already mentioned by working from a total data base of almost 3000 modern proteins, Ivanov and Förtsch (1986) have shown that modern proteins contain many evolutionary relics of the self-ordering process that made ordered primordial proteins possible.

2. Since the emphasis is now so strongly on constructionistic phenomena, it should be said that the declension of highly ordered proteins to almost random proteins has been made possible by studies of amino acid sequence in which the methods are analytical in type (Fox, 1945; Rosmus and Deyl, 1972).

SUMMARY

Oparin emphasized the role of organization in cellular entities, and even of self-organization. On the other hand, he advocated a watery origin of life that was thermodynamically indefensible, he advocated the evolution of cellular order from disorderly precursors, he proposed that unstable first cells would have acquired stability by natural selection, and he proposed that biofunctions would emerge only as the consequence of prolonged evolution. These ideas appear to have been largely derived from his experiments with coacervate droplets. They have been influential in the development of early paradigms.

When instead of using modern polymers to construct protocell models, the steps in evolution were retraced from models of primordial polymers, many kinds of protocellular archetypes, all generically modelled by proteinoid microspheres, were obtained. Such cellular structures are defensible as generated in two steps in the geological realm: the crucial self-ordering of amino acids into thermal proteins and the self-organization of those thermal proteins. The latter organize themselves to cells having stability, membranes, and catalytic activity. The experiments indicate that these properties would have been on hand at the outset, not having needed evolution for their first or later appearance. When the thermal protein contains sufficient basic amino acid such as lysine, the resultant laboratory protocells can themselves catalyze the synthesis of both peptide and phosphodiester bonds. This has been invoked to explain a maximally simple origin of the genetic coding mechanism in cells (Fox, 1981a). Again, these processes relate to the input stage of molecular evolution. Input processes appear to be the most logical for the working of natural selection in modernization of the protocell. The development, however, required internal, forward chemical control at each stage: synthesis of protoprotein, self-organization thereof, synthesis of cellular peptides and polynucleotides, catalysis of energy transfer in metabolism and the rest of the biochemical panoply.

ACKNOWLEDGMENT

This work has been supported by Grant No. NG 10-007-008 of the National Aeronautics and Space Administration. Acknowledgement is made also to numerous co-workers other than those cited in the references.

REFERENCES

Brooke, S. and Fox, S. W., 1977: BioSystems, 9 1.

Calvin, M., 1969: Chemical Evolution, Oxford University Press.

Day, W., 1979: Genesis on Planet Earth, House of Talos Publishers, East Lansing, Michigan.

Dose, K., 1984: In S. W. Fox (ed.) Individuality and Determinism, Plenum Press, New York.

Dyson, F., 1985: The Origins of Life, Cambridge University Press.

Eden, M., 1967: In P. S. Moorhead and M. M. Kaplan (eds.) Mathematical Challenges to the Neo-Darwinian Interpretation of Evolution, Wistar Institute Press, Philadelphia, 5-12.

Fox, S. W., 1945: Advances Protein Chem., 2, 155.

Fox, S. W., 1973: Naturwissenschaften, 60, 359.

Fox, S. W., 1974: Encyclopedia Britannica, 15th edition, Biography of Oparin, Aleksandr Ivanovich.

Fox, S. W., 1980a: Naturwissenschaften, 67, 378.

Fox, S. W., 1980b: Naturwissenschaften, 67, 576.

Fox, S. W., 1981a: In M. Kageyama, K. Nakamura and T. Uchida (eds.), Science and Scientists, Japan Sc. Soc. Press, Tokyo, 39-45.

Fox, S. W., 1981b: J. Chromatogr., 215, 115.

Fox, S. W., 1986a: Amer. Biol. Teacher, 48 (3), 140.

Fox, S. W., 1986b: Quart. Rev. Biol., 61, 375.

Fox, S. W. and Dose, K., 1977: Molecular Evolution and the Origin of Life, Revised Edition, Marcel Dekker, New York.

Fox, S. W. and Harada, K., 1960: J. Amer. Chem. Soc., 82, 3745.

Fox, S. W. and Nakashima, T., 1980: BioSystems, 12, 155.

Fox, S. W. and Windsor, C. R., 1984: Intl. J. Quantum Chem. Quantum Biol. Symp., 11, 103.

Fox, S. W. and Yuyama, S., 1963: J. Bacteriol., 85, 279.

Fox, S. W., McCauley, R. J., Montgomery, P. O'B., Fukushima, T., Harada, K., and Windsor, C. R., 1969: In F. Snell, J. Wolken, G. J. Iverson and J. Lam (eds.), Physical Properties of Biological Membranes, Gordon and Breach, New York.

Fox, S. W., Harada, K. and Hare, P. E., 1976: Geochim. Cosmochim. Acta, 40, 1069.

Francis, S., Margulis, L. and Barghoorn, E. S., 1978: Precambrian Res., 6, 65.

Ho, M. W. and P. T. Saunders (eds.), 1984: In Beyond Neo-Darwinism, and Introduction to the New Evolutionary Paradigm, Academic Press, London.

Huffman, H. M., 1942: J. Phys. Chem., 46, 885.

Ivanov, O. C. and Förtsch, B., 1986: Origins of Life, 17, 35.

Katchalsky, E., 1951: Advances Protein Chem., 6, 123.

Lehninger, A. L., 1975: Biochemistry, Worth and Co., New York.

Miller, S. and Orgel, L. E., 1974: The Origins of Life on Earth, Prentice-Hall, Englewood Cliffs, N.J.

Nakashima, T., 1987: Topics Current Chem., 139, 57.

Nakashima, T. and Fox, S. W., 1980: J. Mol. Evol., 15, 161.

Oparin, A. I., 1957: The Origin of Life on Earth, Academic Press, New York.

Oparin, A. I., 1971: Sub-Cell Biochem., 1, 75.

Oparin, A. I. and Gladilin, K. L.: BioSystems, 12, 133.

Przybylski, A. T. and Fox, S. W., 1986: In F. Gutmann and H. Keyzer (eds.), Modern Bioelectrochemistry, Plenum Press, New York, 377-396.

Rohlfing, D. L. and Fox, S. W., 1969: Advances Catal., 20, 373.

Rosmus, J. and Deyl, Z., 1972: J. Chromatogr., 70, 221.

Shapiro, R., 1986: Origins, Summit Books, New York.

Snyder, W. D. and Fox, S. W., 1975: BioSystems, 7, 222.

Szent-Györgyi, A., 1972: In D. L. Rohlfing and A. I. Oparin (eds.), Plenum Press, New York.

Vegotsky, A., Harada, K. and Fox, S. W., 1958: J. Am. Chem. Soc., 80, 3361.

Woese, C. R., 1980: In H. O. Halverson and K. E. van Holde (eds.), The Origins of Life and Evolution, Alan R. Liss, New York, 65-76.

ADDENDUM

Fox, S., 1987: In memory of Professor Albert Szent-Gyorgyi. Intl. J. Quantum. Chem., Quantum Biol. Symp., 14, 1.

Fox, S. W., Hefti, F., Hartikka, J., Junard, E., Przybylski, A. T., and Vaughan, G., 1987: "Pharmacological activities in thermal proteins: relationships in molecular evolution," Intl. J. Quantum Chem., Quantum Biol. Symp., 14, 347-349.

Fox, S., 1988: "Evolution outward and forward," in M.-W. Ho and S. W. Fox (eds.), Evolutionary Processes and Metaphors, John Wiley & Sons, Chichester, 17-33.

Fox, S., 1988: The Emergence of Life/Darwinian Evolution from the Inside, Basic Books Inc., New York, 208 pages + ix.

Fox, S. W. and Khoury, A., 1988: "The New Evolutionary Paradigm," BioEssays 9, 209-213.

Ho, M.-W. and Fox, S. W., 1988: "Processes and Metaphors in Evolution," in M.-W. Ho and S. W. Fox (eds.), Evolutionary Processes and Metaphors, 1-16.

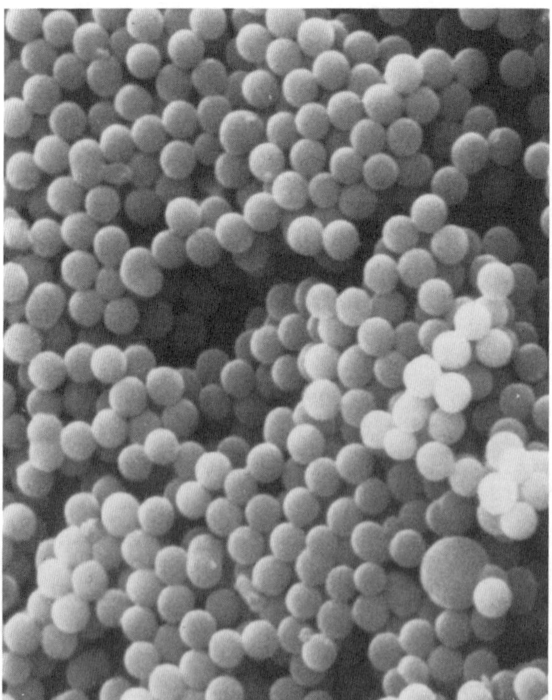

Figure 1. Scanning electron micrograph of proteinoid microspheres. Thanks to Mr. Steven Brooke.

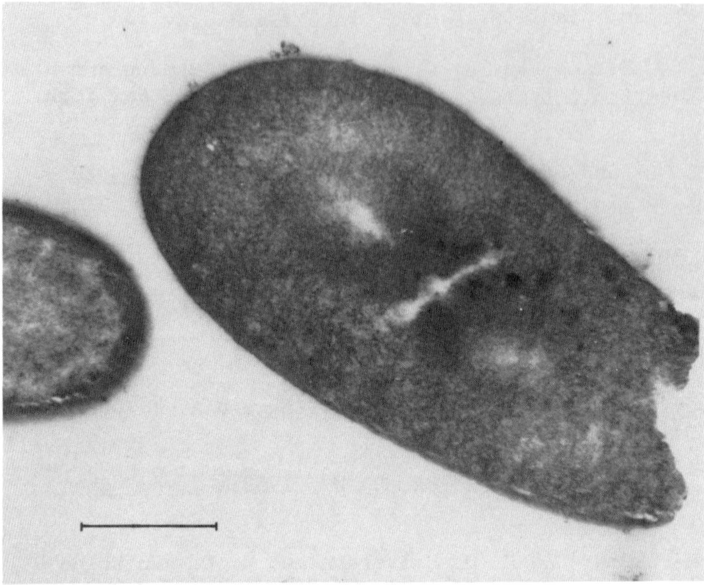

Figure 2. Transmission electron micrograph of proteinoid microsphere, stained with osmium tetroxide and sectioned.

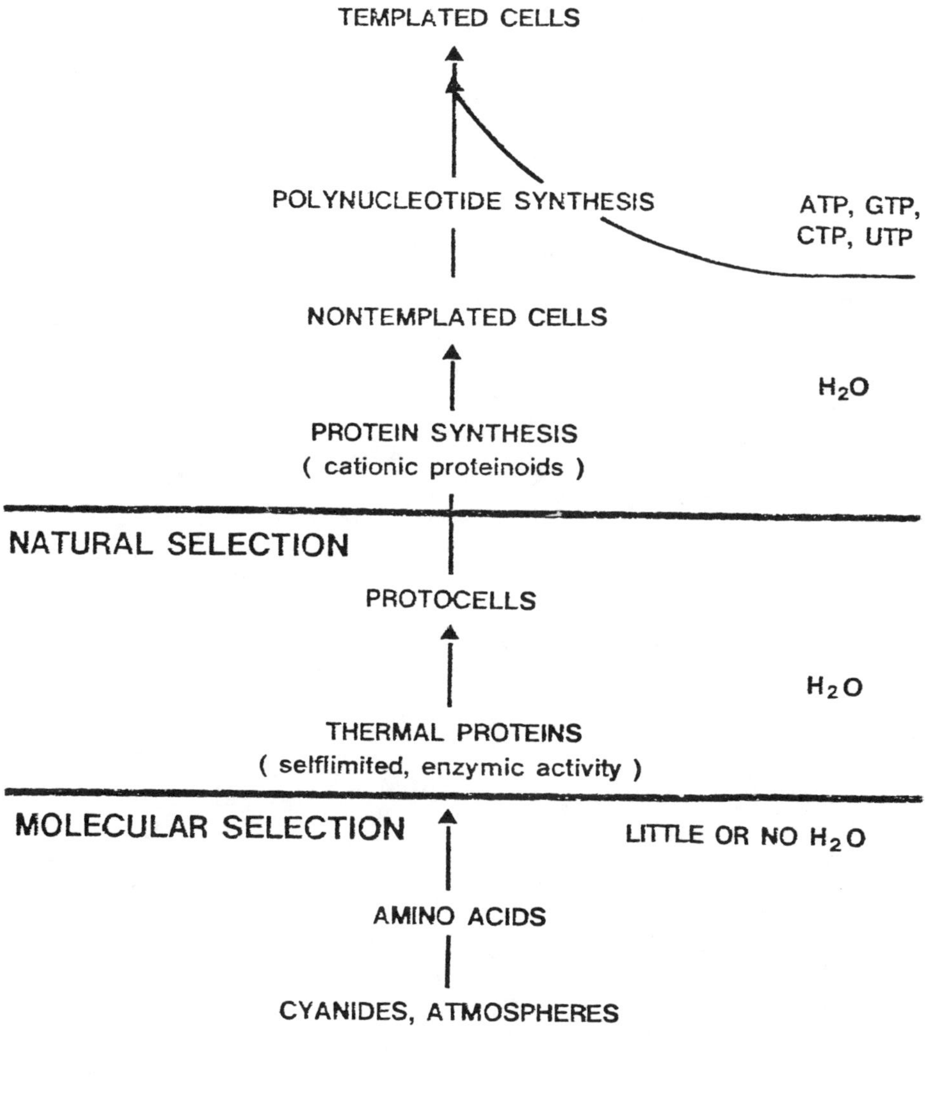

Figure 3. Flowsheet of early molecular and cellular evolution. This demonstrates that molecular selection preceded natural slection, and that informed proteins preceded DNA/RNA.

A STUDY OF PEPTIDE SYNTHESIS BY AMINO ACYL NUCLEOTIDE ANHYDRIDES IN PRESENCE OF COMPLEMENTARY HOMOPOLYNUCLEOTIDES

Sanjay Tyagi and Cyril Ponnamperuma
Laboratory of Chemical Evolution
University of Maryland
College Park, Maryland 20742

ABSTRACT

Alanyl anhydrides of GMP, UMP and AMP were synthesized in order to test the proposition that the complementary homopolynucleotides may enhance peptide synthesis by amino acyl nucleotide anhydrides. We report here that no difference could be detected in the peptide synthesis by these anhydrides in presence and absence of complementary homopolynucleotides. This result may be relevant to prebiotic peptide synthesis.

INTRODUCTION

Spontaneous synthesis of peptides from amino acyl adenylates, the phosphoanhydrides of amino acids and adenosine monophosphoric acid (AMP) could have contributed to the pool of peptides on the prebiotic Earth (Paecht-Horowitz and Katchalsky, 1967; Paecht-Horowitz <u>et al.</u>, 1970; Lacey and Mullins, 1983; and Armstrong et. al., 1978). These compounds appear as necessary intermediates during the linking of amino acids to t-RNA in the contemporary protein biosynthesis and could have existed on the prebiotic Earth (Mullins and Lacey, 1981a and 1981b). The peptide synthesis by amino acyl anhydrides is a general phenomenon -- amino acyl nucleotide anhydrides of a number of amino acids (Krampitz and Fox, 1969) and nucleotides (Tyagi, 1987) are capable of synthesizing peptides.

If this synthesis is to be a model for the appearance of prebiotic peptides, it should be able to evolve into a process which, like current protein biosynthesis, takes instructions from polynucleotides. It is possible to envisage a primitive form of such a system: Amino acyl nucleotide anhydrides could interact with polynucleotides through complementary hydrogen bond interactions and the orientation effect of these interactions could enhance the yield of peptides. This enhancement over a basal rate of peptide synthesis, brought about by polynucleotides, could be considered

ISBN 0-937194-18-2

as a rudimentary form of translation, and the nucleotide part of the anhydride molecule could be considered as a "single nucleotide adaptor."

In order to test this proposition, we have synthesized a set of amino acyl nucleotide anhydrides using alanine as the amino acid, and AMP, GMP and UMP as the nucleotides. Each of these anhydrides was used to carry out peptide syntheses in presence of the complementary homopolynucleotide. If the polynucleotides are able to bring the anhydride molecules close to each other, one could expect an increase in the average degree of polymerization of alanine, over the background peptide synthesis. The pattern of oligopeptides concentrations produced in presence of the polynucleotide was compared with the pattern in its absence.

The physical conditions likely to be favorable for the orientation action mentioned above can be derived from biophysical studies of interactions of nucleotides and polynucleotides. Many purines and their derivatives interact with complementary pyrimidine polynucleotides in discrete integral ratios,closely similar in many respects to those formed between the corresponding pairs of polymers (Howard et al., 1964; Huang and Ts'o, 1966; Davies and Davidson, 1971; and Frazier and Miles, 1982). Purine bases, nucleosides, nucleotides and some nucleotide derivatives like imidazolides have been shown to take part in these interactions (Frazier and Miles, 1982). The conformation of these complexes is similar to the corresponding double helices, except, that one of the strands in the complexes is not continuous but is made up of juxtaposed monomers. These 'monomer-polymer' complexes exhibit characteristic cooperative melting phenomena. For example, 5'-GMP (or 5'-dGMP) and poly(C) form a 1:1 complex which melts at 15°C; adenosine and poly(U) form a 1:2 complex, melting at 18.5°C and 2-amino adenosine and poly(U) form a 1:2 complex, melting at 31°C (Howard et al., 1964; Howard et al., 1966).

The most important aspects of the physical conditions that favor the formation of organized monomer-polymer complexes are temperature and pH (Howard et al., 1964). The most favorable temperature range is 0 to 10°C, depending upon the nature of the nucleotide and the polynucleotides involved. The optimum pH range is 7.5 - 8.5. Salt concentrations have only marginal effects on the organization of these complexes. In the study mentioned above very high concentrations (0.01M to 0.1M) of the monomer and the polymer were employed.

The monomer-polymer interactions can have chemical consequences similar to the one envisaged above. Self-condensation of nucleoside 5'-phosphorimidazolides (Sawai, 1976) is enhanced by the presence of the complementary polynucleotides under the conditions mentioned above (Bridson and Orgel, 1980; Inoue and Orgel, 1982; Frazier and Miles, 1982; Inoue and Orgel, 1983; Joyce _et al._, 1984).

MATERIALS AND METHODS

SYNTHESIS AND PURIFICATION OF AMINO ACYL NUCLEOTIDE ANHYDRIDES

Amino acyl nucleotide anhydrides were prepared by condensation of alanine with AMP, GMP and UMP by adopting a method developed by Armstrong _et al._ (1979) for the synthesis of alanyl adenylate. In this synthesis _Sec_-butyl chloroformate was replaced with _n_-butyl chloroformate as the condensing agent (Tyagi, 1987).

Usually only about 45 percent of the dry weight of the preparations were the amino acyl nucleotide anhydrides, the rest of the material being free amino acid or free nucleotide. Also, on storage, additional unidentifiable impurities would appear in the preparation. A preparative high pressure liquid chromatography (HPLC) method was used to obtain pure anhydrides.

In order to minimize hydrolysis during purification, the mobile phase and the column were chilled by ice. Due to the relatively higher stability of the amino acyl nucleotide anhydrides at acidic pH's (Lewinsohn _et al._, 1967; Lacey _et al._, 1984; and Tyagi, 1987), an acidic elution buffer was chosen. The following is a description for the purification of Ala-UMP; identical procedures were used for the other anhydrides, Ala-GMP and Ala-AMP.

A Whatman ODS-3 M9, 25 cm x 9.5 mm preparative column with Particil-10 C-18 packing was used. A Dupont 830 Liquid chromatograph was used for solvent delivery and UV detection. The mobile phase was 0.001 M KH_2PO_4, adjusted to pH 3.0 with 5 percent phosphoric acid, with a flow rate of 5 ml/min. The dry Ala-UMP preparation (18 mg) was dissolved in distilled water (160 μl) and filtered with Millex-HV_4 filter unit (Millipore Corporation), before loading on the chromatograph. The detector was set at 256 nm and 1.28 AUFS. The chromatograms were recorded with a Varian strip chart recorder.

The effluent from the shaded peak (see Figure 1a, volume 7 to 8 ml) was collected over ice. It was acidified with a 4 μl of 6 N HCl and stored in liquid nitrogen. Solutions of all the anhydrides could be stored in this manner for over a month. If necessary, the samples could be concentrated by careful and fast freeze drying, with little hydrolysis. The purities and concentrations were established by several analytical chromatographic runs just before the use.

ASSAY FOR THE ANHYDRIDES

Traditionally, the amino acyl adenylates have been assayed by hydroxamate assay (Lipmann and Tuttle, 1945; Moldave et al., 1959) which detects the presence of anhydride bonds. The hydroxamate assay, however, is relatively insensitive and is not applicable to sub-millimolar concentrations. A chromatographic assay was developed for this purpose.

The analytical high pressure liquid chromatograph was equipped with either a UV (Dupont 930), or a fluorescence detector (Aminco Fluoromonitor, American Instrument Co.). The UV detector could recognize any species absorbing at the wavelength of 256 nm. The fluorescence detector, in conjunction with "fluorescamine post column derivatization," could recognize the ones having a primary amino group (with the notable exception of the amino groups on the ring system of the nucleotides, Tyagi, 1987).

Using a reverse phase, C-18 column (Supelcosil C-18, 25 cm x 4.6 mm, Supelcosil Inc.), the optimum chromatographic conditions were 0.01 M KH_2PO_4, pH adjusted to 3.0 using 5 percent H_3PO_4; flow rates 1.0 - 0.6 ml/min) were found for the separation of amino acyl nucleotide anhydrides, alanine and peptides. The following description pertains to the assay of Ala-UMP, but an identical method was followed for the other anhydrides: a sample of the Ala-UMP preparation, appropriately diluted with the running buffer and filtered with a Millex HV_4 filter unit, was loaded on to the column. The Ala-UMP peak was identified by the properties it exhibited: it released equimolar amounts of free alanine and UMP upon boiling (Figure 2); showed a positive hydroxamate reaction; and under basic conditions, gave rise to oligopeptides.

The concentration of Ala-UMP was determined by estimating the amounts of alanine and UMP released upon boiling, which in turn could be determined by a comparison with authentic standards of the two compounds (Figure 2). The concentrations obtained from these two independent measures were in close agreement. Linearity of the estimation range was established by using a number of known alanine and UMP standards. Since fluorescence detection gave better correlations, it was used for routine estimations.

Using this chromatographic assay system, it was possible to estimate concentrations as low as 5 μM.

REACTION CONDITIONS

In the following description, tube A contains polynucleotide whereas tube B is a control. All of the concentrations of polynucleotides are in terms of their monomeric units. All the reactions were carried out at 4°C for a week.

Poly(C) and Ala-GMP: The following were placed in tube A and freeze dried: poly(C) (0.05 M, 100 μl), NaCl (1.0 M, 10 μl) and $MgCl_2$ (1.0 M, 10 μl). 2,6-Lutidine buffer, (2.2 M, pH 8.0, 45.5 μl) and water (27.2 μl) were added to both A and B. Then at 4°C, Ala-GMP (0.184 M, 27.3 μl) were added to both tubes. Overall, tube A contained poly(C), Ala-GMP, NaCl and $MgCl_2$ (0.05 M each); tube B contained only Ala-GMP (0.05 M). Both solutions were buffered by 1.1 M lutidine (pH 8.0). A micro-stirrer provided continuous stirring.

At the end of the reaction we confirmed that the mixture maintained it's pH (8.0). The volatile buffer was removed by cycles of addition of water (1 ml) and freeze drying until the smell of lutidine disappeared. Trifluoroacetic acid (100 μl) and water (5 ml) were added to bring the peptides into solution. The precipitated RNA was removed using Milex filter units. Upon analysis, the precipitate showed no significant amounts of peptides. The supernatant was diluted, 1 to 5, by the running buffer of the HPLC.

Chromatographic conditions for the separation of peptides: mobile phase, 0.01 M KH_2PO_4, pH brought to 3.0 by H_3PO_4; column Supelcosil C-18; the flow rate, 0.6 ml/min; and detection fluorescamine post-column-derivation.

Poly(U) and Ala-AMP: The components of tube A were : lutidine buffer (1.5 M, pH 8.0, 75 μl), poly(U) (0.06 M, 62 μl), water (40 μl) and Ala-AMP (0.38 M, 10 μl). Tube B had lutidine buffer (1.5 M, pH 8.0, 75 μl), water (102 μl) and Ala-AMP (0.038 M, 10 μl). The concentrations of poly(U) and Ala-AMP were 0.02 M. The volatile buffer was evaporated as mentioned before. The contents of reaction mixtures were dissolved in 1 ml KH_2PO_4 and 50 μl trifluoroacetic acid. The precipitated RNA from tube A was removed by filtration using Milex filter units. The products were analyzed by HPLC under conditions similar to ones described above for Ala-GMP.

Poly(A) and Ala-UMP: The contents of tube A were: borate buffer (0.02 M, pH 8.0, 100 μl), poly(A) (2.5 mM, 200 μl), water (648 μl) and Ala-UMP (9.6 mM, 52 μl). Tube B had borate buffer (0.2 M, pH 8.0, 100 μl), water (848 μl) and Ala-UMP (9.6 mM, 52 μ). The concentration of Ala-UMP was 0.5 mM. At the end of the reaction period, the reaction mixtures were acidified and the precipitated RNA was removed by filtration. The products were analyzed by HPLC using the following chromatographic conditions: column: Whatman Particil 10, ODS-3 reverse phase C-18; mobile phase: KH_2PO_4 (0.01 M, pH brought to 2.8 by H_3PO_4); flow rate: 1.0 ml/min; and detection: fluorescamine post-column-derivatization.

EFFECT OF THE CONCENTRATION OF POLY(C) ON THE PEPTIDE SYNTHESIS BY ALA-GMP

The concentration of Ala-GMP was kept at 2.5 mM, while the concentration of poly(C) was varied from 0 to 0.0475 M. In a 400 μl reaction mixture, 0, 20, 100 or 380 μl of 0.05 M poly (C) solution were placed. After adding Ala-GMP (0.062 M, 20 μl) and borate buffer (0.2 M, pH 8.0, 100 μl), the volume was increased to 500 μl with water in all tubes. The reaction was carried out at 4°C for a week. The tubes were analyzed after acidification of the sample with 1 N HCl and removal of the precipitated RNA, without any further dilution prior to HPLC.

NINHYDRIN ASSAY

The number of amino terminals in the solutions of the peptides and their hydrolysates was estimated by the ninhydrin method (Spies, 1957). For hydrolysis, 20 μl aliquots of appropriate dilutions were placed in the bottom of thick walled glass tubes with teflon stoppers. Triple distilled, constant boiling, 6 N HCl (1 ml) was added and the tightly closed tubes were heated for 24 hrs at 110°C on a sand bath. After cooling, the samples were frozen and freeze dried. To remove the remaining traces of HCl, 1 ml water was added and evaporated from these tubes by freeze drying. The amount of alanine in these hydrolysates and of free amino terminals in unhydrolyzed samples were estimated by the ninhydrin method. A standard curve was prepared by conducting the ninhydrin reaction on a series of known amounts of alanine.

RESULTS

Ala-GMP, Ala-AMP and Ala-UMP were reacted in the presence and in absence of their respective complementary polynucleotides under conditions which were likely to favor the formation of 'monomer-polymer' complexes. The products of the reactions were analyzed by HPLC. The chromatograms are depicted in figures 3 to 5. The amounts of various peptides were determined by comparison with authentic standards of peptides, Ala, $(Ala)_2$, ..., $(Ala)_6$. The results are listed in Table 1. No significant effect of the polynucleotides on the yields of the polypeptides can be observed.

Change in the concentration of the complementary polynucleotide did not alter this general result. Figure 6 displays some of the chromatograms of the products of reactions where the concentration of poly(C) was changed, while keeping the concentration of Ala-GMP constant. A variety of other conditions were explored in order to see if they lead to a template effect. A number of buffer systems (borate, cacodylate, phosphate and 2.,6-lutidine) were tested; the entire basic pH range, under which the peptide synthesis takes place, was explored; and divalent metal ions were included in the reaction mixtures. No polynucleotide induced enhancement in the yields of the peptides could be seen under these conditions either (data not shown).

In order to confirm these results by an independent analytical methods, the peptide synthesis by Ala-GMP was analyzed in more detail. The peptide products, chromatographed in Figure 3, were subjected to complete acid hydrolysis. The amount of alanine released after hydrolysis was determined by the ninhydrin assay. The ratio of the concentrations of amino terminals after hydrolysis and before hydrolysis is indicative of the average degree of polymerization. This parameter also includes any peptide larger than $(Ala)_6$, not detected by the HPLC technique used. The ratios of concentrations of amino terminals after hydrolysis and before hydrolysis were: in the presence of poly(C), 5.58 (and 5.50 in a replicate) and in absence of poly(C), 5.15. Again, there is no significant difference in the average degree of polymerization in presence and absence of the polynucleotides.

DISCUSSION

The results presented in the previous section do not confirm the proposition that monomer-polymer interactions between amino acyl nucleotide anhydrides and complementary polynucleotides may lead to an enhancement in the degree of polymerization of amino acids. The experiments were performed with a variety of anhydride and polynucleotide combinations and under a variety of physical conditions. If there were to be a significant effect on yields and lengths of peptides, it would have been detected by the two analytical procedures employed. Often, an expected phenomenon is not detected because of the narrow range of physical conditions explored. To broaden the scope of the physical conditions, additional experiments were conducted, varying different parameters that could possible influence the peptide synthesis. No template effect was observed under these conditions either.

The degree and the stability of the monomer-polymer complexes depend upon the monomer-polymer pair. While GMP and poly (C) form double helical, 1:1 complexes, at the same time poly(U) and AMP form triple helical, 1:2 complexes. In case of UMP and poly(A), these complexes are not known. Because of this, the systems of Ala-GMP and poly(C) were most likely to exhibit template effects.

Several factors could make the expected template effects impossible: In the system of amino acyl nucleotide anhydrides and polynucleotides, the organized monomer-polymer complexes may not form at all. If they do form, the stereochemical disposition of the amino acids may not be favorable for the steps of peptide synthesis.

Although the formation of monomer-polymer complexes was a likelihood in the present system, their existence has not been shown in our experiments. The traditional approach towards such a demonstration has been to follow the change in IR spectra of the mixture (Howard, et al., 1964) with temperature. Stretching of the carbonyl (-C=O) bond is sensitive to hydrogen bonding between the complementary nucleotides and can be used to detect the formation of complexes. It would be difficult to conduct such an experiment in the present system because of low the stabilities of the anhydrides at basic pH's. It is essential to use basic pH's because the peptide synthesis takes place only at such pH's. The possibility that the wrong stereochemical disposition of the amino acids is the cause for the lack of a template effect, is even more difficult to test, again due the instability of the anhydrides.

In conclusion, under the conditions studied, the complementary polynucleotides did not influence the peptide synthesis. These results make it doubtful that single nucleotides could have served as adaptors in primeval protein synthesis.

ACKNOWLEDGEMENT

We thank Professor Frederick R. Eirich for his involvement in this work.

REFERENCES

Armstrong, D. W., Senguin, R., McNeal, C. J., Macfarlane, R. D. Fendler, J. H., 1978: J. Am. Chem. Soc., 100, 4605.

Armstrong, D. W., Senguin, R., Saburi, M., Fendler, J. H., 1979: J. Mol. Evol., 13, 103.

Bridson, P. K. and Orgel, L. E., 1980: J. Mol. Biol., 144, 567.

Davis, R. J. H. and Davidson, N., 1971: Biopolymers, 11, 789.

Frazier, J. and Miles, H. T., 1982: J. Mol. Biol., 162, 219.

Howard, F. B., Frazier, J., Lipsett, M. N. and Miles, H. T., 1964: Biochem. Biophys. Res. Commun., 73, 93.

Howard, F. B., Frazier, J., Singer, M. F. and Miles, H. T., 1966: J. Mol. Biol., 16, 415.

Huang, W. M. and Ts'o, P. O. P., 1966: J. Mol. Biol., 16, 523.

Inoue, T. and Orgel, L. E., 1982: J. Mol. Biol., 162, 201.

Inoue, T. and Orgel, L. E., 1983: Science, 219, 859.

Joyce, C. F., Inoue, T. and Orgel, L. E., 1984: J. Mol. Biol., 176, 279.

Krampitz, G. and Fox, S. W., 1969: Proc. Nat. Acad. Sci. USA, 62, 399.

Lacey, J. C. and Mullins, D. W. Jr., 1983: Origins of Life, 13, 3.

Lacey, J. C. Jr., Senaratne, N. and Mullins, D. W. Jr., 1984: Origins of Life, 15, 45.

Lewinsohn, R., Paecht-Horowitz, M., and Katchalsky, A., 1967: Biochim. Biophys. Acta, 140, 24.

Lipmann, F. and Tuttle, L. C., 1945: J. Biol. Chem., 159, 21.

Moldave, K., Castelfranco, P. and Meister, A., ;1959: J. Biol. Chem., 234, 841.

Mullins, D. W. and Lacy, J. C. Jr., 1981a: Z. Naturforsch, 36b, 732.

Mullins, D. W. and Lacy, J. C. Jr., 1981b: Origin of Life, Y. Wolman, ed., D. Reidel Publishing Company, London, 209-215.

Paecht-Horowitz, M., and Katchalsky, A., 1967: Biochim. Biophys. Acta, 140, 14.

Paecht-Horowitz, M., Berger, J. and Katchalsky, A., 1970: Nature, 228, 636.

Sawai, H., 1976: J. Am. Chem. Soc., 98, 7037.

Spies, J. W., 1957, Methods in Enzymology, 3, 467.

Tyagi, S., 1987: An Inquiry into Prebiotic Peptide Synthesis, a Ph.D. Dissertation, submitted to The University of Maryland at College Park, Maryland.

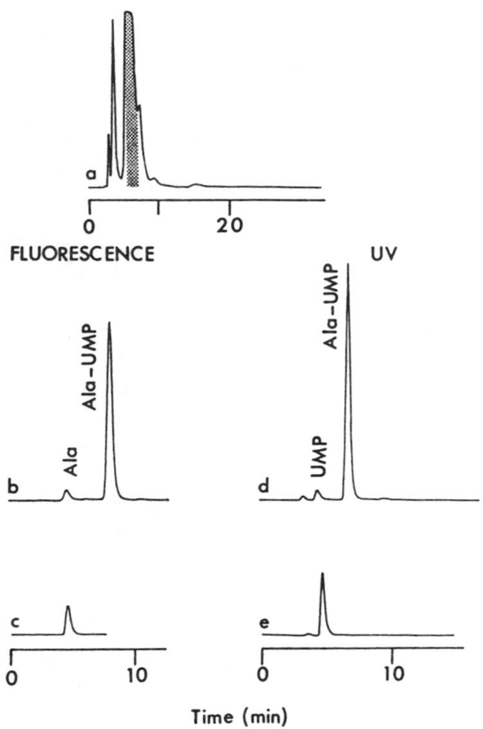

Figure 1. Purification of Ala-UMP by preparative HPLC. The effluent corresponding to shaded area of the preparative chromatogram a was collected. Analytical chromatography (b-e), shows that it does not contain appreciable amounts of free alanine and UMP. Chromatograms b and d are the purified Ala-UMP, and c and e are standards of alanine and UMP respectively.

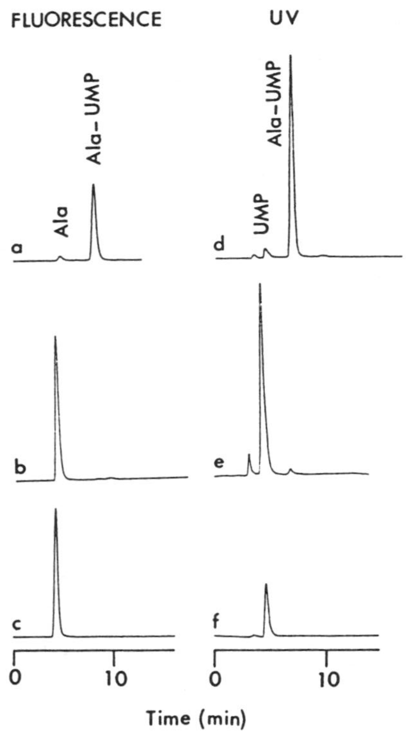

Figure 2. Chromatographic assay of Ala-UMP. The chromatograph could detect compounds with primary amino groups in a-c, whereas in d-f it could detect compounds absorbing at 256 nm. Chromatograms a and d are for a preparation of Ala-UMP, b and e are for the same preparation after 10 min of boiling, and c and f are standards of alanine and UMP. The concentration of Ala-UMP was determined by comparing the area of the alanine peak in b with the one in c or of the UMP peak in e, with the UMP standard in f.

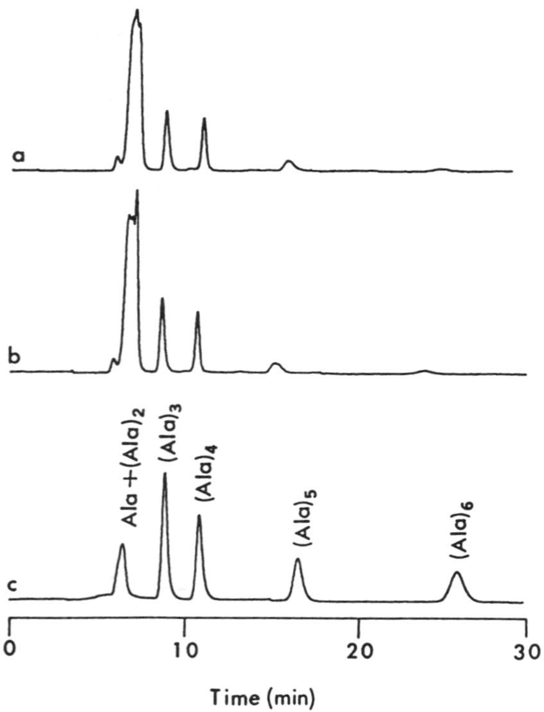

Figure 3. Separation of peptides produced by Ala-GMP in presence and absence of poly(C) under conditions likely to favor the formation of organized "monomer-polymer" complexes. Chromatogram a is in presence of poly(C), b is in its absence, and c is a standard containing equimolar concentrations of the homologous peptides of alanine ranging from monomer to hexamer.

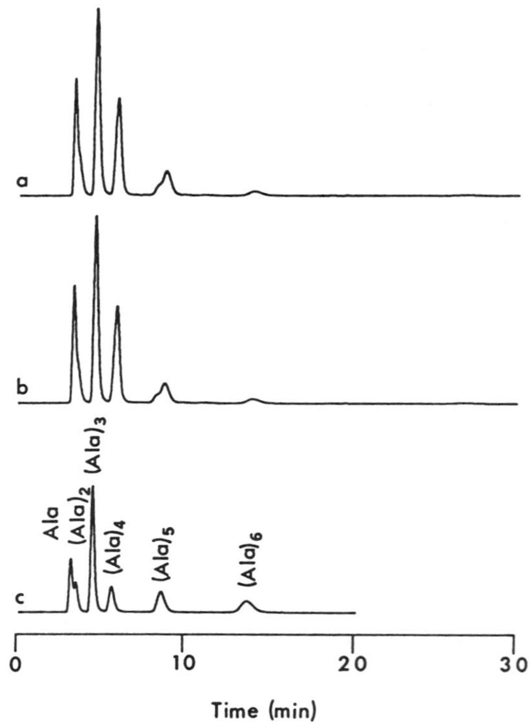

Figure 4. Separation of peptides produced by Ala-UMP in presence and absence of poly(A) under conditions likely to favor the formation of organized "monomer-polymer" complexes. Chromatogram a is in presence of poly(A), b is in its absence, and c is a standard containing equimolar concentrations of the homologous peptides of alanine ranging from monomer to hexamer.

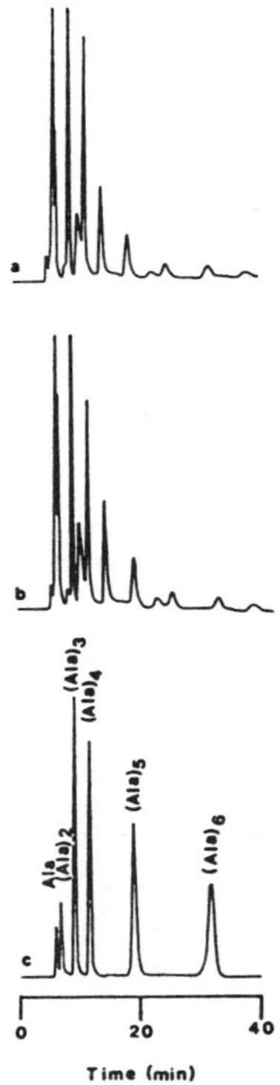

Figure 5. Separation of peptides produced by Ala-AMP in presence and absence of poly(U) under conditions likely to favor the formation of organized "monomer-polymer" complexes. Chromatogram a is in presence of poly(U), b is in its absence, and c is a standard containing equimolar concentrations of the homologous peptides of alanine ranging from monomer to hexamer.

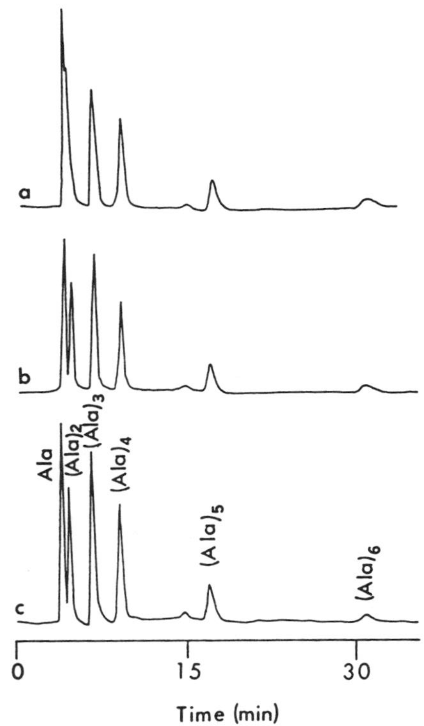

Figure 6. Effect of change in concentration of poly(C) on the peptide synthesis by Ala-GMP. The traces are chromatograms of the peptides produced by Ala-GMP [2.5 base mM] in presence of 0.0475 M poly(C), a; 0.0125 M poly(C), b: and 0 M poly(C), c.

Table 1

The Concentrations of Peptides Produced in Presence and in Absence of Complementary Homopolynucleotide

Anhydride	RNA	Peptide Concentrations (mM)				
		Ala & $(Ala)_2$	$(Ala)_3$	$(Ala)_4$	$(Ala)_5$	$(Ala)_6$
Ala-GMP	poly(C)	19.31	1.01	1.22	0.65	0.20
	poly(C)*	26.20	1.13	1.22	0.63	0.25
	-	20.60	1.18	1.24	0.66	0.24
Ala-AMP	poly(U)	6.33	1.40	0.67	0.69	0.19
	poly(U)*	11.46	-	0.95	0.74	0.25
	-	6.88	1.95	0.68	0.59	0.23
Ala-UMP	poly(A)	0.08	0.07	0.027	0.013	0.003
	-	0.08	0.07	0.024	0.009	0.003

Note: The concentrations of peptides produced by different anhydrides are not comparable with each other because the initial concentration of each anhydride was different.

*Replicate of the experiment in the row above.

ENANTIOSYMMETRY:*
THE FOUNDATION OF PROTEIN ORGANIZATION

Stanley I. Goldberg
Jane M. Crosby
Norberto D. Iusem
Usama E. Younes
University of New Orleans
Lake Front, New Orleans, Louisiana 70148

ABSTRACT

Enantiosymmetry or configurational homogeneity lies at the heart of the uniqueness of protein structure. It plays a major role in determining protein self-assembly, organization and function. Indeed, protein form and function in the contemporary biosphere would be unimaginable if amino acid residues existed in a stereochemically random state instead of in the prevailing configurationally homogeneous condition. Prevital peptides, therefore, had to be assembled with a strong enantiosymmetric bias from the racemic amino acids generally presumed to be present on the primitive Earth: otherwise the number of stereoisomers would have been impossibly high. Stereo-random formation of even a modest-sized decapeptide, for example, would give 2^{10} or 1,024, configurational isomers of which only two (the all L and the all D) would be enantiosymmetric.

The idea that only enantiosymmetric prevital peptides were qualified as candidates for participation in the chemical evolution of life is examined in the light of available experimental evidence.

*An enantiosymmetric peptide is made up of chiral amino acid residues possessing the same configuration. The term enantiosymmetric replaces the previously misapplied term "isotactic" which is one of a set of stereodescriptors reserved for polymers formed from substituted vinyl monomers (Chemical Abstracts Service).

ISBN 0-937194-18-2

INTRODUCTION

At the core of Wald's 1957 proposal for a biologically-based explanation of configurational homogeneity in the biosphere is the requirement for amino acids to have formed enantiosymmetric peptides. Wald suggested that the helical shapes of growing peptides could have served as the stereoselectivity agents, promoting the incorporation of only D-amino acids into already formed all D-helical chains, and L-amino acids into all L-helical chains. While this part of the proposal suffers from the fact that, even for the onset of helicity, peptides are required to contain about eight to twelve amino acid residues in a state of complete enantiosymmetry (see discussion in Goldberg, et al., 1987), this is not to say that in larger peptides stable secondary structures could not be important factors in the conformationally-dependent components of chain growing processes. The inherent idea that enantiosymmetry in prevital peptides could have served as the basis of stereochemical amplification mechanism has since been pursued (Goldberg and Younes, 1977; Blair and Bonner, 1980; Bonner et al., 1981; Kricheldorf and Mang, 1983; Kricheldorf, et al., 1985; Goldberg et al., 1987). Carrying this one step further, we should like to suggest that only enantiosymmetric prevital peptides were qualified for the evolutionary journey to contemporary life.

Since enantiosymmetric proteins dominate the present biosphere, perhaps peptide enantiosymmetry was a required condition in the beginning as well. If true, then a major simplification is automatically granted to the problem of accounting for the evolutionary development of protein form and function. But, what evidence is available in support of this view? (see Scheme I)

There is in fact a body of observations which seems to suggest that it may be surprisingly easy to find reaction conditions to allow or promote enantiosymmetric assembly of amino acids. Furthermore, the accepted mechanism of the formation of the amide or peptide bond appears to be consistent with this view. In this report we shall review some of that evidence, starting with an examination of the tetrahedral mechanism for nucleophilic substitution on the acyl carbon (March, 1985).

THE TETRAHEDRAL MECHANISM

Consideration of this mechanism reveals its potential to support or allow enantiosymmetric growth of a peptide chain. The general features of the mechanism are shown in Scheme II, where A represents the acylating component, with O-L standing for the various leaving groups used for carboxyl activation. The reacting amino acid B adds to the carbonyl carbon of A to give the tetrahedral intermediate C, which must be deprotonated to give D, before O-L can be lost to form the peptide bond in E.

Applying this mechanism to the problem of enantiosymmetric growth means that L-A will prefer to combine with L-B to form L,L-E, rather than combine with D-B to form L,D-E. [Of course, the required mirror image process -- the combination of D-A and D-B to form D,D-E -- will go on to an equal extent.] In order to be able to understand how such a system would favor enantiosymmetric growth, it would be necessary to be able to evaluate the large number of rate constants involved in the four major diastereomerically related pathways (and their mirror images) represented in Scheme III, where B and B′ are blocking groups.

It is easy enough to see the difficulty of evaluating the relative contribution of each pathway, owing to the complexity involved. The situation is further complicated in that each major pathway may also include a number of energetically similar conformations -- particularly with the lower molecular weight components -- requiring evaluations as well. But some simplification occurs by invoking the principles of stereoelectronic control (Deslongchamps, 1983). For example, the orbital orientation shown in 3 is required for the loss of O-L to give the Z-AMIDE, which is preferred presumably over its E diastereomer. This implies that deprotonation of 2 must take place as shown. Indeed, the details of the base catalysis in this system, although not as yet studied experimentally, appear to be important factors. But overall, anything -- temperature, solvents, etc. -- that affects the many conformational equilibria of these pathways can be expected to have a bearing on the net stereochemical outcome, that of enantiosymmetric or non-enantiosymmetric growth.

While we might despair over our inability to deal with all these conformational questions, the irony is that it is precisely because of the system's susceptibility to respond to so many factors that there should be many sets of reaction conditions capable of yielding the net result of enantiosymmetry. Supporting this idea are a fair amount of available experimental data.

SUPPORTING DATA AND DISCUSSION

The Table I shows some examples of dipeptide stereoselectivities, measured under differing conditions of solvents, reaction temperatures, activating groups and blocking groups. The important point is that variations in one or more of these parameters lead to stereoselectivities ranging from enantiosymmetric to non-enan tiosymmetric growth in each case. Moreover, reaction variations of this type, in polymerizations of valine (Akaike et al., 1975; Blair and Bonner, 1980; Kricheldorf and Mang, 1983) have given rise to polyvalines indicated to possess net stereochemistries ranging between enantiosymmetric and non-enantiosymmetric.

These data, while not constituting rigorous proof, do seem to support the idea that peptide bond-forming systems are stereochemically rather sensitive to a fair number of reaction parameters; they are, therefore, capable of being tuned or adjusted to provide biases in favor of enantiosymmetric peptide growth. This implies the possibility of yet-to-be-discovered reaction systems that could serve as evolutionary prototypes, producing complete, or nearly complete, enantiosymmetric peptides under simulated prebiotic conditions.

REFERENCES

Akaike, T., Aogaki, Y. and Inoue, S., 1975: Biopolymers, 14, 2577.

Blair, N. E. and Bonner, W. A., 1980: Origins of Life, 10, 255.

Bonner, W. A., Blair, N. E. and Dirbas, F. M., 1981: Origins of Life, 11, 119.

Cervinka, O. and Budilova, J., 1967: Collect. Czech. Chem. Commun., 32, 2383.

Deslongchamps, P., 1983: Stereoelectronic Effects in Organic Chemistry, Pergamon, New York.

Goldberg, S. I. and Younes, U. E., 1977: Presented at the 173rd ACS Meeting, New Orleans, March 20-25; See Abstracts of Papers, ORGN 005.

Goldberg, S. I., Crosby, J. M., Iusem, N. D. and Younes, U. E., 1987: J. Amer. Chem. Soc., 109, 823.

Kricheldorf, H. R. and Mang, T., 1983: Int. J. Biol. Macromol., 5, 258.

Kricheldorf, H. R., Au, M. and Mang, T., 1985: Int. J. Peptide Protein Res., 26, 149.

March, J., 1985: Advanced Organic Chemistry, 3rd ed., Wiley, New York, 290-295.

Otvos, L., Tomoskozi, I. and Mohacsi, T., 1970: Tetrahedron Lett., 1995.

Sokolowska, T. and Biernat, J. F., 1966: Roczniki Chem., 40, 1665.

Steinman, G., 1967: Experienta, 23, 177.

Wald, G., 1957: Ann. N.Y. Acad. Sci., 57, 352.

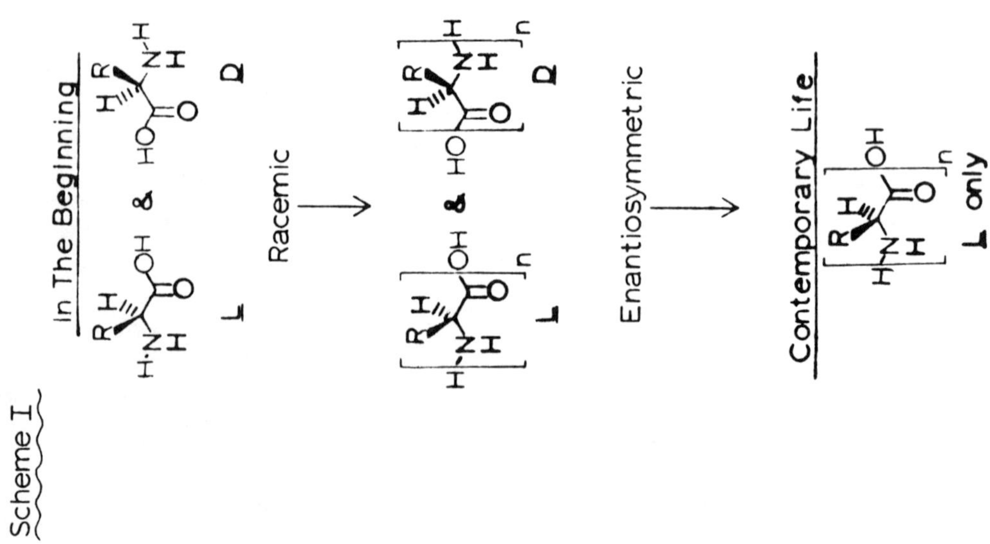
Scheme I
In The Beginning
L
D
Racemic
L
D
Enantiosymmetric
Contemporary Life
L only

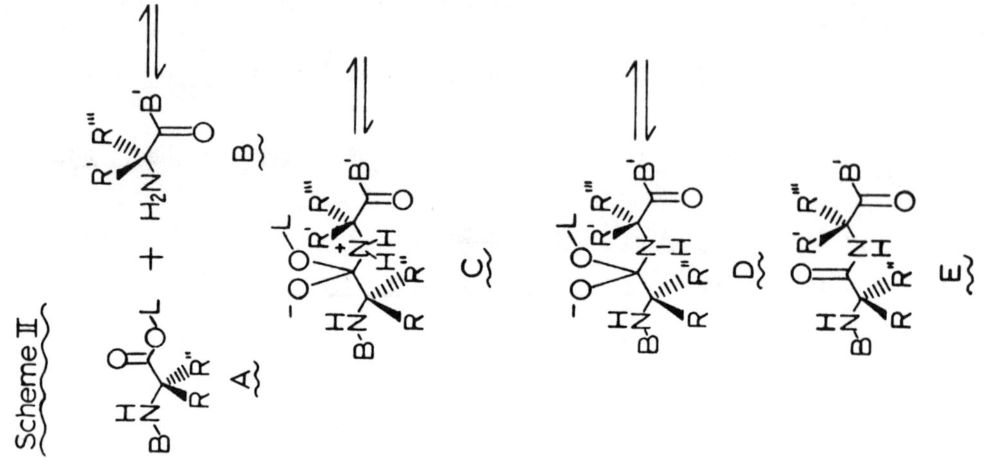
Scheme II
A
B
C
D
E

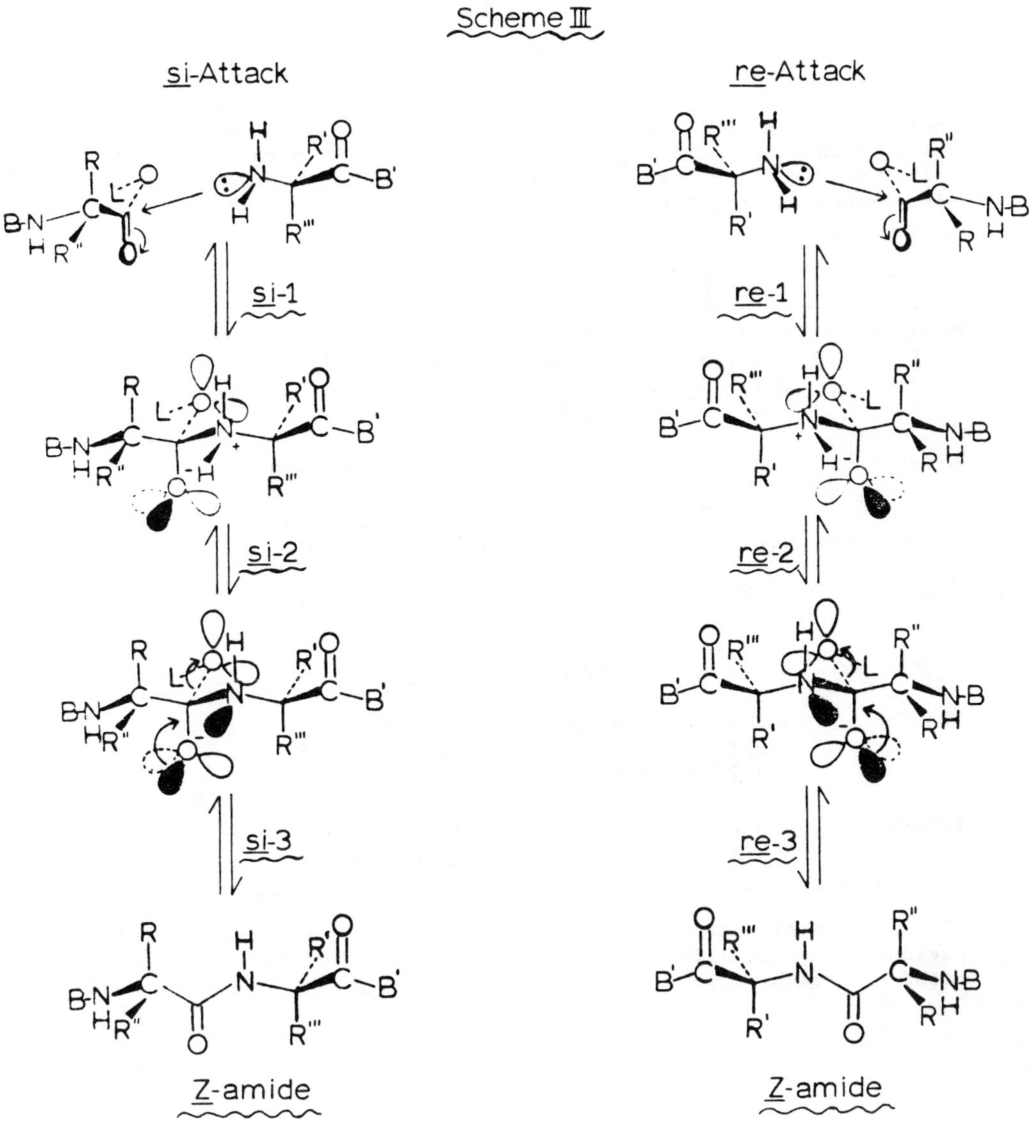
Scheme III
si-Attack
re-Attack
si-1
re-1
si-2
re-2
si-3
re-3
Z-amide
Z-amide

TABLE I

Some examples of variations in enantiosymmetric to non-enantiosymmetric formation ratios (L-L/L-D) of dipeptides, which may be attributed to differences in solvents, reaction temperatures, activating groups and blocking groups.*

DIPEPTIDE	L-L/L-D	REFERENCE
Ala-Ala	0.87/1.0	Otvos *et al.*, 1970
"	1.1-1.9/1.0	Kricheldorf and Mang, 1983
"	0.7-1.8/1.0	Kricheldorf *et al.*, 1985
"	1.4/1.0	Goldberg *et al.*, 1987
Ala-Phe	0.5-2.3/1.0	Sokolowska and Biernat, 1966
"	<1.0/1.0	Cervinka and Budilova, 1967
"	1.1-1.3/1.0	Kricheldorf and Mang, 1983
"	1.0-1.6/1.0	Kricheldorf *et al.*, 1985
Ala-Leu	<1.0/1.0	Cervinka and Budilova, 1967
"	1.0/1.0	Steinman, 1967
"	1.7-2.3/1.0	Kircheldorf and Mang, 1983
"	1.1-1.2/1.0	Kircheldorf *et al.*, 1985
Phe-Phe	0.6-1.9/1.0	Sokolowska and Biernat, 1966]
"	1.6-1.8/1.0	Kircheldorf and Mang, 1983
"	0.3-2.1/1.0	Kircheldorf *et al.*, 1985
Phe-Val	0.9-2.1/1.0	Sokolowska and Biernat, 1966
"	0.8-1.1/1.0	Kircheldorf *et al.*, 1985

*The details of these differences are too extensive for a concise presentation here. They may be seen by consulting the references cited with each entry.

MOLECULAR EVOLUTION OF NUCLEIC ACID POLYMERASES

A. Lazcano, V. Valverde, J. Fastag
Escuela Nacional de Ciencias Biologicas, IPN
Apdo. Postal 4-870, 06400 Mexico, D.F., Mexico

P. Gariglio
Centro de Investigacion y Estudios Avanzados, IPN
Apdo. Postal 14-740, 07000 Mexico, D.V., Mexico

C. Ramirez
University of Victoria
P.O. Box 1700, V8W 242, Victoria, B.C., Canada

J. Oro
University of Houston
Houston, Texas 77004

ABSTRACT

Although the chemical nature of the earliest genetic material is unknown, it is widely accepted that RNA preceded DNA as a cellular informational macromolecule. This implies that an RNA-dependent RNA polymerase must have been one of the earliest enzymes. Following our previous work, we argue that an important vestige of the ancestral RNA replicase is found in the contemporary eubacterial β'-subunit of the DNA-dependent RNA polymerase and its homologues among archaebacterial, eukaryotic and viral RNA polymerases. This hypothesis is supported by the highly conserved nucleotide sequence found for the genes coding for the largest subunits of different eukaryotic and viral RNA polymerases. We show the existence of homology between the viral-coded components of the replicases of both the Qβ virus and the MS2 virus. The ubiquity among viral RNA-dependent polymerases of a segment consisting of a Asp-Asp sequence flanked on both sides by hydrophobic amino acids had been reported previously by Kamer and Argos (1984). We have extended this observation to include the E.coli polynucleotide phosphorylase and several cellular DNA-dependent RNA polymerases. We also discuss the possible relationship between these sequences and a somewhat comparable segment found in nucleotide binding enzymes, ATPases, and yeast inorganic pyrophosphatase by Baltscheffsky et al. (1987).

ISBN 0-937194-18-2

The hypothesis that a number of contemporary cellular and viral RNA polymerases contain a polypeptide from times before the appearance of double-stranded DNA is further supported by the in vitro RNA-directed RNA polymerase activity of normally DNA-dependent RNA polymerases. The possible evolutionary relationship between the ancestral RNA polymerase and other proteins involved in bacterial DNA replication, such as DNA polymerase and DNA primase, is also discussed.

INTRODUCTION

The existence of an intracellular genetic apparatus able to store, express and, upon reproduction, transmit to its progeny information capable of undergoing evolutionary change is an essential requirement of all contemporary cells (Hartman _et al._, 1985). Replication schemes in which genetic information is stored in proteins, clay minerals, etc., have been suggested, but there is no evidence supporting their existence (Schwartz, 1987). Template-directed polymerization reactions including "nucleic acid-like" molecules in which ribose and deoxyribose have been substituted by glycerol or other molecules (Schwartz and Orgel, 1985) support the idea that both RNA and DNA were preceded by other simple linear molecules in which genetic information was coded in nitrogen bases (Schwartz, 1987). Ribonucleic acid may have been preceded by a polymer made of flexible, acyclic, perhaps prochiral nucleotide analogues of non-enzymatic origin (Joyce _et al._, 1987).

Although the precise chemical nature of the molecules that were used as carriers of genetic information in the earliest forms of life is not known, it is generally accepted that RNA was used by living systems prior to DNA as an informational macromolecule (Follmann, 1981; Lazcano _et al._, 1986). This hypothesis implies that RNA polymerase must be one of the oldest enzymes (Lazcano, 1986). We have argued elsewhere that an important vestige of the ancestral RNA-dependent RNA polymerase is found in the contemporary eubacterial DNA-dependent RNA polymerase β'-subunit and its homologues among eukaryotic, archaebacterial and viral units (Lazcano _et al._, 1987). The purpose of this paper is to review this suggestion and to develop it further by reporting on studies of additional characteristics of the eubacterial β'-subunit and its cellular and viral homologues, including codon usage and replicase activity.

HOMOLOGUES TO THE β'-SUBUNIT ARE FOUND IN ALL CELLS

Genes coding for RNA polymerases are present in many viruses and in all cells (Figure 1). In spite of their diversity, there is now evidence available that suggests a common origin for most of them. Immunochemical comparison studies between different archaebacterial DNA-dependent RNA polymerases with the _Escherichia coli_ and yeast enzymes have demonstrated that the eubacterial β'-subunit is strongly conserved and is homologous to the largest component of the three eukaryotic RNA polymerases, to the largest subunit of the enzyme from halophilic and methanogenic bacteria, and to the second largest polypeptide from thermoacidophilic archaebacterial RNA polymerases (Zillig _et al._, 1985).

These conclusions have been further confirmed by nucleotide sequence studies of the genes coding for the β'-subunit of _E.coli_ and the largest polypeptides of different eukaryotic DNA-dependent RNA polymerases. It has been shown that the eubacterial RNA polymerase β'-subunit is homologous to the largest subunits of 1) yeast RNA pol II and III (Allison _et al._, 1985); 2) _Drosophila melanogaster_ RNA pol II (Biggs _et al._, 1985); 3) mouse RNA pol II (Corden, in Biggs _et al._, 1985); and 4) vaccinia-virus RNA pol. These results are further supported by immunochemical studies that have shown that antibodies for the subunit IIo (240-kd) of the calf thymus DNA-dependent RNA pol II also recognize the _E.coli_ RNA pol β'-subunit (Llaca _et al._, 1987).

Further evidence of the presence of the genes coding for proteins homologous to the eubacterial RNA polymerase β'-subunit in eukaryotic cells is provided by their identification in the liverwort (Ohyama _et al._, 1986) and tobacco (Shinozaki _et al._, 1986) chloroplast genomes. We have found no regions homologous to the nucleotide sequence of the gene coding for the _E.coli_ RNA pol β'-subunit in animal mitochondrial genomes. We interpret these negative results to mean that the genes coding for the mitochondrial transcriptional apparatus have already migrated to the nucleocytoplasmic genome. This explanation is consistent with the endosymbiotic theory on the origin of eukaryotes (Margulis, 1981).

IS β' A CATALYTIC SUBUNIT ?

We have argued elsewhere that the evolutionary conservation of β' and its homologues among cellular and viral RNA polymerases is due to its catalytic role in the elongation phase of RNA synthesis (Lazcano, 1986; Lazcano _et al._, 1987). This idea is supported by 1) the streptolydigin sensitivity of the β'-subunit of _Bacillus subtilis_ RNA polymerase (Doi et al., 1980) and 2) the Zn-atoms that the bacterial RNA polymerase requires for catalytic activity (Mildvan and Loeb, 1979) which are bound only to the β'-subunit of the B.subtilis enzyme (Halling et al., 1977); at least one Zn-atom is present in the E.coli β'-subunit (Wu et al., 1977; Miller et al., 1979).

Further support for the hypothesis that the eubacterial β'-subunit is involved in RNA elongation can be derived from the study of the largest subunits of the eukaryotic DNA-dependent RNA polymerases which are homologous to it (cf. Section II). The highly conserved nucleotide sequences of the genes coding for the largest subunits of different eukaryotic RNA polymerases and the E.coli β'-subunit suggest that these polypeptides perform comparable functions in RNA elongation (Lazcano et al., 1987). Direct involvement in RNA elongation has been shown for the largest subunits of 1) yeast RNA pol I (Martial et al., 1975) and RNA pol II (Ruet et al., 1980); 2) Drosophilia melanogaster RNA pol II (Coulter and Greenleaf, 1985); 3) rat liver RNA pol I and II (Martial et al., 1985); and 4) rat hepatoma RNA pol I (Rose et al., 1983).

VIRAL RNA POLYMERASES

As shown in Figure 2, the genomes of the RNA Qβ- and MS2 bacteriophages are organized in a similar way. Moreover, their replicases have the same structure (Figure 3). These observations prompted us to compare the primary structure of the phage MS2-coded polypeptide of its replicase (Fiers et al., 1976) with that of the Qβ-coded polypeptide of the Qβ-replicase (Mekler, 1981). Details of the computer program used for studying these consequences will be published elsewhere. As shown in Figure 4, large portions of these polypeptides are particularly well matched, suggesting that both are derived from a common ancestor.

Primary structure comparison of RNA-dependent DNA polymerases and RNA polymerases from several plant, animal and eubacterial viruses led Kamer and Argos (1984) to discover a highly conserved fourteen-residue segment that consists of an Asp-Asp pair flanked by hydrophobic amino acid residues. It has been hypothesized that this segment may correspond to the RNA-dependent polymerase active site and/or to the nucleic acid recognition region for these polymerases (Kamer and Argos, 1984). As shown in Figure 5, we have found that this segment is also present 1) once in the viral-coded component of the Qβ-replicase; 2) once in the E.coli RNA pol β'; 3) twice in the E.coli RNA pol β- subunit; 4) once in the E.coli polynucleotide phosphorylase (Regnier et al., 1987). Similar segments can be identified in the largest subunits of 1) the Drosophila melanogaster RNA pol II; 2) the yeast RNA pol II and II; and 3) the vaccinia virus DNA-dependent RNA pol.

CODON USAGE

The codon usage for the _E.coli_ RNA pol β'-subunit and its homologues among cellular and viral RNA polymerases is shown in Table 1. Also shown is the codon usage in the genes coding for the viral-coded components of the MS2 and the Qβ replicases, and for the gene coding for the _E.coli_ polynucleotide phosphorylase (Regnier _et al._, 1987).

The patterns of codon usage seen for the _E.coli_ RNA pol β'-subunit and its homologues are similar for those seen for genes expressed at a very high level (Grantham _et al._, 1981; Crawford, 1982). The pattern of choices between synonymous codons shows considerable differences between the viral, the prokaryotic and the eukaryotic genomes (Table 1). This observation is consistent with previous comparisons on code usage among other bacterial and eukaryotic genes (Grantham _et al._, 1981). It is interesting to note that although no sequence similarity has been detected between the genes coding for the _E.coli_ RNA pol β'- subunit (Ovchinnikov _et al._, 1982) and the polynucleotide phosphorylase enzyme (Regnier _et al._, 1987), codon usage is similar in both cases (Table 1).

IS THE β'-SUBUNIT A REPLICASE?

The hypothesis that the eubacterial RNA pol β'-subunit and its archaebacterial, eukaryotic and viral homologues are vestiges of the ancestral RNA-dependent RNA polymerase is strongly supported by the observation that cellular DNA-dependent RNA polymerases can use RNA molecules and polyribonucleotides as templates. We interpret this replicase activity as vestigial from times before the evolutionary development of double-stranded DNA as an informational macromolecule. The evidences that support our contention that the catalytic site involved in RNA elongation is located on the eubacterial β'-subunit and its homologues also imply that this polypeptide is directly involved in RNA synthesis when reading RNA templates (Lazcano _et al._, 1987).

The two evidences of replicase activity of the eubacterial RNA pol β'-subunit and its homologues are:

1) eubacterial DNA-dependent RNA polymerases can use synthetic polyribonucleotides, viral RNA genomes and different cellular RNA molecules as templates in the absence of DNA and Mg^{++} and in the presence of Mn^{++} (Adman and Grossman, 1965; Biebricher and Orgel, 1973; Llaca _et al._, 1987).

2) viroids are small single-stranded covalently closed RNA molecules with a highly base-paired rodlike structure, which are found as pathogens of plant cells (Diener, 1982). Viroids do not code for their own polymerases, and the absence of viroid-specific DNA in infected cells implies that only RNA intermediates are formed during viroid replication (Riesner and Gross, 1985). The observations that confirm the suggestion that viroid synthesis is catalyzed by a DNA-dependent RNA pol reading a RNA template include a) the inhibition of *in vivo* viroid replication by an intracellular concentration of 10^{-8} M of α -amanitin (Muhlbach and Sanger, 1979); and b) in vitro viroid replication catalyzed by tomato or wheat DNA-dependent RNA pol II (Rackwitz et al., 1981) and by *E.coli* RNA pol (Rohde et al., 1982).

ADAPTING TO DOUBLE-STRANDED DNA

We have argued elsewhere that cellular double-stranded DNA-based genetic systems were selected for by neo-Darwinian mechanisms to stabilize earlier cells whose reproduction was based on RNA replication (Lazcano *et al.*, 1986) mediated by an RNA-replicase (Lazcano *et al.*, 1987). This assumption eliminates the need of primordial genes coding for ribonucleotide reductases, DNA polymerases, DNA topoisomerases, DNA-dependent ATPases that couple the unwinding of double-stranded DNA with the hydrolysis of DNA, etc. (Reanner, 1982; Lazcano, 1986). As noted above, changes in the template specificity of both cellular and viral RNA polymerases can be easily achieved in the presence of Mn++ ions (Goodman *et al.*, 1984). This suggests that the evolutionary transition from the ancestral RNA-dependent RNA replicase into the present DNA-dependent RNA polymerase required only minor modifications of the enzyme active site (Lazcano *et al.*, 1987).

The functional similarities between the template-directed enzymatic synthesis of both, RNA and DNA leads us to suggest that the genes coding for some of the proteins involved in DNA replication are the result of a process of differentiation from the ancestral RNA-dependent RNA polymerase gene. This may be the case of prokaryotic DNA polymerases. The possible evolutionary relationship between the eubacterial RNA polymerase β'-subunit and DNA polymerase is supported by 1) the sequence similarity between the β'-subunit and the DNA-binding domain of the *E.coli* DNA polymerase I (Allison *et al.*, 1985); and 2) the *in vitro* RNA-directed DNA synthesis catalyzed by the *E.coli* DNA polymerase I, using viroids as templates (Rohde *et al.*, 1982). A similar explanation for the origin of DNA primases has been discussed elsewhere (Lazcano *et al.*, 1987).

CONCLUSIONS

The classification of RNA-containing coliphages has led to the recognition of three to five different groups. The f2, R17 and MS2 viruses have been included in Group I, while the Qβ virus is a member of Group III (Weissmann et al., 1973). This taxonomical distribution is supported by 1) the differences in the sizes of the plaques formed by the Qβ- and the MS2 viruses (Overby et al., 1966); 2) the lack of serological cross-reaction among the two groups (Weissman et al., 1973); 3) the unique preference of the MS2- and the Qβ-replicases for their homologous RNA molecules as templates, i.e., neither replicase can function with the other's RNA (Haruna and Spiegelman, 1965); and 4) the competition hybridization results that had been interpreted to imply the lack of homology between the RNA genomes of the MS2- and the Qβ-virus (Weisman and Ochoa, 1967). However, the considerable similarity we discuss in this paper between the viral-coded components of the Qβ-replicase and the MS2-replicase suggest a common origin for both coliphages. Codon usage in the viral genes coding for these polypeptides support this conclusion (Table 1). This result is in agreement with similar suggestions by Konigsbert et al. (1970), and by Adams and Cory (1970) based on the comparison of the genes coding for the coat proteins of these two viruses.

In this paper, we have argued that the eubacterial RNA polymerase β'-subunit and its homologues are vestiges of the ancestral RNA-dependent RNA replicase. As discussed elsewhere (Lazcano et al., 1987), the original protoenzyme must have been a relatively simple, small, catalytic polypeptide with binding sites for the RNA template, the ribonucleotides and the metallic cofactors. It is, of course, tempting to identify the Asp-Asp sequence flanked by hydrophobic amino acids described in Section III with such ancestral forms of the enzyme.

It has been suggested that the Asp-Asp segment flanked by hydrophobic amino acids may be a possible active site, or a nucleic acid recognition region (Kamer and Argos, 1984). These possibilities are consistent with the distribution of this segment in all the polymerases we have discussed in this paper. Additional support for this interpretation is provided by 1) the presence of this sequence in the E.coli RNA pol β'- and β'-subunits, which are directly involved in RNA elongation and initiation, respectively; and 2) the Zn atoms of the E.coli RNA pol which may be located in a hydrophobic region (Miller et al., 1979). Furthermore, inactivation studies of rat liver RNA pol I and II and yeast RNA pol I by pyridoxal 5'-phosphate has shown that at least one lysyl residue is present in the active site located in the largest subunits of these enzymes (Martial et al., 1975). One such highly conserved lysyl residue is present near the Asp-Asp sequence found in the yeast

RNA pol II and III and in the largest subunits of the vaccinia virus enzyme (Figure 5).

On the other hand, Baltscheffsky et al. (1987) have shown in yeast that there are important similarities between amino acid sequences of soluble inorganic pyrophosphatase and some nucleotide binding polypeptides which include adenylate kinase and the α- and the β- subunits of eubacterial ATP synthases. These sequences are rich in hydrophobic amino acids, and precede an Asp-residue which appears to be involved in the binding of inorganic pyrophosphate or ATP. The similarity of these sequences with the Asp-Asp segment flanked by hydrophobic amino acid discussed in Section III, suggest that this may be a highly conserved region common to all nucleotide binding proteins and may be related to the most primitive form of the RNA-dependent RNA polymerase. Experiments now in progress will attempt to answer this and related questions.

ACKNOWLEDGMENT

We are indebted to Drs. Lynn Margulis and Ricardo Amils for many useful discussions, and to Professor Emile Zuckerkandl for his comments on the importance of homology studies; to Drs. M. Grumberg-Manago, H. Baltscheffsky and G. F. Joyce, who made available to us their results prior to publication; to Dr. Blanca Lilia Barron for pointing out to us the role of plant RNA polymerases in viroid replication; to Dr. E. Domingo for providing us with the Qβ virus genome; to Drs. H. Ozeki and M. Siguira for providing us with the nucleotide sequences of the liverwort and tobacco chloroplast genomes, and to Monsieur Cesar Hernandez for several useful references. This research has been partially supported by an OEA, PRDC y T (1986-1987) Grant to P.G., of the University of Victoria, a Fellowship to C.R., and by NASA Grant NGR 44-005-002 to J.O.

REFERENCES

Adams, J. M. and Cory, S., 1970: Nature, 227, 570-574.

Adman, R. and Grossman, L., 1965: Fed. Proc., 24, 602.

Allison, L. A., Moyle, M., Shales, M. and Ingles, C. J., 1985: Cell, 42, 599-610.

Baltscheffsky, H., Alauddin, M., Falk, G. and Lundin, M., 1987: Acta Chem. Scand., in press.

Biebricher, C. K. and Orgel, L. E., 1973: Proc. Natl. Acad. Sci. USA, 70, 934-938.

Biggs, J., Searles, L. L. and Greenleaf, A. L., 1985: Cell, 42, 611-621.

Broyles, S. S. and Moss, B., 1986: Proc. Natl. Acad. Sci. USA, 83, 3141-3145.

Coulter, D. E. and Greenleaf, A. L., 1985: J. Biol. Chem., 260, 13190-13198.

Crawford, I. P., 1982: In R. Milkman (ed.), Perspectives on Evolution, Sinauer Associates, Sunderland, Massachusetts, 149-163.

Diener, T. O., 1982: Ann. Rev. Microbiol., 36, 239-258.

Fiers, W., Contreras, R., Duerinck, F., Haegaman, G., Isenrentant, D., Merregaert, J., Min Jou, W., Molemans, F., Raeymaekers, A., Van den Berghe, A., Volckaert, G. and Ysebaert, M., 1976: Nature, 260, 500-507.

Follmann, H., 1981: Chemie und Biochemie der Evolution, Quelle und Meyer, Publisher, Heidelberg.

Goodman, T. C., Nagel, L., Rappold, W., Klotz, G. and Riesner, D., 1984: Nucleic Acid Res., 12, 6231-6246.

Grantham, R., Gautier, C., Gouy, M., Jacobzone, M. and Mercier, R., 1981: Nucleic Acid Res., 9, r43-r74.

Halling, S. M., Burtis, K. C. and Doi, R. H., 1978: Nature, 272, 837-839.

Hartman, H., Lawless, J. G. and Morrison, P. (eds.), 1985: Search for the Universal Ancestors, NASA SP-477, Washington.

Haruna, I. and Spiegelman, S., 1965: Proc. Natl. Acad. Sci. USA, 54, 579-587.

Joyce, G. F., Schwartz, A. W., Miller, S. L. and Orgel, L. E., 1987: Proc. Natl. Acad. Sci. USA, in press.

Kamer, G. and Argos, P., 1984: Nucleic Acid Res., 12, 7269-7282.

Konigsberg, W., Maita, T., Katze, J. and Weber, K., 1970: Nature, 227, 271-273.

Lazcano, A., 1986: In L. Margulis, R. Guerrero and A. Lazcano (eds.), The Origin of Life and the Evolution of Cells, Treballs Soc. Catalana Biol., 39, in press.

Lazcano, A., Guerrero, R., Margulis, L. and Oro, J., 1986: Abstracts of the Fifth ISSOL Meeting, Berkeley, California, July 21-25, 1986, 334-335.

Lazcano, A., Fastag, J., Garioglio, P., Ramirez, C. and Oro, J., 1987: J. Mol. Evol., submitted.

Liaca, V., Silva, E., Lazcano, A., Rangel, L. M., Gariglio, P. and Oro, J., 1987: This volume.

Margulis, L., 1981: Symbiosis in Cell Evolution, Freeman Co., San Francisco, California.

Martial, J., Zaldivar, J., Bull, P., Venegas, A. and Valenzuela, P., 1975: Biochemsitry, 14, 4907-4911.

Mekler, P., 1981: Ph.D. Thesis, University of Zurich.

Mildvan, A. S. and Loeb, L. A., 1979: CRC Crit. Rev. Biochem., 6, 219-244.

Miller, J. A., Serio, F. G., Howard, R. A., Bear, J. L., Evans, J. E. and Kimball, A. P., 1979: Biochim. Biophys. Acta, 579, 291-297.

Muhlbach, H. P. and Sanger, H. L., 1979: Nature, 278, 185-188.

Ohyama, K., Fukusawa, H., Kohchi, T., Shirai, H., Sano, T., Sano, S., Umesono, K., Shiki, Y., Takeuchi, M., Chang, Z., Aota, S., Inokuchi, H. and Oseki, H., 1986: Nature, 322, 572-574.

Ovchinnikov, Yu. A., Monastyrskaya, G. S., Gubanov, V. V., Guryev, S. S., Salomatina, I. S., Shuvaeva, T. M., Lipkin, V. M. and Sverdlov, E. D., 1982: Nucleic Acid Res., 10, 4035-4044.

Overby, I. R., Barlow, G. H., Doi, R. H., Jacob, M. and Spiegelman, S., 1966: J. Bacteriol., 91, 442-448.

Reanney, D. C., 1982: Ann. Rev. Microbiol., 36, 47-61.

Regnier, P., Grunberg, Manago, M. and Portier, C., 1987: J. Biol. Chem., 262, 63-68.

Riesner, D. and Gross, H. J.: Ann. Rev. Biochem., 54, 531-564.

Rohde, W., Rackwitz, H. R., Boege, F. and Sanger, H. L., 1982: Biosci. Rep., 2, 929-939.

Rose, K. M., Maguire, K. A., Wurpel, J. N. D., Stetler, D. A., Marquez, E. D., 1983: J. Biol. Chem. 258, 12976-12981.

Ruet, A., Sentenac, A., Fromageot, P., Windsor, B. and Lacroute, F., 1980: J. Biol. Chem., 255, 6450-6455.

Schwartz, A. W., 1987: Adv. Space Res., in press.

Schwartz, A. W. and Orgel, L. E., 1985: Science, 228, 585-588.

Shinozaki,K., Ohme, M., Tanaka, M., Wakasugi, T., Hayashida, N., Matsubayashi, T., Zaita, N., Chunwongse, J., Obokata, J., Sugita, M., Deno, H., Kamogashira, T., Yamada, K., Kusuda,J., Takaiwa, F., Kato, A., Tohdoh, N., Shimada, H., Sugiura, M., 1986: EMBO Jour., 5, 2043-2049.

Weissmann, C., Billeter, M. A., Goodman, H. M., Hindley, J. and Weber, H., 1973: Ann. Rev. Biochem., 42, 303-328.

Wu, C., Wu, F. and Speckhard, D. C., 1977: Biochemistry, 16, 5449-5454.

Zillig, W., Schnabel, R. and Stetter, K. O., 1985: Curr. Topics Microbiol. and Immunol., 114, 1-18.

ADDENDUM

Argos, P., 1988: Nucleic Acid Res., 16, 9909-9916.

Ishihama, A., and Nagata, K., 1988: CRC Critical Rev. Biochem., 23, 27-76.

Lazcano, A., Fastag, J., Gariglio, P., Ramirez, C., and Oro, J., 1988: J. Mol. Evol., 27, 365-376.

Leffers, H., Gropp, F., Lottspeich, F., Zillig, W., and Garrett, 1989: J. Mol. Biol., 206, 1-17.

Wang, T. S.-F., Wong, S. W., and Korn, D., 1989: The FASEB Jour. 3, 14-21.

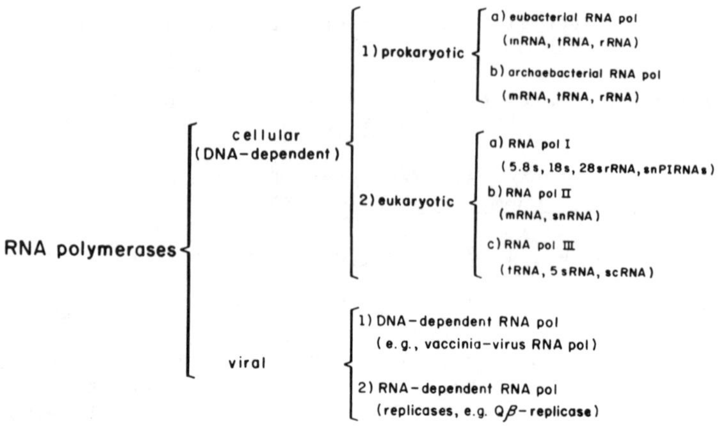

Figure 1. Major types of RNA polyerases. Neither DNA primases nor plant cell RNA-dependent RNA polymerases are included (based on Lazcano et al., 1987).

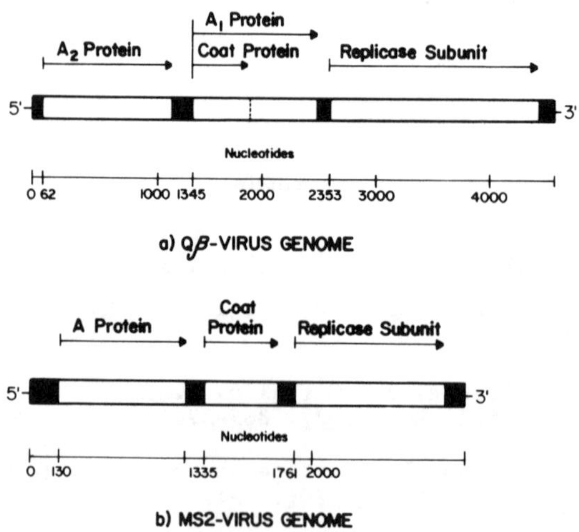

Figure 2. Schematic representation of the genetic organization of the a) Qβ virus and b) MS2 virus. Non-translated intercistronic regions are indicated in black. A2 and A are maturation proteins.

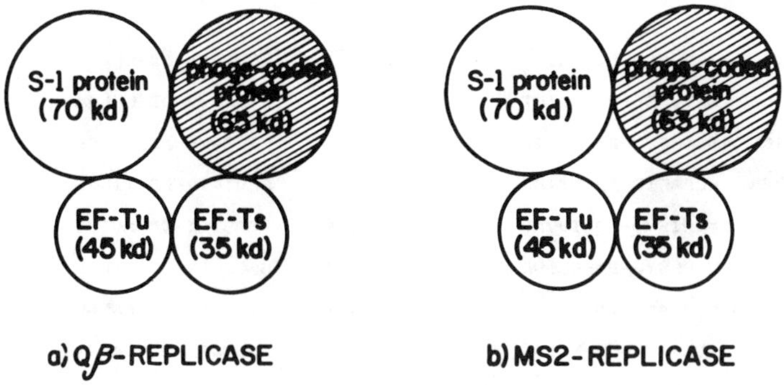

Figure 3. Subunit structure of the a) Qβ-replicase and b) MS2-replicase. EF-Tu and EF-Ts are <u>E.coli</u> ribosomal proteins.

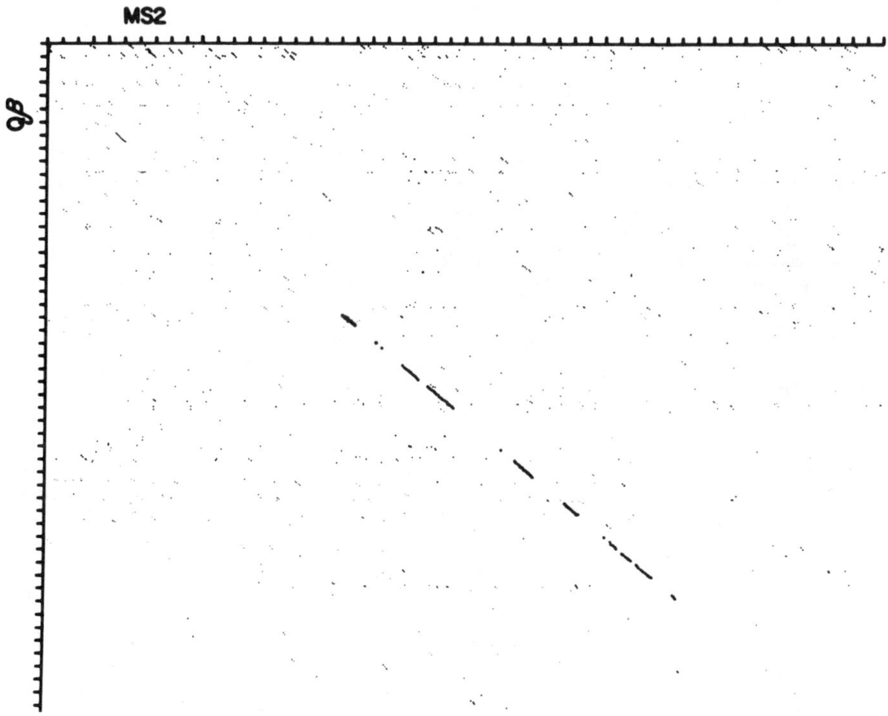

Figure 4. Comparison matrix of amino acid sequences of the Qβ-coded component (Y-axis) and the MS2-coded component (X-axis) of their viral replicases. Sequence similarity begins in the Qβ component amino acid residue at position 209.

TMV	IKGAFCG-DD-SLLYF	RSV	LCMLHYM-DD-LLLAA
AMV	NFVVASG-DD-SLIGT	MuLV	LILLQYV-DD-LLLAA
BMV	DCAIFSG-DD-SLIIS	CaMV	KFCCVYV-DD-ILVFS
SNBV	RCAAFIG-DD-NIIHG	Qβv	SEVTVYG-DD-IILPS
FMD	YTMISYG-DD-IVVAS	E.coli β'	SGASVGI-DD-MVIPE
EMC	VKVLSYG-DD-DLLVA	E.coli β	GSGILSK-DD-IIDVM
Polio	LKMIAYG-DD-VIASY	E.coli β	LIPFLEH-DD-ANRAL
CPMV	IGLVTYG-DD-NLISV	E.coli pnp	AAVMVSM-DD-TAVFV
ATLV	CTILQYM-DD-ILLAS	yeast RNA pol II	NESQRGE-DD-LTFKL
MSV	GTIGIYG-DD-IICPS	yeast RNA pol III	DSPASNE-DD-LTVKL
FLU	NASCAAM-DD-FGLIP	Drosophila RNA pol II	FGAAKNQ-DD-LTHKL
HBV	CLAFSYM-DD-VVLGA		

Figure 5. Alignment for several viral and cellular polymerase regions around the Asp-Asp sequence. (Symbols: V=virus; TM, tobacco mosaic; BM, brome mosaic; AM, alfalfa mosaic; SNB, sindbis; CPM, cowpea mosaic; FM, foot and mouth disease; EM, encephalomyocarditis; RS, Rous sarcoma; HB, hepatitis B; Mul, Maloney Murine Leukeumia; ATL, adult T-cell leukemia; MS, phage MS2; FL, influenza; CaM, cauliflower mosaic; Qβ, Qβ phage. Sequences for yeast and <u>Drosophila</u> RNA polymerases refer to the largest subunit, and pnp to <u>E.coli</u> polynucleotide phosphorylase.

Table 1. Codon usage for the E.coli RNA pol β'-subunit gene, the genes for the largest subunit of Sacharomyces cerevisae RNA pol II and III (S.c.II and S.c.III), and of Drosophila (Drosoph.), the 147-kda subunit of the vaccinia RNA pol, and liverwort chloroplast (L.rpo C1 and C2). Codon usage is also shown for E.coli polynucleotide phosphorylase (pnp), and the viral-coded components of the $Q\beta$- and the MS2-replicase ($Q\beta$ rep and MS2 rep, respectively).

aa	Codon	β' Freq.	β' % tot	$Q\beta$ rep Freq.	$Q\beta$ rep % tot	MS2 rep Freq.	MS2 rep % tot	pnp Freq.	pnp % tot	Vaccinia Freq.	Vaccinia % tot	S. cerev II Freq.	S. cerev II % tot	S. cerev III Freq.	S. cerev III % tot	Drosophila Freq.	Drosophila % tot	L. rpoC1 Freq.	L. rpoC1 % tot	L. rpoC2 Freq.	L. rpoC2 % tot
Ala	GCU	28	1.99	15	2.55	15	2.75	25	3.52	26	2.02	38	2.20	45	3.08	1	.21	16	2.34	15	1.08
	GCC	11	.78	13	2.21	7	1.28	9	1.27	14	1.09	22	1.27	14	.96	14	2.97	2	.29	4	.29
	GCA	32	2.27	10	1.70	8	1.47	16	2.25	14	1.09	28	1.62	26	1.78	4	.85	14	2.05	19	1.37
	GCG	52	3.70	9	1.53	12	2.20	23	3.23	6	.47	9	.52	13	.89	6	1.27	1	.15	3	.22
Arg	CGU	75	5.33	14	2.38	11	2.02	32	4.50	4	.31	45	2.61	25	1.71	9	1.91	6	.88	7	.51
	CGC	24	1.71	6	1.02	10	1.83	13	1.83	1	.08	0	0.00	1	.07	13	2.76	2	.29	0	0.00
	CGA	0	0.00	4	.68	7	1.28	0	0.00	8	.62	1	.06	6	.41	4	.85	15	2.19	14	1.01
	CGG	0	0.00	5	.85	7	1.28	0	0.00	0	0.00	1	.06	0	0.00	4	.85	0	0.00	2	.14
	AGA	0	0.00	7	1.19	5	.92	0	0.00	25	1.94	27	1.56	38	2.60	1	.21	18	2.63	12	.87
	AGG	0	0.00	7	1.19	2	.37	0	0.00	6	.47	9	.52	14	.96	4	.85	1	.15	1	.07
Asn	AAU	1	.07	13	2.21	12	2.20	4	.56	52	4.04	41	2.38	54	3.70	8	1.70	33	4.82	113	8.15
	AAC	47	3.34	13	2.21	8	1.47	16	2.25	28	2.18	31	1.80	25	1.71	10	2.12	3	.44	11	.79
Asp	GAU	34	2.42	24	4.07	18	3.30	19	2.67	57	4.43	80	4.63	61	4.18	20	4.25	19	2.78	33	2.38
	GAC	47	3.34	14	2.38	14	2.57	24	3.38	16	1.24	22	1.27	28	1.92	9	1.91	1	.15	3	.22
Cys	UGU	7	.50	10	1.70	5	.92	0	0.00	15	1.17	18	1.04	21	1.44	2	.42	7	1.02	9	.65
	UGC	8	.57	5	.85	2	.37	1	.14	4	.31	3	.17	3	.21	10	2.12	4	.58	4	.29
Gln	CAA	3	.21	4	.68	7	1.28	7	.98	28	2.18	53	3.07	41	2.81	1	.21	42	6.14	63	4.55
	CAG	47	3.34	6	1.02	8	1.47	25	3.52	9	.70	13	.75	6	.41	18	3.82	2	.29	2	.14
Glu	GAA	83	5.90	7	1.19	9	1.65	47	6.61	65	5.05	84	4.87	75	5.14	3	.64	44	6.43	68	4.91
	GAG	26	1.85	21	3.57	13	2.39	12	1.69	19	1.48	34	1.97	13	.89	22	4.67	2	.29	6	.43
Gly	GGU	85	6.04	19	3.23	18	3.30	33	4.64	31	2.41	86	4.98	76	5.21	13	2.76	16	2.34	21	1.52
	GGC	28	1.99	10	1.70	7	1.28	29	4.08	3	.23	10	.58	13	.89	19	4.03	2	.29	7	.51
	GGA	0	0.00	3	.51	8	1.47	0	0.00	28	2.18	13	.75	6	.41	6	1.27	26	3.80	37	2.67
	GGG	2	.14	7	1.19	10	1.83	0	0.00	7	.54	5	.29	10	.68	0	0.00	0	0.00	5	.36
His	CAU	4	.28	4	.68	4	.73	4	.56	15	1.17	28	1.62	24	1.64	6	1.27	14	2.05	27	1.95
	CAC	17	1.21	7	1.19	6	1.10	7	.98	6	.47	7	.41	6	.41	9	1.91	3	.44	4	.29
Ile	AUU	17	1.21	14	2.38	7	1.28	13	1.83	60	4.66	60	3.48	58	3.97	10	2.12	41	5.99	84	6.06
	AUC	75	5.33	9	1.53	13	2.39	39	5.49	37	2.87	26	1.51	17	1.16	19	4.03	2	.29	5	.36
	AUA	0	0.00	13	2.21	12	2.20	0	0.00	25	1.94	11	.64	20	1.37	1	.21	15	2.19	58	4.18
Leu	UUA	3	.21	8	1.36	8	1.47	0	0.00	36	2.80	50	2.90	59	4.04	1	.21	45	6.58	98	7.07
	UUG	1	.07	11	1.87	5	.92	2	.28	27	2.10	41	2.38	33	2.26	9	1.91	3	.44	8	.58
	CUU	3	.21	8	1.36	8	1.47	3	.42	19	1.48	20	1.16	14	.96	4	.85	23	3.36	32	2.31
	CUC	7	.50	16	2.72	15	2.75	0	0.00	13	1.01	5	.29	3	.21	3	.64	1	.15	0	0.00
	CUA	0	0.00	8	1.36	8	1.47	1	.14	15	1.17	16	.93	18	1.23	2	.42	6	.88	5	.36
	CUG	125	8.88	9	1.53	7	1.28	45	6.33	12	.93	11	.64	13	.89	14	2.97	1	.15	1	.07
Lys	AAA	62	4.41	7	1.19	9	1.65	29	4.08	65	5.05	63	3.65	58	3.97	8	1.70	52	7.60	154	11.11
	AAG	25	1.78	17	2.89	15	2.75	5	.70	49	3.81	29	1.68	46	3.15	28	5.94	2	.29	3	.22
Met	AUG	35	2.49	7	1.19	10	1.83	20	2.81	29	2.25	45	2.61	37	2.53	19	4.03	11	1.61	18	1.30
Phe	UUU	9	.64	13	2.21	12	2.20	2	.28	33	2.56	38	2.20	27	1.85	5	1.06	29	4.24	77	5.56
	UUC	26	1.85	15	2.55	16	2.94	17	2.39	26	2.02	20	1.16	24	1.64	8	1.70	1	.15	6	.43
Pro	CCU	3	.21	15	2.55	10	1.83	2	.28	16	1.24	39	2.26	25	1.71	3	.64	19	2.78	16	1.15
	CCC	0	0.00	3	.51	4	.73	0	0.00	7	.54	8	.46	8	.55	10	2.12	2	.29	1	.07
	CCA	10	.71	2	.34	3	.55	3	.42	11	.85	73	4.23	26	1.78	10	2.12	12	1.75	21	1.52
	CCG	45	3.20	9	1.53	9	1.65	26	3.66	10	.78	2	.12	7	.48	8	1.70	2	.29	2	.14
Ser	UCU	24	1.71	17	2.89	7	1.28	17	2.39	31	2.41	50	2.90	41	2.81	2	.42	17	2.49	38	2.74
	UCC	27	1.92	8	1.36	12	2.20	6	.84	11	.85	26	1.51	20	1.37	7	1.49	1	.15	6	.43
	UCA	1	.07	5	.85	6	1.10	0	0.00	15	1.17	39	2.26	16	1.10	0	0.00	6	.88	24	1.73
	UCG	5	.36	12	2.04	10	1.83	1	.14	4	.31	19	1.10	9	.62	9	1.91	0	0.00	1	.07
	AGU	0	0.00	7	1.19	4	.73	0	0.00	22	1.71	18	1.04	11	.75	3	.64	10	1.46	21	1.52
	AGC	14	1.00	7	1.19	9	1.65	8	1.13	5	.39	9	.52	4	.27	1	.21	0	0.00	2	.14
Thr	ACU	23	1.63	7	1.19	4	.73	13	1.83	23	1.79	27	1.56	19	1.30	3	.64	13	1.90	40	2.89
	ACC	46	3.27	6	1.02	12	2.20	25	3.52	10	.78	18	1.04	13	.89	9	1.91	0	0.00	4	.29
	ACA	1	.07	9	1.53	7	1.28	0	0.00	23	1.79	49	2.84	29	1.99	7	1.49	11	1.61	35	2.53
	ACG	7	.50	9	1.53	7	1.28	3	.42	14	1.09	19	1.10	6	.41	6	1.27	0	0.00	4	.29
Trp	UGG	9	.64	7	1.19	9	1.65	2	.28	10	.78	13	.75	10	.68	3	.64	9	1.32	16	1.15
Tyr	UAU	7	.50	11	1.87	6	1.10	2	.28	43	3.34	26	1.51	26	1.79	4	.85	22	3.22	51	3.68
	UAC	27	1.92	12	2.04	16	2.94	11	1.55	22	1.71	30	1.74	14	.96	5	1.06	4	.58	3	.22
Val	GUU	53	3.77	19	3.23	9	1.65	34	4.78	25	1.94	70	4.06	44	3.01	4	.85	19	2.78	29	2.09
	GUC	7	.50	5	.85	10	1.83	7	.98	15	1.17	20	1.16	28	1.92	15	3.18	0	0.00	3	.22
	GUA	32	2.27	4	.68	6	1.10	13	1.83	29	2.25	16	.93	13	.89	3	.64	8	1.17	26	1.88
	GUG	17	1.21	3	.51	7	1.28	16	2.25	8	.62	12	.70	20	1.37	12	2.55	4	.58	4	.29
	Total	1407	100.00	589	100.00	545	100.00	711	100.00	1287	100.00	1726	100.00	1460	100.00	471	100.00	684	100.00	1386	100.00

THREE MODELS OF A MINIMAL CELL

Gail Raney Fleischaker
Boston University
Boston, Massachusetts 02215

ABSTRACT

What we require in reconstructing the _origins_ of life is a set of rigorous criteria by which we can recognize _life_. That statement implies the possibility of distinguishing life from non-life, and this paper offers the criteria of autopoiesis by which that categorical distinction can be made. In developing the criteria, three models of a minimal cell will be presented:

1. the _minimal contemporary cell_, a description of the simplest living entity observable among all organisms today, that is, the simplest known phospholipoprotein- bound entity capable of DNA-directed, ATP-driven, enzyme-mediated metabolic processes common to all life as we know it;

2. the _minimal Terran cell_, a theoretical construction of the simplest possible membrane-bound entity capable of self-assembly and metabolism within the constraints of our planet; and

3. the _minimal universal cell_, a theoretical construction of the organizing principle definitive of life wherever it may be found.

I submit that _autopoiesis_ is that life-defining principle. Its criteria are the self-boundedness, self-generation, and self-perpetuation of an entity as a consequence of that entity's dynamic interchange with its surroundings.

INTRODUCTION

The problem to be addressed here can be put quite starkly: in pursuit of the _origins_ of life, we have yet to define _life_. Clearly we must have a definition of life before we can possibly identify life at its origin. Neither an intuitive sense, nor even a practiced eye alone will suffice. What is needed is a set of rigorous criteria by which life can be recognized.

ISBN 0-937194-18-2

My aim in this paper is to bring the criteria of autopoiesis, the principle of organization of the living, to the problem of the origins of life. Before developing those criteria, however, I want to point briefly to four underlying philosophic issues. First is the implication, in demanding criteria for life itself, that a distinction between life and non-life is even possible. I argue here not only that a categorical distinction is possible but that the principle of autopoiesis is unequivocal in making that distinction.

The other three issues serve as operating assumptions and can be stated simply. First, life has a single common ancestor. While recent life is surely more complex than life at its origin, it is continuous with that origin. Second, the complexity of life observed through evolutionary time is the increasing complexity of organization -- the proliferation of pathways and sub-pathways productive of cellular components. Third, in its minimal description, life is processes, not parts alone. It is clear that metabolic pathways require molecular components, yet the components and the pathways are complements of one another and have no role apart from each other: while physical structures are required, the chemical pathways are not reducible to those structures.

There are two different avenues for approaching the origins of life. One is by chemical synthesis of the molecular types presumably present on the ancient prebiotic Earth. The second is by theoretical analysis of contemporary biological phenomena with extrapolation of the simplest possible living system, the minimal cell. In the best of all possible scientific worlds, the two approaches will meet in a reciprocal characterization of the first cell, of life at its origin (Morowitz, 1985).

With these assumptions made explicit, it must be evident that my approach to the criteria-of-life problem is by the second avenue, by a system analysis of contemporary life. What distinguishes life from non-life is the kind of relationships among the structural components of a living system. Those relationships are maintained through a set of dynamic and systematic processes required in all living cells. The organization of those life processes is amenable to systems analysis. The intent of the system-analytic approach in reducing life to a minimal description is to expose the <u>logic of biological organization</u> at its simplest level. Extrapolation to the simplest possible living system will identify those cell processes minimally required for life. Generalization of those irreducible processes, then, will provide a set of criteria definitional of life itself.

THE MINIMAL CONTEMPORARY CELL

A recent example of the minimal cell model (Margulis, 1981), based on the now-classic analysis of biological self-replicating systems (Morowitz, 1967), is shown in Figure 1. Minimally-required cell structures are indicated graphically, and the several classes of cell components (DNA, RNAs, enzymes, membrane, and cytoplasm) are listed along with the related function and the types of biochemical compounds implied in the synthesis of each component group. This cell could be described as a volume of cytoplasmic solvent, capable of DNA-directed, ATP-driven, and enzyme-mediated metabolism, enclosed within a phospholipoprotein membrane capable of active energy and material transport. Because the cells from which the model is drawn are available for our direct observation, the minimal cell -- its structural components and their functions -- can be stipulated in considerable detail. The model serves both as a _minimal_ description for all living cells and as a _complete_ description for the simplest of all living cells, such as the free-living microaerophilic bacterium, _Thermoplasma acidophilum_ (Searcy, 1982).

While the list of the groups of components in this minimal cell is short, the molecules are too complex and the cell functions are too elaborate to represent anything comparable to life at its origin: the list itself may be minimal, but only in the context of the components' functions in contemporary life.

One could argue about how many enzymes, amino acids, or proteins are required for the proper functioning of the minimal contemporary cell, but that would be only quibbling. That a minimal number of enzymes is required is because enzymes themselves are required. Reducing the number of enzymes may disrupt the system, but it will not change the fact of its being an enzyme-dependent system -- that is, it may well render the cell non-functional, but it cannot make the cell any less contemporary, nor any more minimal.

THE MINIMAL TERRAN CELL

The primordial cell will have differed from the contemporary cell not in metrics but in levels of organization: it will have fewer kinds of molecular components and will require fewer pathways and fewer hierarchical levels in any pathway. A model of that primordial cell, of life at its origin, will be the simplest conceivable entity capable of the simplest possible life on Earth: the minimal Terran cell. Unlike the minimal contemporary cell, rich in observed detail, the minimal Terran cell is a lean, theoretical model by necessity: because we cannot know the atmospheric composition and surface conditions of the planet at the time when life arose, we must refrain from detailing specific structures and components. Indeed, in modeling the minimal Terran cell, we are entitled to stipulate only two things beyond the logical requirement of its system-defining membrane and its simplest possible organization: 1) the original cell will have been formed within the physical and thermodynamic constraints of our planetary chemistry, and 2) the elements composing its cellular structure will have been the same as those found in contemporary cells. (The second stipulation, derived from the underlying assumption of continuity with an original common ancestor, has the effect of ruling out any theory of "genetic takeover" in evolution from the original cell.)

The minimal Terran cell model will serve as a general profile against which any particular candidate postulated as the original cell can be compared. And while the general model must be agnostic as to molecular structure and biochemistry of components, any particular candidate will be specifiable in detail. That is, to be viable, a candidate must present itself as a detailed and operational cell system -- as an entity fully capable of function prior to the evolutionary acquisition of the particular components and metabolic pathways required in the contemporary cell. A candidate can be extrapolated from the minimal contemporary cell by reducing contemporary organizational requirements. Such system simplification can be accomplished by deletion of an entire molecular type, by dismantling an entire network of relationships within that cell system: if the resulting system still has integrity -- if metabolic functions are conceivable even in the absence of the deleted components -- its organization will necessarily be at a simpler level. If the molecular activities of the postulated system are consistent with empirical research findings, the candidacy of that system as the original cell is at least possible.

To be plausible, the candidate must present itself in evolutionary continuity with the kinds of molecules and synthetic pathways of the contemporary cell. It is not enough, for example, to hypothesize that a particular molecule was capable of a particular function in the ancient system: the evolution of both, the contemporary molecule and its contemporary role, from that ancient system must also be postulated.

AN ANCESTRAL CELL . . .

Because proteins are such complex molecules and their synthesis requires so many steps, the are unlikely components of a fundamentally simple cell system. A cell system prior to the evolutionary acquisition of enzymes, an "eobiont" (Pirie, 1953), can be extrapolated from the minimal contemporary cell by systematic deletion of all proteins. Structural components lost as a result of that wholesale deletion are: 1) all structural proteins (e.g., membrane transporters and pumps) and all enzymes (e.g., polymerases, isomerases, and reductases as well as metabolic enzymes); 2) all structures and molecules specific to protein synthesis (i.e., ribosomes, mRNAs, and rRNAs); and 3) DNA and tRNA, genetic molecules whose synthesis in the contemporary cell is protein-dependent. The respective activities displayed by these components within the contemporary cell system are: 1) mediator of energy transport and substrate metabolism; 2) catalyst in protein synthesis; and 3) director of system activity and replicator of the system-informational molecules.

To warrant the candidacy of a protein-independent system for the original cell, it must be shown that these activities are possible in the absence of protein. Evidence for the first is the permeability to water, small ions, and small inorganic molecules of liposomes made from biological lipids (Hargreaves and Deamer, 1978); the osmotic activity and maintenance of ionic gradients (Deamer and Oro, 1980) and macromolecular encapsulation (Deamer and Barchfeld, 1982) in bilayer vesicles. These research findings support the possibility of active membrane transport in the absence of proteins.

In the contemporary cell system, RNA and DNA are involved in protein catalysis and genetic replication. Since the synthesis of both RNA and DNA is enzyme-dependent, the deletion of protein from the contemporary cell would mean the loss of both nucleic acids. Both system roles would be lost as well, unless one of the nucleic acids could assume the two roles and could itself be synthesized in a protein-independent cell. Since protein-synthesis is known to take place in the absence of DNA but not in the absence of RNA (Lazcano, 1986), it is reasonable to examine the possibility of RNA's playing the dual role. The activities of RNA and DNA in contemporary cell systems are of two clearly distinct kinds: enzymatic activity (resulting from the steric, three-dimensional molecular structure) and informational replication (resulting from the two-dimensional molecular structure) (Gilbert, 1986). The two activities must be shown as possible in the absence of protein. The possibility of protein-independent enzymatic activity is supported in recent findings of RNA enzymatic activity (Chech, 1985). The possibility of RNA self-replication is supported in recent findings of DNA- and protein-independent production of RNA (Sharp, 1985). These two findings taken together lend credence to the suggestion that RNA could have been capable of assuming both enzymatic and informational roles in the ancient cell.

The hypothesis of a protein-independent cell as an ancestor to contemporary life is given plausibility in a recent summary of the evidence for RNA as the original cell genome (Lazcano, 1986), in several recent scenarios (Alberts, 1986; Gilbert, 1986), and in a recent explanation for the evolutionary transition from RNA to DNA in early cells (Lazcano et al., in press).

. . . AND A CANDIDATE FOR THE ORIGINAL CELL

The force of the minimal Terran cell model lies in its insistence on a single question: how can the cell system at hand be further simplified? The three system roles -- metabolic mediation, protein catalysis, and system-information replication -- must be examined and the necessity for each in an ancestral eobiont system must be questioned. 1) Energy capture and transduction and metabolic transformation are essential to the living system and will be found intact even in the absence of mediating enzymes. It is reasonable to expect that in a fundamentally simple system the metabolic processes would be direct and there would be no need for mediation. 2) In the absence of proteins, there would be no role for protein catalysis or synthesis. 3) Where there is a central director of system activity, a mechanism for replication of system information is required. Yet while system activity itself is required in every living cell, in the simplest possible cell there may be no need for centralization of that activity. With no centralization of system activity, there is no role for the system-informational molecules, and thus no need of a mechanism for their replication.

In sum, then, the minimal Terran cell is the simplest possible cell to exhibit life on Earth. The chemical compounds comprising the membrane and the metabolic solvent must be bracketed for the minimal Terran cell model, but will be specified in detail in any candidate for the single common ancestor of life on Earth. The membrane is the only required cell structure and metabolism is the only required cell function (see Figure 2). The Terran cell membrane, a selectively permeable self-boundary, would have established system identity and enabled the sequestering of system products against external chemical gradients. Metabolism, the Terran mechanism of internal transformations, would have enabled the whole system to survive the constant replacement of its parts. Such Terran life would have arisen in self-assembly driven by ionic forces, hydrogen bonding, and hydrophobic forces -- an inevitable effect of our planetary chemistry.

THE MINIMAL UNIVERSAL CELL

From minimal Terran life, we can generalize to the processes that are definitional of life. True of all life as we know it, yet removed from the contingencies of Terran existence, those processes will display the organizing principle of life -- autopoiesis -- and will provide a definition of living systems not restricted to the specific molecules or structures with which we are familiar in contemporary life. In identifying living <u>systems</u>, autopoiesis provides a literally "meta-physical" view of life. The criteria of <u>autopoiesis</u> (Gk. "self" + "formation") (Maturana, 1981; Schwegler and Tarumi, 1986; Varela, <u>et al.</u>, 1974) are the systematic criteria by which any living system may be recognized in our laboratories, on other planets, and under other conditions yet unknown.

The minimal universal cell (Figure 3) is a schematic embodiment of the autopoietic criteria. A boundary structure is a system-logical requirement -- and the membrane is a bio-logical requirement -- for a system's being separated and contained as distinct from its surroundings. The first autopoietic criterion is <u>self-boundedness</u>, the self-assembly of the living system's boundary and component structures. The illustration is not meant to imply any particular boundary structure but to indicate only that the membrane is permeable to its surroundings and is assembled by processes internal to the cell. The second autopoietic criterion is <u>self-generation</u>, the transport and export of matter and energy across the cell membrane and away from thermodynamic equilibrium. The requirement of dynamic interchange with the cell's surroundings ensures the in-gathering of energy sufficient to start up and drive the system processes. That is, the electrochemical ionic forces of the self-assembled molecular structures of the original cell would have been in themselves enough to account for the thermodynamic driving force of the cell system. The third autopoietic criterion is <u>self-perpetuation</u>, the recursive organization of processes internal to the cell sufficient for transformation of raw materials and constant replacement of constituent parts, so that the whole ongoing system tends away from dissipation.

In short, the principal of autopoiesis identifies as living a system which arises as a consequence of its own activities within a boundary of its own making. Surely, if we were to observe a stable, membrane-bound entity undergoing metabolic transformations away from thermodynamic equilibrium, we would instinctively affirm that entity as alive. The principle of autopoiesis goes beyond instinctive affirmation by providing operational criteria for life by which we may test any system. Of the three autopoietic criteria, self-assembly is a structural requirement and accounts for system identity and structural integrity. The second and third criteria are organizational requirements, providing for a system that is thermodynamically open and informationally self-determined. Self-generation accounts for the energetics capable of starting up the system process and ensuring its coherence in space; self-perpetuation accounts for the internal processes of metabolic transformations capable of ensuring the system's continuity in time. In combination, these three criteria are both necessary and sufficient to define life (Maturana and Varela, 1980).

Absent from the autopoietic list are three commonly-held criteria: genetic replication, growth, and reproduction. While those three are characteristic of particular life (contemporary Terran), they are not definitional of universal life. While genetic replication is unarguably efficient in securing structural stability and in directing the system away from dissipation, it will be found only in life systems complex enough to require centralization of system activity and, thus, faithful copying of the system-informational molecules. Thus genetic replication accomplishes greater prolongation of system parts, but is not in itself a criterion of life: as a mechanism productive of non-dissipation, it is already subsumed in the autopoiesis of an entity.

Much the same case can be made against growth and reproduction as criteria of life. Metabolism is a mechanism accomplishing synthesis and degradation of system parts. The rate by which those two sets of processes differ is either metabolically negative (not life-sustaining), or metabolically positive (life-sustaining) (see Figure 4). It is the least differential rate, ensuring the necessarily constant replacement of system components, that is sufficient for self-perpetuation of the living system. When the rate of synthesis outpaces the rate of degradation, and there is a physical accumulation of the system's structural components, there is morphological expansion ("growth") up to some critical point of accumulation, and cell division and multiplication ("reproduction") beyond that critical point. Thus neither growth nor reproduction is a criterion of life: as a consequence of positive metabolism, each is subsumed in the autopoiesis of an entity.

CONCLUSION

The three models of a cell depicted here present the biological cell system in its molecular components (the minimal contemporary cell), in its operational roles (the minimal Terran cell), and in its principle of organization (the minimal universal cell). The minimal universal cell has utility for origins-of-life research in providing operational criteria by which any living system may be recognized. The minimal Terran cell has utility in providing a model for the simplest living system of the ancient Earth. An evolutionary continuity from the fundamentally simple cell to the minimal contemporary cell, the simplest living system observable on Earth today, must be postulated for any candidate for the original ancestor of life.

These minimal cell systems are amenable to different analyses, theoretical and experimental, that should be mutually informative: the results of the system-theoretic analysis must be consistent with the details of the experimental analysis. As a systems philosopher, I offer the scientist a logic for the origin of life, and I look to the scientist for the details from which that logic must be born.

ACKNOWLEDGEMENT

I gratefully acknowledge conversations with Antonio Lazcano and thank him for his probing questions and patience.

REFERENCES

Alberts, B. M., 1986: The Function of the Hereditary Materials: Biological Catalyses Reflect the CEU's Evolutionary History, American Zoologist, 26, 781-796.

Chech, T. R., 1985: Self-Splicing RNA: Implications for Evolution, International Review of Cystology, 93, 3-22.

Deamer, D. W. and Barchfeld, G. L., 1982: Encapsulation of Macromolecules by Lipid Vesicles Under Simulated Prebiotic Conditions, Journal of Molecular Evolution, 18, 203-206.

----- and Oro,J., 1980: Role of Lipids in Prebiotic Structures, BioSystems, 12, 167-175.

Gilbert, W., 1986: Origin of Life: The RNA World, Nature, 319, 618.

Hargreaves, W. R. and Deamer, D. W., 1978: Origin and Early Evolution of Bilayer Membranes, Light Transducing Membranes - Structure, Function, and Evolution, D. W. Deamer, ed., Academic Press, New York, New York, 23-59.

Lazcano, A., 1986: Prebiotic Evolution and the Origin of Cells, Origin of Life and Evolution of Cells, L. Margulis, R. Guerrero, and A. Lazcano, eds.; Treballs de la Societat Catalana de Biologia, 39, 73-103.

Lazcano, A., Guerrero, R., Margulis, L. and Oro, J.: The Evolutionary Transition from RNA to DNA in Early Cells, Journal of Molecular Evolution, in press.

Margulis, L., 1981: Symbiosis in Cell Evolution, W. H. Freeman and Company, San Francisco, California.

Maturana, H. R., 1981: Autopoiesis, Autopoiesis--A Theory of Living Organization, M. Zelery, ed., North Holland, New York, New York, 21-33.

Maturana, H. R. and Varela, F. J., 1980: Autopoiesis and Cognition - The Realization of the Living, H. D. Reidel Publishing Co., Dordrecht, Holland.

Morowitz, H. J., 1967: Biological Self-Replicating Systems, Progress in Theoretical Biology, 1, 35-58.

-----, 1985: The Origin of Life, What Darwin Began, L. R. Godfrey, ed., Allyn and Bacon, Inc., Boston, Massachusetts, 243-257.

Pirie, N. W., 1953: Ideas and Assumptions about the Origin of Life, Discovery, 14, 238-242.

Schwegler, H. and Tarumi, K., 1986: The 'Protocell': A Mathematical Model of Self-Maintenance, BioSystems, 19, 307-315.

Searcy, D. G., 1982: Thermoplasma: A Primordial Cell from a Refuse Pile, Trends in Biochemical Science, 7, 183-185.

Sharp, P. A., 1985: On the Origin of RNA: Splicing and Introns, Cell, 42, 397-400.

Varela, F. J., Maturana, H. R. and Uribe, R., 1974: Autopoiesis: The Organization of Living Systems, A Characterization and a Model, BioSystems, 5, 187-196.

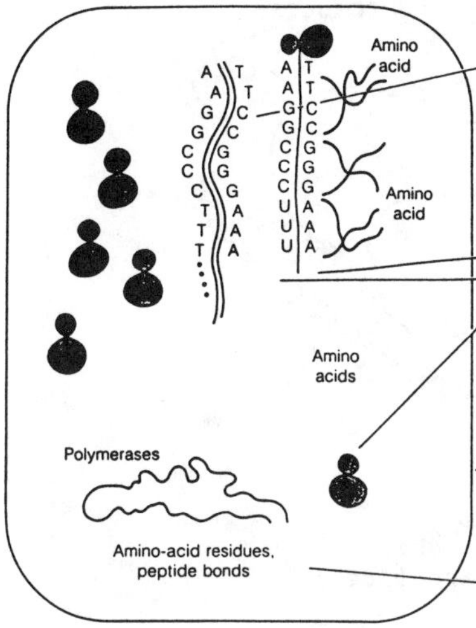

CELL COMPONENTS	FUNCTION	CHEMICAL COMPOUNDS IMPLIED
DNA	Replication	Deoxyribose, guanine (purine derivative), adenine (purine derivative), thymine (pyrimidine derivative), cytosine (pyrimidine derivative), phosophoric acid
RNA (transfer, messenger, ribosomal, etc.)	Protein synthesis	Ribose, guanine, adenine, uracil (pyrimidine derivative), cytosine, phosphoric acid, amino acid acylating enzymes
Enzymes	Catalysis	Proteins, peptides, amino acids, nucleotide coenzymes
Membrane	Maintenance of internal cellular environment	Lipids, saturated and mono-unsaturated fatty acids, glycerol, proteins
Ground cytoplasm	Solvent system, food	All of the above in water, K^+, metabolites (acetate, lactate, etc.)

Figure 1. Minimal contemporary cell (from Symbiosis in Cell Evolution by L. Margulis, copyright 1981, W. H. Freeman and Company, used with permission).

CELL COMPONENTS	FUNCTION	CHEMICAL COMPOUNDS IMPLIED
Membrane	Maintenance of internal cellular environment	[Lipids, saturated and mono-unsaturated fatty acids, glycerol, proteins]
Ground cytoplasm	Solvent system, food	[All of the above in water, K^+, metabolites (acetate, lactate, etc.)]

Figure 2. Minimal Terran cell

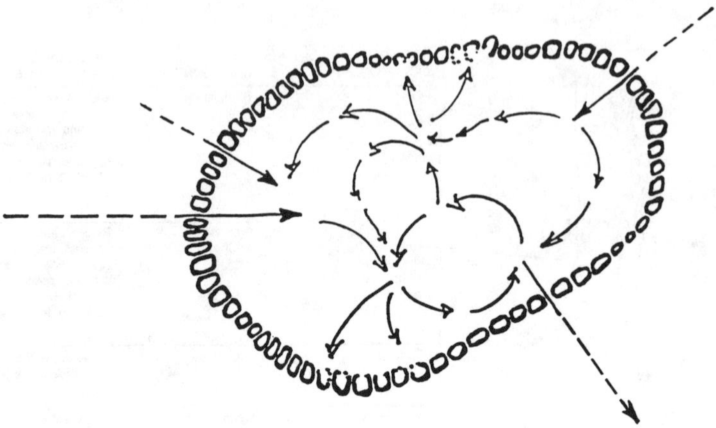

Figure 3. Minimal universal cell

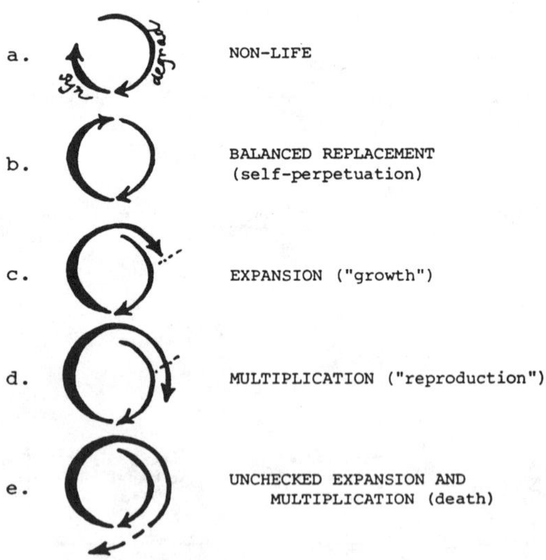

Figure 4. Consequences of differential rates of metabolic synthesis and degradation.

IN SEARCH OF THE ANCESTRAL RNA POLYMERASE: AN EXPERIMENTAL APPROACH

V. Llaca, E. Silva, A. Lazcano
Escuela Nacional de Ciencias Biologicas, IPN
Apdo. Postal 4-870, 06400 Mexico, D.F., Mexico

L. M. Rangel, P. Gariglio
Centro de Investigacion y Estudios Avanzados, IPB
Apdo. Postal 14-740, 07000 Mexico, D. F., Mexico

J. Oro
University of Houston
Houston, Texas 77004

ABSTRACT

The hypothesis that contemporary prokaryotic RNA polymerases were initially RNA-dependent, i.e., were replicases, is strongly supported by the observation that cellular DNA-dependent RNA polymerases can use different types of RNA molecules and ribonucleotides as templates. These RNA molecules include RNA viral genomes, viroids, cellular RNAs (tRNA, mRNA, rRNA) and synthetic homo- and heteropolyribonucleotides. The enzyme-mediated DNA-directed RNA synthesis requires a DNA template, ribonucleoside triphosphates, and Mg^{++}. Based on the fact that the template-specificity of both cellular and viral RNA polymerases is reduced in the presence of Mn^{++}, replicase activity has been observed with eubacterial DNA-dependent RNA polymerase. Here we report some preliminary results of our studies on the RNA-directed synthesis of RNA using *E.coli* RNA polymerase. Optimal replicase activity is observed at 1 mM Mn^{++}, KCl (ionic strength) lower than 3 mM, and slightly basic conditions (pH 8.0 to 8.5). Attempts to observe replicase activity with Zn^{++} and no Mn^{++} have been unsuccessful. RNA-dependent RNA synthesis was found to be resistant to actinomycin-D, and is slightly enhanced at 4mM spermidine.

We have also reported here additional evidence of the evolutionary conservation of polypeptides homologous to the eubacterial RNA pol β' subunit by immunochemical studies that have shown that antibodies for the largest subunit (IIo) of calf thymus DNA-dependent RNA pol II also recognize the *E.coli* RNA pol β' subunit.

PREBIOLOGICAL SELF ORGANIZATION OF MATTER
C. Ponnamperuma and F.R. Eirich (Eds.)

ISBN 0-937194-18-2

INTRODUCTION

It has been recently suggested that the first genetic material may have been a polymer made of flexible, acyclic perhaps prochiral nucleotide analogues of non-enzymatic origin which eventually gave way to cellular RNA genomes (Joyce et al., 1987). The existence of cellular systems whose reproduction was based on the replication of a RNA genome has been discussed by a number of authors (cf. Lazcano, 1986). Accordingly, RNA replication must have been achieved by an RNA-dependent RNA polymerase. It has been suggested that vestiges of this enzyme can still be identified as the eubacterial RNA pol β' subunit and its homologues among archaebacterial, eukaryotic and viral RNA polymerases (Lazcano et al., 1987 a, b). This hypothesis is supported by the direct involvement of eukaryotic and eubacterial DNA-dependent RNA polymerases in RNA-directed RNA syntheses under in vitro conditions (Biebricher and Orgel, 1973) and during viroid replication (Lazcano et al., 1987a, b).

The purpose of this paper is to make a preliminary report of our attempts to develop these ideas using an experimental approach. We present results of our studies on a) the ubiquity of polypeptides homologues to the eubacterial RNA pol β' subunit by showing that polyclonal antibodies directed against the largest subunit of calf thymus DNA-dependent RNA pol II recognize the eubacterial β' polypeptide; and b) based on previous work by Biebricher and Orgel (1973) and others, we have established the optimal physicochemical conditions for achieving an RNA-directed RNA synthesis using the E.coli DNA-dependent RNA polymerase.

MATERIALS AND METHODS

REAGENTS

Tryptone, yeast extract, kanamycin, Tris-HCl, sodium ethyldiaminetetraacetate (Na2EDTA), lysozyme, phenylmethanesulphonyl fluoride (PMSF), acetic acid, methanol, glycine, bovine serum albumin, Triton X-100, Coomassie stain, salmon DNA, poly (A,C) and unlabelled ATP were purchased from Sigma Chemical Company. Disodium hydrogen phosphate 12-hydrate, tetrasodium pyrophosphate 10-hydrate, sodium hydroxide, phenol, chloroform, methanol, acrylamide, silver nitrite, citric acid, magnesium chloride, manganese chloride, zinc chloride and ammonium sulphate were obtained from Merck. Sodium dodecyl sulphate, cellulose (Cellex 410), Bio-Gel A 5m, nitrocellulose sheets, protein A, iodogen and TEMED were obtained from Bio-Rad. Glycerol, polymin-P 50% and proteinase-K were purchased from BRL. 2-mercaptoethanol, glycine and sodium deoxycholate were purchased from BDH Chemicals Co. The Scintillation liquid column was obtained from ICN. Dithiothreitol, calf-thymus DNA, unlabelled CTP, GTP and UTP were obtained from Calbiochem Co.. Spermidine was obtained from Aldrich Co. Ethanol, sodium chloride, hydrogen chloride and isoamyl alcohol were purchased from J. T. Baker S. A. de C. V., and 3H-UTP was obtained from New England Nuclear, 125-I from Amersham, and anti-RNA polymerase from Dr. Michael H. Dahmus.

STRAIN AND GROWTH CONDITIONS

E.coli RNase III- was grown to the late log phase in 10% tryptone, %5 yeast extract, 10% NaCl and 0.25 μg/ml kanamycin, previously filtered by Millex-Ha (millipore) 0.45 filer, at 37°C with continuous stirring; 68 g of cells (wet weight) were harvested by centrifugation at 8500 g during 10 min, in a GSA rotor of a Sorvall 2B centrifuge. They were then frozen and stored at -80°C until needed.

BUFFERS

All the buffers were prepared with double-distilled water. Stock solutions were 2.0 M Tris-HCl (pH 7.9), 0.2 M Na2EDIA (pH 7.9), 4 M NaCl and 0.1 M Dithiothreitol. The buffers were 1) Grinding buffer: 0.05 M Tris-HCl, 5% (v/v) glycerol, 2 mM Na2EDTA, 0.1 mM DTT, 1 mM 2-mercaptoethanol, 0.233 M NaCl, 130 μg/ml, lysozyme and 23 μl/ml phenylmethanosulfonyl fluoride (JPMSF); 2) TGED buffer: 0.01 M Tris-HCl, 5% (v/v) glycerol, 0.1 mM Na2EDTA and 0.1 mM DTT which contained different NaCl concentrations depending on the purification step (Burgess and Jendrisack, 1975); and 3) Dialyzing (storage) buffer: 0.01 M Tris-HCl, 50% (v/v) glycerol, 0.1 mM Na2EDTA and 0.1 mM DTT (NaCl was not added).

DNA-CELLULOSE COLUMN

A DNA-Cellulose column was prepared according to Litman (1968). DNA was from calf-thymus; 500 mg of total DNA were suspended in 10 mM Tris-HCl (pH 7.9) and Na2EDTA 1 mM to obtain a solution of 2 mg/ml. This solution was mixed in previously washed and dried Cellex 410. The mixture was exposed to ultraviolet light (10^5 ergs/mm^2) during 45 minutes to fix the DNA to the cellulose. The resultant paste was then dried and lyophilized, and then resuspended and washed several times in the same buffer before packing it into a column (Pharmacia). The buffer used for washing the paste was scanned in a DU-7 Beckman spectrophotometer (250-310 nm) to confirm the DNA fixation to the cellulose.

RNA POLYMERASE PURIFICATION

The E.coli RNA polymerase purification (Figure 1) was performed at 4°C following the method described by Burgess and Jendrisack (1975); 68 grams of frozen E.coli (ww) were broken and placed in a waring blender containing 150 ml of grinding buffer, and blended at low speed for 2 minutes. The mixture rested for 20 minutes, and then 2.5 ml of sodium deoxycholate were added blending for 30 sec at low speed, and 30 sec at high speed after 20 min. The mixture was centrifugated at 10 000 g for 45 min and the supernatant was then collected; 10 ml of polymin-P 10% were added to it and stirred for 5 min. The mixture was centrifugated 15 min at 8 500 g. The pellet was drained and resuspended in the blender in 200 ml of TGED + 0.5 M NaCl with low stirring during 5 minutes, and then centrifugated at 8 500 g for 30 min. The supernatant was then collected, and its proteins were precipitated with 0.35 g/ml ammonium sulphate, and centrifugated at 10 000 g for 45 minutes. The drained pellet was resuspended in TGED until its conductivity was equivalent to TGED + 0.15 M NaCl. This mixture was passed through a 40 cm DNA-Cellulose column prepared as described above. The column was equilibrated with TGED + 0.15 M NaCl. The mixture was eluted with a 200 ml linear salt gradient (0.15 M to 1 M NaCl). RNA polymerase assays were made, and the reactions in which activity was detected were pooled and passed through a Bio-Gel A 5 m column (Bio-Rad) equilibrated with TGED + 0.5 M NaCl. RNA polymerase assays were repeated. The fractions with a very high activity were pooled (aliquote A2) and separated from the fractions with moderate activity (aliquote B2). Both aliquotes were dialyzed against a storage buffer for one day and stored at -70°C until needed. Protein determination was made following the procedure described by Bradford (1976). The purity of the enzyme at each step was determined by acrylamide gradient (5 to 17.5%) electrophoresis in the presence of SDS (Laemmli, 1970) treated with the $AgNO_3$-stain technique described by Wray (1981).

RNA POLYMERASE ASSAYS

RNA polymerase activity was measured in 50 μl of reaction mixture. The reaction mixture used for activity determination in each step of RNA polymerase purification was 40 mM Tris-HCl (pH 7.9), 5 mM KCl, 8 mM $MgCl_2$, 0.2 mM dithiothreitol, 0.33 mM CTP, 0.33 mM GTP, 0.33 mM ATP, 70 mM UTP 1 μCi 3H-UTP, 1 μg DNA and 10 μl of fraction sample.

The reaction mixture in the determination of optimal conditions varied according to each experiment and is described in the figure captions. The solutions were incubated at 37°C, and the reaction was stopped with 2 μl of 500 mM Na2EDTA. The mixture was bound to Whatman DE-81, and washed several times with 0.8% tetrasodium pyrophosphate, 10% disodium hydrogen phosphate, 0.1% sodium dodecyl sulphate. Dry filters were treated with scintillation liquid, and counted in a Miniaxi Beta 260 Tricarb 4000 (United Technological Packart) liquid scintillation counter. Incorporation of UMP was estimated from the counts per minute and 3H-UTP incorporation.

ASSAYS WITH SPERMIDINE

To measure the effect of spermidine, different concentrations of this polycation (0.0 mM, 0.5 mM, 2 mM and 4 mM) were prepared in a total volume of 145 μl of 10 mM Tris-HCl (pH 7.0), 1 mM NaCl. Samples of 25 μl at each concentration were incubated with 4 μl of 500 μg/μl salmon DNA. or E. coli total RNA at room temperature for 45 minutes to associate the template to spermidine. The assays were then made under the same conditions as described above.

PROTEIN TRANSFER AND WESTERN BLOTTING

Electrophoretic protein transfer from gradient polyacrylamide gels to nitrocellulose was made according to the method described by Towbin et al. (1979) and modified by Burnette (1981). Additional modifications are described below. Total protein from HeLa cells and E. coli RNA polymerase subunits were separated in acrylamide gradient (5 to 17.5%) gels in the presence of sodium dodecyl sulphate in the buffer system of Laemmli (Laemmli, 1970). Each gel was placed in close contact with a nitrocellulose sheet (anode side) in a Bio-Rad Transfer Chamber containing a transfer buffer made of 25 mM Tris-Base, 20% methanol, 192 mM glycine. The transfer was made in 5.5 hours at 500-980 mA at 6°C with cooling to avoid overheating. Following the method described by Garcia-Carranca et al. (1986), the nitrocellulose was blocked with 10 mM Tris (pH 7.4), 150 mM NaCl, 0.5% Triton X-100, 0.1% SDS, 3% bovine serum albumine (BSA) during 15 hours. It was then incubated in 3 mM Tris (pH 7.4), 100 mM NaCl, 0.3% Triton X-100, 0.07% SDS, 0.7% BSA and minor amounts of Anti-RNA polymerases II. The nitrocellulose was washed 5 times with the same buffer without BSA; subsequently, the paper was incubated with the same buffer, with BSA and 125-I-labelled protein A, which was prepared according to Tolan et al. (1980). After 6 washes, the nitrocellulose was dried and kept with a XK-1 Kodak plate for several days to detect labelled proteins. The observation of unlabelled proteins was achieved by staining the nitrocellulose sheet after exposure by the method of Schaffner and Weissmann (1973) as modified to Coomassie stain by Burnette (1981). Molecular markers were included in the gels.

RESULTS

IMMUNOASSAYS WITH ANTI-RNA POLYMERASE II

In order to study the distribution of polypeptides homologous to the E.coli RNA pol β' subunit, Western blots of its SDS-polyacrylamide gel electrophoretic pattern on nitrocellulose sheets were challenged with polyclonal antibodies directed against the calf-thymus DNA-dependent RNA polymerase IIo subunit. A homology character of these two polypeptides is indicated by the positive reaction shown in Figure 2.

REPLICASE ACTIVITY OF THE EUBACTERIAL RNA POLYMERASE

As shown in Figure 3A, RNA-directed RNA synthesis was observed in the E.coli DNA-dependent RNA polymerase, if no KCl was added, Mn++ was used instead of Mg++, and an RNA template (poly A,C or E.coli total RNA) was employed. The absence of any possibly undetected DNA template was confirmed by running assays in the presence of actinomycin-D (0.04 μg/ml). Results are shown in Figure 3B.

In order to achieve an optimal replicase activity, assays were made using different Mn++ concentrations and ionic strengths (Figure 4A), different pH's (Figure 4B), and spermidine concentrations (Figure 5A). Attempts to substitute Mn++ by Zn++ were unsuccessful (Figure 5B). The replicase activity at optimal Mn++ concentration (1 mM Mn++, 0.0 mM KCl) is inhibited by Zn++ (Figure 8); with no Mn++ some replicase activity is observed at 1 mM Zn++. Optimal physicochemical conditions required by the E.coli RNA polymerase when reading either DNA or RNA (poly A,C or E.coli RNA) are summarized in Table 1.

DISCUSSION AND CONCLUSIONS

Homology between the E.coli RNA poly β' polypeptide and the subunit IIo of calf thymus RNA polymerase II is inferred by the cross-reaction of the antibody against the eukaryotic component with the eubacterial subunit (Figure 2). This result is consistent with previous observations of similar amino acid sequences between the largest component of animal RNA polymerases and the E.coli RNA pol β' subunit (Biggs et al., 1985; Corden, in Biggs et al., 1985). The eukaryotic RNA pol II subunit IIo is a highly phosphorylated polypeptide (Corden et al., 1985). The positive reaction shown in Figure 2 supports the hypothesis that extensive polypeptide chain phosphorylation and other modifications of the nuclear DNA-dependent RNA polymerases largest subunits are secondary adaptations (Lazcano et al., 1987a) that were selected for after the origin of eukaryotic cells (Margulis, 1981).

It is known that under appropriate conditions (absence of a DNA template, Mn++ instead of Mg++) both the Gram-positive (Fox et al., 1964; Adman and Grossman, 1965; Steck et al., 1968) and the Gram-negative (Gomatos et al., 1964; Nigoyi and Stevens, 1965; Biebricher and Orgel, 1973) eubacterial DNA-dependent RNA polymerases will read RNA templates that include tRNA, rRNA, homo-polyribonucleotides and viral RNA genomes. Our obser-ation that RNA polyerase-mediated RNA synthesis is improved, if poly (A,C) is used as a template (Figure 3A) is in agreement with previous observations that have shown that the enzyme reads RNA templates with little secondary structure more efficiently (Biebricher and Orgel, 1973). This suggests to us that the ancestral cellular RNA genomes (Lazcano et al., 1986) may have had only small double-stranded regions.

The results of our studies on the replicase activity of DNA-dependent RNA polymerases are consistent with previous observations on the changes in the template specificity of cellular and viral nucleic acid polymerases in the presence of Mn++ ions (Biebricher and Orgel, 1973; Goodman et al., 1984). Attempts to observe replicase activity of DNA-dependent RNA polymerases in the presence of Mg++, Co++, Fe++ (Biebricher and Orgel, 1973), or Zn++ (Figure 5B) were unsuccessful.

It is well known that DNA-directed RNA synthesis is strongly inhibited by actinomycin-D, an antibiotic that presumably intercalates into the double-stranded DNA template between the base-paired dinucleotide sequence dG-dC (Polya, 1977). Actinomycin-D is known to inhibit E.coli RNA polymerase-mediated viroid replication (Rohde et al., 1982). This may be due to the double helix-like structure of viroids. Other RNA-dependent RNA syntheses are not affected (Sanger and Knight, 1963). It is likely that in our model system with total E.coli RNA replicase the activity was not inhibited by actinomycin-D because the enzyme is reading preferentially RNA molecules with little secondary structure.

Replicase activity of normally DNA-dependent RNA polymerases has been interpreted as vestigial from before the appearance of double-stranded DNA molecules as carriers of cellular genetic information (Lazcano, 1986; Lazcano et al., 1987 a,b). Table 1 shows that the most significant factors for achieving replicase activity in a DNA-dependent RNA polymerase are 1) the presence of Mn++ instead of Mg++; and 2) a very low ionic strength. The change in template specificity reported here supports the hypothesis that the evolutionary transition from a RNA genome to the use of double-stranded DNA molecules required only minor modifications of the active site of an ancestral RNA-dependent RNA polymerase (Lazcano et al., 1987 a,b). The optimal physicochemical conditions for a RNA-directed RNA synthesis of the E.coli RNA polymerase, summarized in Table 1, will be used in future attempts to establish the replicase activity of the isolated eubacterial RNA pol β' subunit.

ACKNOWLEDGMENT

This research has been partially supported by an OEA, PRDC y T (1986-1987) Grant to P.G., and NASA Grant NGR 44-005-002 to J.O.

REFERENCES

Adman, R. and Grossman, L., 1965: Fed. Proc., 24, 602.

Allison, L. A., Moyle, M., Shales, M. and Ingles, C. J., 1985: Cell, 42, 599-610.

Biebricher, C. K. and Orgel, L. E., 1973: Proc. Natl. Acad. Sci. USA, 70, 934-938.

Biggs, J., Searles, L. L. and Greenleaf, A. L., 1985: Cell, 42, 611-621.

Bradford, M. J., 1976: Anal. Biochem., 72, 248-252.

Burgess, R. and Jendrisak, J., 1975: Biochemistry, 14, 4634-4638.

Burnette, W. N., 1981: Anal. Biochem., 112, 195-203.

Corden, J. L., Cadena, D. L., Ahearn, J. M. and Dahmus, M. E., 1985: Proc. Natl. Acad. Sci. USA, 82, 7934-7938.

Fox, C. F., Robinson, W. S., Haselkorn, R. and Weiss, S. B., 1964: J. Biol. Chem., 239, 186-195.

Garcia-Carranca, A., Miguel, F., Dahmus, M. E. and Gariglio, P., 1986: Arch. Biochem. Biophys., 251, 232-238.

Gomatos, P.J., Krug, R. M. and Tamm, I., 1964: J. Mol. Biol., 9, 193-207.

Goodman, T. C., Nagel, L., Rappold, W., Klotz, G. and Riesner, D., 1984: Nucleic Acid Res., 12, 6231-6246.

Joyce, F. G., Schwartz, A. W., Miller, S. L. and Orgel, L. E., 1987: Proc. Natl. Acad. Sci. USA, in press.

Laemmli, L. M., 1970: Nature, 227, 680-685.

Lazcano, A., 1986: In L. Margulis, R. Guerrero and A. Lazcano, eds., The Origin of Life and the Evolution of Cells, Treballs Soc. Castalana Biol., 39, in press.

Lazcano, A., Guerrero, R., Margulis, L. and Oro, J., July 21-25, 1986: Abstracts of the Fifth ISSOL Meeting, Berkeley, California, 334-335.

Lazcano, A., Fastag, J., Gariglio, P., Ramirez, C. and Oro, J., 1987a: J. Mol. Evol., submitted.

Lazcano, A., Valverde, V., Fastag, J., Gariglio, P., Ramirez, C. and Oro, J., 1987b: This volume.

Litman, R. M., 1968: J. Biol. Chem., 243, 6222-6233.

Margulis, L., 1981: Symbiosis in Cell Evolution, Freeman Co., San Francisco, California.

Nigoyi, S. K. and Stevens, A., 1965: J. Biol. Chem., 240, 2587-2592.

Polya, G. M., 1977: The Ribonucleic Acids, P. R. Stewart and D. S. Letham, eds., Springer-Verlag, Heidelberg, 9-41.

Rohde, W., Rackwitz, H. R., Boege, F. and Sanger, H. L., 1982: Biosci. Rep., 2, 929-939.

Sanger, H. L. and Knight, C. A., 1963: Biochem. Biophys. Res. Comm., 13, 455-461.

Schaffner, W. and Weissmann, K., 1973: Anal. Biochem., 56, 502-514.

Steck, T. L., Caicuts, M. S. and Wilson, R. G., 1968: J. Biol. Chem., 243, 2769-2778.

Tolan, D. R., Lambert, J. M., Boileau, G., Fanning, T. G., Kenny, J. W., Vassos, A. and Traut, R. R., 1980: Anal. Biochem., 103, 101-109.

Towbin, H. T., Staehelin, I. K. and Gordon, J. H., 1979: Proc. Natl. Acad. Sci. USA, 76, 4350-4354.

Wray, W. T., Boulikas, T., Wray, V. P. and Hancock, R., 1981: Anal. Biochem., 118, 197-203.

ADDENDUM

Lazcano, A., Fastag, J., Gariglio, P., Ramirez, C., and Oro, J., 1988: J. Mol. Evol., 27, 365-376.

Loeb, L. A., and Mildvan, A. S., 1981: In G. L. Eichorn and L. G. Marzilli, eds., Metal Ions in Genetic Information Transfer, Elsevier, New York, 125-142.

Mildvan, A. S., and Loeb, L. A., 1981: In G. L. Eichorn and L. G. Marzilli, eds., Metal Ions in Genetic Information Transfer, Elsevier, New York, 103-123.

Table 1. E.coli RNA POLYMERASE-MEDIATED RNA SYNTHESIS

	DNA template (salmon)	RNA template (E.coli)
Mn++	---	1.0 mM
Mg++	8.0 mM	---
Zn++	---	---
KCl	100.0 mM	---
pH	7.9	8.0 - 8.5
spermidine	4.5 mM	4.0 mM

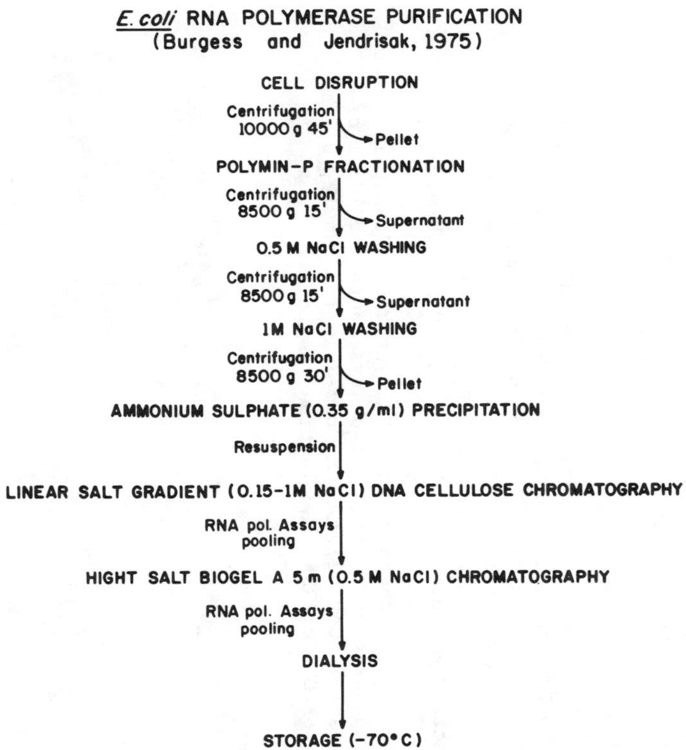

Figure 1. An outline for the purification of E.coli RNA pol (Burgess and Jendrisak, 1975).

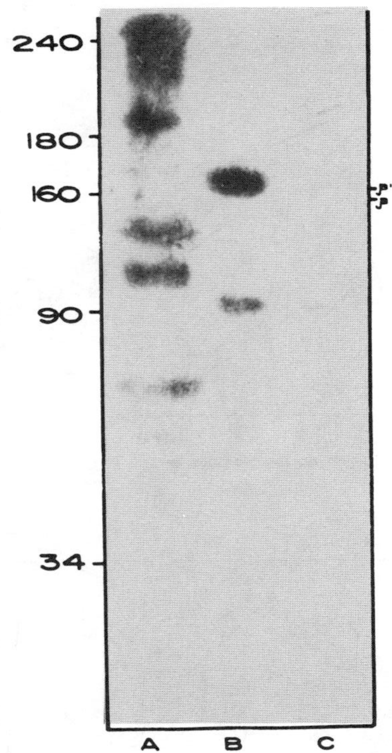

Figure 2. Homology between the subunit IIo of the calf thymus RNA pol II and the E.coli RNA pol β' subunit. Component patterns of the enzymes obtained by SDS-polyacrylamide gel electrophoresis transferred to nitrocellulose filters were challenged with antibody raised against the IIo subunit. Cross-reaction was visualized by 125-I-labelled protein A (see Materials and Methods). Track A, HeLa RNA pol II; Track B, E.coli RNA pol.

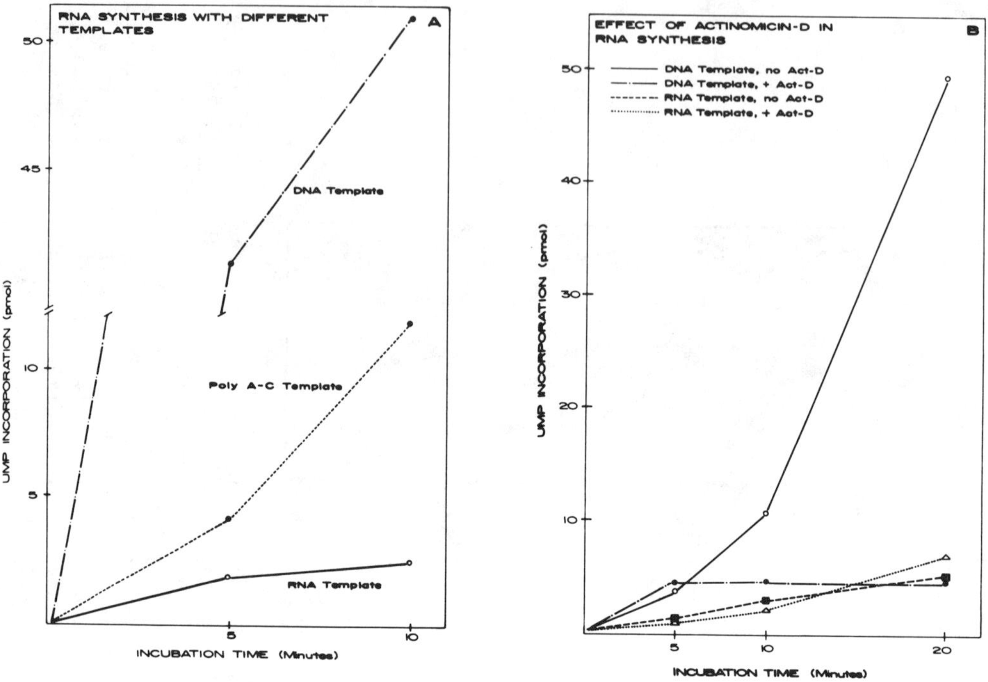

Figure 3

A) <u>E.coli</u> RNA polymerase-mediated RNA synthesis. Activity was measured by UMP incorporation for various intervals, in a 50 μl reaction mixture containing 10 μl of purified enzyme aliquote A2 (see Materials and Methods), 50 mM Tris-HCl, 22 mM $MnCl_2$, 0.2 mM dithiothreitol, 0.22 mM CTP, 0.33 mM GTP, 0.33 mM ATP and 70 μM UTP. The reaction mixture was labelled with 2 μCi 3H-UTP.

B) Effect of actinomycin-D on RNA synthesis. RNA polymerase activity was measured as in 3A. All assays contained 10 μl of purified enzyme (aliquot B2), 50 mM Tris-HCl, 1 mM MnCl2, 0.2 mM DTT, and 1 μg of DNA or RNA. Nucleotides as in 3A.

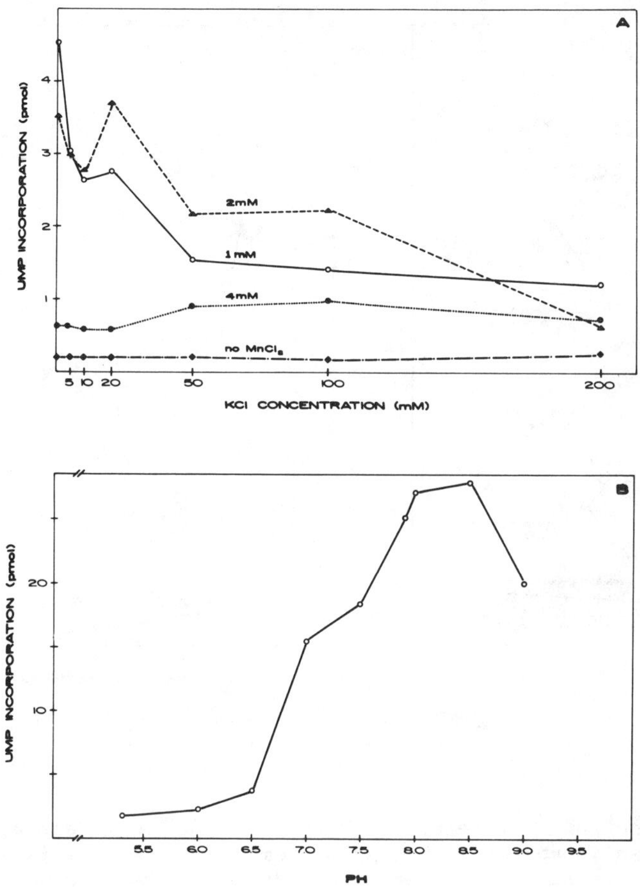

Figure 4

A) Replicase activity of E.coli RNA polymerase with different concentrations of Mn++ and KCl. UMP incorporation (pmol in 50 μl) was measured after 10 minutes in 50 μl of 50 mM Tris-HCl, 0.2 mM DTT, 0.33 mM CTP, 0.33 mM GTP, 0.33 mM ATP, 70 μM UTP, 2 μCi 3H-UTP, with 10 μl of purified RNA polymerase aliquote B2 (see Materials and Methods) and 1 μg of RNA template.

B) Replicase activity of E.coli RNA polymerase at different pH values. The reaction mixture included 1 mM $MnCl_2$, 0.2 mM dithiotreitol, 0.33 mM CTP, 0.33 mM GTP, 0.33 mM ATP, 70 μM UTP, 2 μCi 3H-UTP, 10 μl purified enzyme (aliquote A2), and 40 mM of an acetate or Tris-based buffer with different pH values.

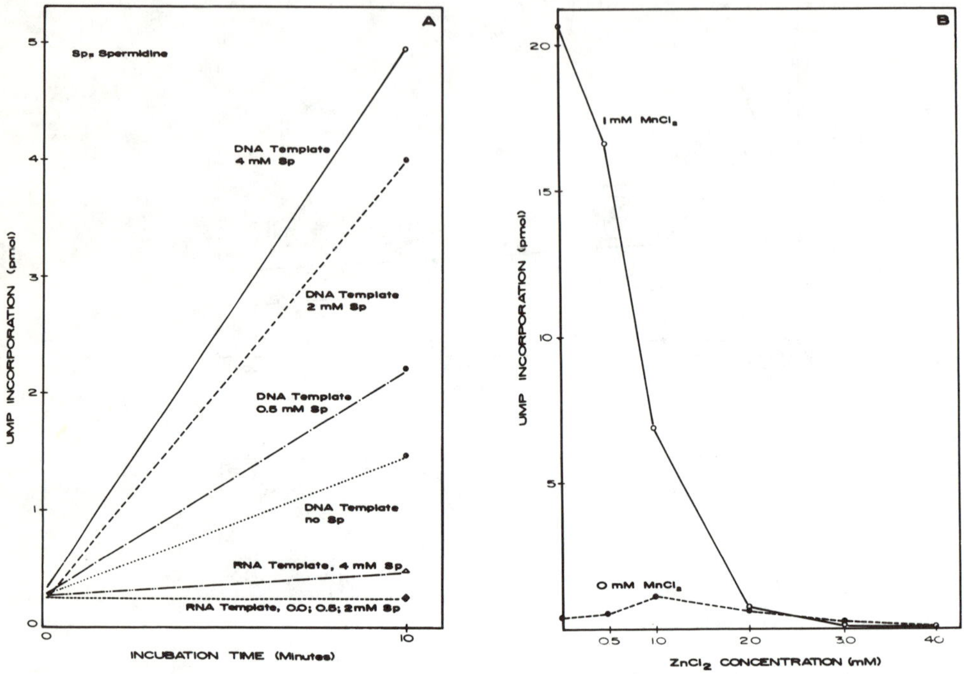

Figure 5

A) Effect of spermidine on RNA polymerase replicase activity. Spermidine was incubated with 2 μg of template (see Materials and Methods), before adding the reaction mixture (50 mM Tris-HCl), 0.2 mM dithiotreitol, 0.33 mM CTP, 0.33 mM GTP, 0.33 mM ATP, 70 μM UTP and 2 μCi 3H-UTP, which contained 10 μl of purified RNA polymerase (aliquote B2).

B) Replicase activity of RNA polymerase in the presence of Zn++. UMP incorporation was measured after 10 minutes of incubation in 50 μl of 50 mM of Tris-HCl, 0.2 mM dithiothreitol, 0.33 mM CTP, 0.33 mM GTP, 0.33 mM ATP and 2 μCi 3H-UTP, with 10 μl of purified enzyme aliquote A2 (see Materials and Methods).

FROM ORIGINS OF LIFE TO EVOLUTION OF MICROBIAL COMMUNITIES: A MINIMALIST APPROACH

Lynn Margulis
University of Massachusetts
Amherst, Massachusetts 01003

and

Ricardo Guerrero
University of Barcelona
Barcelona, Spain

ABSTRACT

The problem of "the origins of life" is limited to the problem of the origin of the earliest autopoietic (self-forming, self-sustaining information) systems. Neither the "prebiotic chemistry," nor the "minimal biological system" approaches have solved this problem. Here we distinguish common imperative operations of the morphologically least complex cells (e.g., control of Na^+, K^+, H^+, PO_4^{3-}, Mg^{2+}, Ca^{2+} ions; replication of DNA, transcription and translation of DNA to RNA and protein, and the binding of these functions within a dynamic lipoprotein membrane). Attempting to reduce these functions of the least complex cells known to a minimum, we propose that the emergence of all metabolic pathways not directly related to reproduction of the minimal autopoietic system is a product of bacterial evolution. We emphasize the importance of the early Archean environment for the evolution of bacteria and their communities.

INTRODUCTION

In contrast to living beings, the inanimate world does not retain, store and utilize messages for thousands of millions of years. The accumulation of information during the course of Earth's

----

*This article is an expanded version of the work published as "Not 'origin of life' but 'evolution in microbes'" in <u>Treballs de la Societat Catalana de Biologia</u>, vol. 39, pp. 105-112 (1986).

ISBN 0-937194-18-2

history is a property of life. Two essential features of life's information storage system are the ability to replicate the messages and to duplicate the entire set of molecules and structures produced by these messages, i.e., the capacity to reproduce cells. The problem of the origin of life is the problem of the origin of this self-maintaining information system, one that apparently has been in continuous existence for over 3000 million years.

Elso S. Barghoorn, the father of pre-Phanerozoic paleobiology, was fond of saying that he did not study the origin of life, an idea shrouded in mystery, but rather that he was interested in the antiquity of life (Margulis and Stolz, 1984). Even though present mysteries may become knowledge in the future, current scientific attempts to resolve the problem of the origin of life usually deal with tentative extrapolations from the possibilities inherent in certain limited chemical reactions. Extrapolations, or logical inferences from "prebiotic chemical systems," have not yet explained what actually occurred on the early Earth. Nor is there any body of current scientific knowledge that allows us to predict the course of success in the production of an autopoietic system in the laboratory.

We can divide the scientists studying the origin of life into two groups: those attempting to mimic the simplest structures and functions with autopoietic capacity (a chemical system approach), and those trying to dissect the essential structures and functions of living cells to analyze the minimal conditions for life (a biological system approach).

The present gap between the experimental gains of the former and the maximal permitted simplification of the latter is enormous, indicating that we are still very far from knowing how life began its history on Earth.

The main purpose of this article is to reduce the gap epistemologically in the form of an inquiry into the essential mechanisms and structures for the earliest living forms. What were the minimal requirements for the multiplication and growth of the earliest beings in their ancient habitats? We proceed as follows: 1) We distinguish many features of cells, both their components and operations that evolved subsequent to the origin of life itself. Such components and capabilities have been characteristic of life since the Archean Aeon, but they were not essential for the reproduction of the earliest cells. 2) We attempt to reduce to a minimum those common imperative operations of the simplest cells, the bacteria. We propose that the complexity of metabolism is a product of bacterial evolution. And 3) we emphasize the great importance of the environment for Archean evolution of bacteria. Only when we consider the subtleties and gradients of the environmental parameters which surround a given bacterial population is it possible to un derstand the changes in number and in physiological state of the different cells comprising that population. Early life maintained itself through interaction of populations that formed the bacterial communities, i.e., by cycling the chemical elements and using external energy sources. Throughout the history of life since its origin, different bacterial populations, provided with environmental sources of energy, carbon, other elements and reducing power, were organized into microbial ecosystems.

We present in approximate chronological order increasingly organized steps in the evolution of life. We begin with the first autopoietic systems assuming they had the structure and metabolism of simple fermentative, obligately anaerobic, bacterial cells. We proceed to the essential structures and functions of the first eukaryotic cells, the *sine qua non* for the Phanerozoic explosive radiation of animals and plants. We see that many of the characteristics thought to be the results of animal and plant evolution had already appeared in microorganisms by the turn of the Cambrian Era.

AUTOPOIESY

> "One important result of evolution may be seen in the increasing intensification of autopoietic life in the present by means of including the experience of the past and anticipation of the future"
>
> Erich Jantsch, 1980

The problem of the origins of life has become the problem of autopoiesy, that is: how did the minimal autopoietic entity come into being? The minimal autopoietic thing, we think now, is a bacterial cell -- or more probably because it is the nature of bacterial cells to multiply by division, a population of bacterial cells. But what is an autopoietic system, or, as Jantsch puts it, "autopoietic life."

Autopoietic entities are self-making and self-maintaining. They have boundaries -- cell membranes, cuticles, skin -- that clearly mark their relation with the outside. They expend matter and energy, add and take away bits of themselves; they maintain their identity. Autopoietic entities respond to external communications by self-preserving activities: they heal their membranes, rewax their cuticles, grow back and cover lesions in their skins. Within a range of limits of food, light, temperature and other external changing factors, autopoietic entities are recognizably themselves.

In 1973, Francisco Varela and Humberto Maturana developed their concept of autopoiesis, recognizing that it fits with what we expect of life (Maturana and Varela, 1980). "Reproduction and mutation and the capacity to transmit mutation" alone, although popular, is not good enough as a definition of life since many sterile animals and old women who will not reproduce and mutate are certainly still alive. Indeed, they are autopoietic entities that behave in ways that insure the persistence of themselves.

Viruses, in the absence of their host cells, are not autopoietic entities. They lack dynamic borders maintained by active interchange of parts. They respond passively to their environment and its change. Viruses behave in an autopoietic way only when they are inside of autopoietic entities such as any kind of host cell. Hence, the minimal autopoietic entity is still a collection of bacterial cells. The origin of life is the problem of the origin of a population of metabolizing membrane-bounded spherical entities less than a micrometer in diameter -- namely bacterial cells.

SEMES

> "For nothing is born in the body in order that we may be able to use it, but rather, having been born, it begets a use."
>
> Lucretius (99-55 B.C.)
> On the Nature of Things (Book IV)

When the living world was thought to be divided into the two kingdoms, Animals and Plants, the problem of the origins of life was conceived of as the problem of the origins of plants. Since it was patently obvious that animals cannot exist without the benefit of food generated by photosynthesis and since, in the limited dichotomizing mind of man, Animal <u>or</u> Plant were considered the only forms of life -- the problem of the "origin of life" was the same problem as "the origin of the primitive plant."

We know now that the "origin of plants" is an enormous complexity presenting problems that include events at many different levels at very different times during the history of life on Earth. For example, the "origin of plants" includes the problems of the origins of prokaryotes (bacterial cells), photosynthesis, eukaryotic cells, symbiotic cyanobacteria that became chloroplasts, cell motility, mitosis, sexual reproduction, and others. Now that the living world is much more appropriately divided into prokaryotes and eukaryotes, we begin to define the "origins of life" as the problem of the origins of the first spherical heterotrophic, fermenting anaerobic bacteria. Yet the old errors continue as many authors consider the origin of life as a single epiphanous event, trying to explain too much at once. In this article we can begin to distinguish the problem of the origin of life from a whole series of other issues, often confused, that clearly belong to the domain of the evolution of microbial life.

The very first bacterial-like cells were probably much less complex than even the simplest cells of extant life. A recent interesting hunch is that the first cells were RNA-cells, composed of RNA and protein within a lipid membrane, i.e., of membrane-associated catalytic and replicating RNA (Alberts, 1986). If this view is correct, even DNA metabolism was a product of microbial evolution. In any case, following the original analysis of Morowitz (1967), we can recognize 0.5 μm diameter spheres, which like modern mycoplasmas lack a peptide-glycan cell wall, as common ancestors of all life on Earth. From such autopoietic entities, from their DNA, the universal features of all present-day autopoietic entities evolved, including messenger RNA and its triplet code, acylating enzymes, lipoprotein membranes and many features (semes) which we attempt to list below.

Semes are useful units to deal with in a discussion of evolution: they are features or characteristics of organisms that have clearly identifiable selective advantages. They are the characteristics with which evolutionists traditionally deal: fruits, uteri, eyes, oxygen respiration, swim bladders, nitrogen fixation, etc. Many genes are required to make a seme. Not to repeat what has been written elsewhere, we simply outline, in rough order of appearance, those new semes, i.e., neosemes (Hanson, 1977) that clearly evolved inside microbes and therefore are not intrinsic to the origin of life.

A seme can be thought of as an inherited trait or feature, the product of many genes, of clear selective advantage to the organisms that carry it. The appearance of new semes depends on the net accumulation of genetic material. Among extant organisms we know that genetic material, in the form of lengths of DNA, can be newly acquired by many mechanisms including replication of genomes in the absence of cell division, gene duplication, phage or plasmid transfer, polyploidization, organellar replication and even hereditary symbiosis (Margulis and Bermudes, 1985). Of course, genes can be modified and lost by complementary mechanisms, but the net loss or alteration by mutation of genetic material is of less consequence to the appearance of neosemes than the net gain.

Newly appearing semes are of great interest to evolutionists who also recognize changes in semes. For discussion of semes (a concept first proposed, in German, by Remane, then later developed in English by Hanson) see Hanson (1977). These authors have identified and described, for example, increase and decrease of semes (hypersemes, hyposemes), change in preexisting semes (aposemes) and loss through time of semes in populations of organisms.

By the distribution of, and variation in, semes it is possible to reconstruct the group of organisms in which they first appeared (Hanson, 1977; Margulis, 1981). Using this evolutionary practice, we summarize here distinguishable stages in the origin and evolution of the first, necessarily microbial, autopoietic systems. Our summary is in the form of a list of semes, in roughly what we think may have been the chronological order, beginning with the earliest semes to have appeared. The entries themselves have been developed elsewhere, e.g., in the references given in parentheses.

LIST OF SEMES APPEARING BEFORE, OR DURING, THE EMERGENCE OF THE FIRST PROKARYOTES ANCESTRAL TO SUBSEQUENT LIFE

> "The chief purpose of models is not to be right or wrong, but to give us a place to store our data."
> Robert Garrels, 1984
> (in Sagan, 1985, p. 241)

1) RNA polymerase and other proteins, as well as the accompanying nucleoside triphosphate (ATP, GTP) metabolisms, are the _sine qua non_ of the Earth's autopoietic system ["RNA-cells" (Lazcano, 1987)]. Due to the segregation of the autopoietic system from the environment, lipid metabolism involved in the production of cell membrane must have also occurred in these earliest common ancestors. The central role and universal distribution of acetate suggests that from the beginning all life on Earth had a capacity for acetate metabolism.

2) Ribosome-mediated protein synthesis evolved inside cells, according to the analysis of Lazcano (1987).

3) DNA-mediated messenger RNA, deoxyribose metabolism and other steps in DNA synthesis also evolved in RNA-cells (Lazcano, 1987).

4. DNA polymerase, ligase, restriction endonucleases and, in general, editing and repair systems (apparently universally distributed) evolved in common ancestors of all life.

5) Ion channels for sodium, potassium and calcium, the basis of internal ion control, also distributed ubiquitously in living organisms must have been present in our common ancestors. Since ion-pumping forms an essential component of autopoiesy, it probably

evolved prior to or during the emergence of the first common prokaryotic ancestors. Membrane control of ions is an intrinsic feature of the self-maintaining system of any cell.

LIST OF SEMES APPEARING AFTER THE FIRST COMMON PROKARYOTE

"Omne vivum e cellula (all life comes from the cell)"

Werner Schwemmler, 1984

1) The evolution of transcription, the process that copies lengths of RNA from DNA in a precise order given by the rules of base pairing probably evolved inside bacteria (Zillig _et al._, 1985). Studies of the enzymes involved in RNA synthesis suggest that there are at least two classes of different ways of performing RNA synthesis in bacteria: archaebacterial and eubacterial. The transcription apparatus in these groups of bacteria apparently diverged early in the history of life on Earth. That in these groups of bacteria intervening sequences are present, e.g., in archaebacterial DNA, suggests that eukaryotes merely elaborated on a genetic organization already present in their prokaryotic ancestors.

2) DNA splicing and ligating enzymes, topoisomerases and other requisites for DNA repair and recombining, presumably evolved in common prokaryotic ancestors in response to ultraviolet light and other mutagenic threats (Margulis and Sagan, 1986a).

3) Murein, pseudomurein, and peptidoglycan cell walls clearly evolved in different ways in different lineages of prokaryotes (Kandler, 1985). Because cell walls differ in composition and construction and are even absent in some bacteria, it is clear that walls are not a necessary component of all autopoietic systems.

4) Fermentation of small organic molecules (i.e., the metabolism characteristic of bacteria that are entirely intolerant of molecular oxygen) is probably the kind of metabolism that requires the smallest quantity of genetic material (fewer enzymes) to manage it.

5) Desiccation-resistant spores clearly evolved in bacteria that ferment small carbon compounds in the absence of oxygen (such as _Clostridium_). An ancient origin of these useful structures is likely.

6) Rotary motors, the cell wall "wheels" of flagellated bacteria, are present in a number of types of bacteria that ferment

carbon compounds in the absence of oxygen, suggesting the early appearance of this most useful seme.

7) Flagellated bacteria are often chemotactic, displaying sensitivities (moving toward and away from) to various organic compounds found in their surroundings. The appearance of locomotory behavior sensitive to chemical stimuli probably first occurred in flagellated bacteria where it is still found today. Ancestors to flagellated bacteria also evolved before free oxygen appeared in the Earth's atmosphere.

8) The first production of hard minerals inside the cytoplasms of cells (biomineralization) is associated with microaerophilic, motile, magnetotactic bacteria. Biomineralization, too, probably had an early origin in bacteria (Westbroek and de Jong, 1983).

9) Isopentyl pyrophosphate metabolism, which is nearly universally distributed, is an absolute prerequisite for chlorophyll reaction center photosynthesis, and must therefore have preceded any photosynthetic system.

11) Photoheterotrophy: light-sensitive membranes (rhodopsin) and photophosphorylation of ATP presumably is far less complex than photoautotrophy. Since it therefore requires a less extensive genetic system for its administration, it probably evolved in bacteria prior to photoautotrophy.

12) Anoxygenic photosynthesis evolved in several lineages of bacteria which thus must have had light-sensitive membranes and the chlorophylls, carotenoids and porphyrins which comprise them.

13) Since sulfate reduction (sulfate respiration) is even today limited to anaerobic bacteria, it clearly evolved in this group of organisms, probably from a simplification of the electron chain of anoxygenic photosynthesis (Trüper, 1982).

14) Chemolithotrophy based on sulfide, ammonia, hydrogen and methane oxidation are bacterial modes of metabolism characteristic of different lineages of bacteria and requiring gaseous oxygen. Limited to bacteria, they obviously evolved in bacteria.

15) Oxygenic photosynthesis evolved in the ancestors to cyanobacteria and <u>Prochloron</u>.

16) Differentiation, i.e., linear development based on multicellularity such as that present in cyanobacteria which form heterocysts, evolved in prokaryotes where it still can be found (Saier and Jacobson, 1984).

17) Predation, the attachment, penetration and killing of one type of bacterium by another (e.g., Vampirococcus, Daptobacter) evolved in bacteria (Guerrero et al., 1986; Guerrero et al., 1987).

18) Nitrate respiration, present in facultative anaerobic bacteria and unknown in eukaryotes, evolved in obligate anaerobic bacteria.

19) Aerobic (free oxygen) respiration probably evolved in those bacteria that already contained cytochrome chains for the respiration of nitrate.

20) Myxobacterial and actinobacterial multicellularity (Starr et al., 1981) involving the filamentous habit, exospore and cyst formation, evolved in various lineages of bacteria.

21) DNA-coated protein, HT-A histone-like protein and acid tolerance (Searcy and Delange, 1980) evolved in the ancestors to Thermoplasma, probably some archaebacteria.

22) Since methylotrophs are capable of steroidogenesis (Bloch, 1985) this process, too, evolved in bacteria, albeit an extremely limited group of bacteria.

23) Stable associations of prokaryotes for the origins of organelles (Margulis, 1981) are known in modern bacteria (e.g., Pelochromatium) and can be considered preadaptations to the origins of those intracellular organelles of eukaryotes that demonstrate genetic continuity.

24) Internal membranes for DNA segregation (including nuclear membrane and outer organellar membranes) probably evolved concurrently with intracellular associations of bacteria (Margulis and Sagan, 1986a).

This list, clearly oversimplified, can be augmented by a second list (section V) derived from the appearance in the Proterozoic Aeon of the eukaryotic level of cell organization (Vidal and Knoll, 1983).

What are the implications of this list for the problem of the "origin of life"? Well-developed stromatolites and microfossils have been found associated with rocks more than 3000 million years old -- both in Africa (the Swaziland Formation, the Barberton Mountains) and in northwest Australia. Because we now have a great deal of experience interpreting stromatolites and microfossils (Walter, 1976; Schopf, 1984) we infer that, in the lower Archean (around 3500 million years ago) a diversity of well-developed communities of microbes already covered the surface of the Earth. Stromatolites, the lithified remains of microbial communities, are primarily formed by cyanobacteria, although other kinds of bacteria may participate in their formation (Fig. 1). These rocks can be thought of as skyscrapers made by communities of microbes in the early Proterozoic Eon, one 2000 million years ago. They were built by the accumulation and precipitation of sediment due to microbial communities. In a few places on Earth today, investigators can still identify microbial communities comparable, in principle, to those which dominated the early Earth. We infer that life originated and left remains much earlier than previously suspected and that the complexity of life in the subvisible microcosm, those communities below the limit of resolution of the light microscope, is impressive.

Even today in certain isolated places bacterial communities have persisted in ancient lifestyles. An example of the diversity, complexity and importance of microbial communities becomes evident when we investigate in detail the nature of what superficially seems like "dirty water." We refer for example to the microbial communities developing seasonally in the small Lake Ciso, near Lake Banyoles, northeast Spain. These communities are interpreted to be much like those that appeared shortly after life originated. They are supported by photoautotrophic (purple and green sulfur) bacteria and display all the major aspects of the community ecology we tend to associate with eukaryotic ecosystems (e.g., primary production, consumption, predation, cell motility, etc.) (Margulis *et al.*, 1986).

What appears to be reddening water that develops over a few weeks is a phenomenon of ecological succession, the growth of communities of bacteria (Fig. 2). The red color is primarily due to two types of phototrophic bacteria: large, motile purple cells belonging to the genus *Chromatium*, and small non-motile green sulfur bacteria belonging to the genus *Chlorobium*. Many other types of organisms are present as well in this biota which is almost exclusively prokaryotic throughout most of the year. Direct evidence amassed for the existence of the remains of anaerobic bacterial communities, based on bacterial autotrophy, reaffirms our contention that much of what we tended to regard as the "origin of life" is really the evolution of early life following the existence of prokaryotic ecosystems. The major aspects of community life still extant were well developed on Earth before the appearance of *any* eukaryotes, that is, prior to the evolution of protoctists, animals, plants and fungi. We can now begin to identify the semes that evolved after the appearance of the first eukaryote.

LIST OF SEMES APPEARING AFTER THE FIRST EUKARYOTES

> "It often seems to be taken for granted that the transition [between the prokaryotic and the eukaryotic cells] took place in such a way that one procaryotic cell has developed into a eucaryotic one. It may, however, be just as fruitful to discuss the possibility that one eucaryotic cell has evolved from a number of procaryotic cells, for example, that it originates from something like a coenocytic relationship (that is, a relationship in which the cells are brought into contact without intervening cell walls) between procaryotic cells."
>
> J. Goksøyr, 1967

1) Since intracellular motility is limited to eukaryotes, it evolved probably in the earliest eukaryotic forms. One of the earliest intracellular motility systems is that of undulipodia (e.g., cilia) which may have evolved by symbiotic acquisition of spirochetes (Margulis and Bermudes, 1985).

2) The various eukaryotic intracellular motility systems (tubulin, actin-myosin based processes of pinocytosis, phagocytosis) are unknown in prokaryotes -- they may have originated by merger of the original spirochete with its bacterial host. In any case, this is a eukaryotic seme.

3) The polyphyletic acquisition of mitochondria from respiring eubacteria seems to be established as a process that occurred in eukaryotic cell evolution (Gray, 1983; Obar and Green, 1985).

4) The appearance of histone proteins as components of nucleosomes in the origin of chromatin occurred in the evolution of several lineages of protists (Searcy and Delange, 1980). Such nucleosome histones did not appear in the ancestors of dinomastigotes.

5) The evolution of specialized cell organelles (e.g., extrusomes, hydrogenosomes, methanogenosomes) occurred in eukaryotes ancestors to modern protists (Van Bruggen *et al.*, 1985; Van Bruggen, 1986).

6) Studies of protist cell division reveal that mitotic microtubules, kinetochors and mitosis evolved in this group of organisms (Margulis and Sagan, 1986b).

7) Symbiotic plastid acquisition occurred polyphyletically in many groups of protists and lead to the origins of various kinds of algae (Gray, 1983).

8) Colonial morphogenesis, characteristic of several groups of protoctists (myxomycota, labyrinthulida) evolved in this group of organisms (Margulis and Sagan, 1986b).

9) Cannibalism, diploidization and higher polyploidization can be understood as preadaptation for the origin of meiotic sex in protoctists (Margulis and Sagan, 1986a).

10) The reduction of polyploidization, a preadaptation for the origin of meiosis, evolved in several protoctist lineages (Margulis and Sagan, 1986a).

11) Alternation of haploid and diploid generations and the origin of embryogenesis are polyphyletic phenomena making their appearance in the transition from protoctists to animals and plants (Margulis and Sagan, 1986a).

This list, as presented here, is immensely over-simplified. Yet by the last entry we have arrived at Mendelian genetic systems and the origin of animals. All of the seme acquisition listed here occurred in the pre-Phanerozoic, that is, prior to 580 million years ago, before the so-called "Cambrian explosion of fossils." When analyzed in this way, it becomes clear that the Cambrian appearance in the fossil record of well-preserved hard parts simply made the previously subvisible highly visible, analogous to the way abandoned automobiles will be the manifestations in the fossil record of the far less fossilizable hominoid cerebral cortex.

CONCLUSIONS AND PERSPECTIVES

The subject of the origin of life, or at least the origin of more manifest living beings, has fascinated mankind since the appearance of the earliest civilizations. However, only in this century have we begun to understand enough biochemistry, genetics and paleobiology to develop a focused idea of the possible nature and antiquity of the first living entities. In this article we have seen that the salient characteristic of life is autopoiesy. We believe that the first autopoietic systems were prokaryotic cells, i.e., bacteria. The very earliest cells were probably of the fermentative, obligately anaerobic type. From very early in the history of life, such fermenters were most likely accompanied by primary producers with a photomixotrophic metabolism, i.e., organisms capable of utilizing chemical compounds in their environment and sunlight as a source of energy. Even today we still have abundant living representatives of both types of obligate anaerobes, for example some members of the communities in Lake Ciso. Certainly, bacteria have evolved such that extant fermentative and photomixotrophic bacteria possess many characteristics which were not requisite for the earliest functioning of their semes (e.g., motility, or the ability to cope with measurable quantities of molecular oxygen in the water). During the first steps in the evolution of life many characteristics, structures and functions developed. When the living world was divided into animals and plants these characteristics (e.g., oxygenic photosynthesis and respiration) came to be considered necessary conditions for the origin of any life.

During the Archean Aeon and at least half of the Proterozoic, life was exclusively prokaryotic. In the great Monera kingdom (i.e., bacteria) the essential genetic and metabolic mechanisms first appeared and diversified (Margulis, 1982). On different occasions, various sorts of prokaryotic cells associated to form communities which evolved into permanent symbioses. Such hereditary symbioses eventually developed into the different types of eukaryotic cells (i.e., protoctists, fungi, animals and plants). Prokaryotic associations that became eukaryotes include microbial descendants that we recognize as the protoctista of today. The prokaryotic mechanisms of DNA duplication, recombination of genes and (in the case of symbiosis) of entire genetic systems formed new autopoietic entities. These entities, bacterial and nucleated cells and their descendants, were subjected to immense selection pressures from the physical environment of the Earth's surface as well as from other autopoietic entities. With time the microorganisms (always as components of communities) expanded, recombined and reformed until they became the familiar macroscopic biota. We recognize today that nearly every major biochemical pathway present in animals and plants had already appeared in the cells comprising the Archean and Proterozoic microbes. No longer do we have to insist on a single solution to a large number of problems hiding under the phrase "origins of life." Rather, as soon as we understand the origin of the first membrane-bounded autopoietic entity capable of reproduction by division, we will have solved, in principle, the scientific problem of the origin of life.

ACKNOWLEDGMENT

This work was supported by NASA Life Sciences (NGR-004-025) and the Richard A. Lounsbery Foundation. We are grateful to Professor Andrew Fabian of the Darwin College Lecture Series (Cambridge University, England), Gail Fleischaker, Lorraine Olendzenski and Dorion Sagan for aid with the manuscript, and to Geraldine Kline for expert typing services.

REFERENCES

Alberts, B. M., 1986: "The Function of the Hereditary Materials: Biological Catalyses Reflect the Cell's Evolutionary History," Amer. Zool., 26, 781-796.

Bloch, K., in D. E. Vance and J. E. Vance (eds.), 1985: Biochemistry of Lipids and Membranes, Benjamin Cumming, Menlo Park, 1.

Goksøyr, J., 1967: "Evolution of Eucaryotic Cells," Nature, 214, 1161.

Gray, M. S., 1983: "Bacterial Ancestry of Chloroplasts and Mitochondria," BioScience, 33, 693-699.

Guerrero, R., Esteve, I., Pedros-Alio, C. and Gaju, N., 1987: "Predatory Bacteria in Prokaryotic Communities: The Earliest Trophic Relationships," Annals of the New York Academy of Sciences, 503, 238-250.

Guerrero, R., Pedros-Alio, C., Esteve, I., Mas, J., Chase, D. and Margulis, L., 1986: "Predatory Prokaryotes: Predation and Primary Consumption Evolved in Bacteria," Proc. Natl. Acad. Sci. USA, 83, 2138-2142.

Hanson, E., 1977: The Origin and Early Evolution of Animals, Wesleyan University Press, Middleton, Conecticut.

Jantsch, E., 1980: The Self-Organizing Universe, Pergamon Press, New York.

Kandler, O., 1985: "Evolution of the Systematics of Bacteria," in K. H. Schleifer and E. Stackebrandt (eds.), Evolution of Prokaryotes, Academic Press, London, 335-361.

Lazcano, A., 1987: "Prebiotic Evolution and the Origin of Cells" in L. Margulis, R. Guerrero and A. Lazcano (eds.) Trebalis de La Societat Catalana de Biologia 39, 73-103.

Margulis, L., 1981: Symbiosis in Cell Evolution, W. H. Freeman Co., San Francisco.

Margulis, L., 1982: Early Life, Science Books, Boston.

Margulis, L. and Bermudes, D., 1985: "Symbiosis as a Mechanism of Evolution: Status of Cell Symbiosis Theory," Symbiosis, 1, 101-123.

Margulis, L. and Sagan, D., 1986a: Origins of Sex, Yale University Press, New Haven.

Margulis, L. and Sagan, D., 1986b: Microcosmos. Four Billion Years of Evolution from our Microbial Ancestors, Summit Books, New York.

Margulis, L. and Stolz, J. F., 1984: In Memoriam Elso S. Barghoorn, Origins of Life 15, 1-3.

Margulis, L., Chase, D. and Guerrero, R., 1986: "Microbial Communities," BioScience, 36, 160-170.

Maturana, H. and Varela, F., 1980: Autopoiesis and Cognition: The Realization of the Living, D. Reidel, Boston.

Morowitz, H., 1967: "Biological Self-Replicating Systems," Progress in Theoretical Biology, 1, 35-58.

Obar, R. and Green, J., 1985: "Molecular Archaeology of the Mitochondrial Genome," J. Molecular Evolution, 22, 243-251.

Sagan, D., (ed.), 1985: Planetary Biology and Microbial Ecology 1984: The Global Sulfur Cycle (NASA Technical Memorandum 87570).

Saier, M. and Jacobson, G., 1984: The Molecular Basis of Sex and Differentiation: A Comparative Study of Evolution, Mechanism, and Control in Microorganisms, Springer-Verlag, New York.

Schopf, J. W. (ed.), 1984: The Earth's Earliest Biosphere, Princeton University Press, Princeton, New Jersey.

Schwemmler, W., 1984: Reconstruction of Cell Evolution: A Periodic System, CRC Press, Boca Raton, Florida.

Searcy, D. G. and Delange, R. J., 1980: "Thermoplasma Acidophilum Histone-like Protein: Partial Amino Acid Sequences Suggestion of Homology to Eukaryotic Histones" Biochimica biophysica Acta, 609, 197-200.

Starr, M. P., Stolp, H., Trüper, H. G., Balows, A. and Schlegel, H. G. (eds.), 1981: The Prokaryotes, vols. 1 and II, Springer-Verlag, Berlin.

Trüper, H. G., 1982: "Microbial Processes in the Sulfur Cycle through Time," in H. D. Holland and M. Schidlowski (eds.), Mineral Deposits and the Evolution of the Biosphere, Springer-Verlag, Berlin.

Van Bruggen, H., 1986: "Methanogenic Bacterial Endosymbionts of Sapropelic Bacteria," Stichting Studentenpers, Nijmegen, The Netherlands.

Van Bruggen, J. J. A., Stumm, C. K., Zwart, K. B. and Vogels, G. D., 1985: "Endosymbiotic Methanogenic Bacteria of the Sapropelic Amoeba, Mastigella," FEMS Microbiology Ecology 31, 187-192.

Vidal, G. and Knoll, A. H., 1983: "Proterozoic Plankton," Geological Society of America Memoires, 161, 265.

Walter, M. R. (ed.), 1976: Stromatolites, Elsevier Scientific Publishing Company, Amsterdam.

Westbroek, P. and de Jong, E. W. (eds.), 1983: Biomineralization and Biological Metal Accumulation, D. Reidel, Boston.

Zillig, W., Schnabel, R. and Stetter, K. O., 1985: "Archaebacteria and the Origin of Eukaryotic Cytoplasma," Current Topics of Microbiology and Immunology, 114, 1.

ADDENDUM

Lazcano, A., Guerrero, R., Margulis, L., and Oro, J., 1988: "The Evolutionary Transition From RNA to DNA in Early Cells," Journal of Molecular Evolution, 27, 283-290.

Loomis, W. F., 1988: Four Billion Years: An Essay on the Evolution of Genes and Organisms, Sinauer Associates, Inc., Sunderland, Massachusetts.

Margulis, L., Corliss, J. O., Melkonian, M., and Chapman, D. J. (eds.), 1989: Handbook of Protoctista, Jones and Bartlett Publishers, Inc., Boston.

Rambler, M. B., Margulis, L., and Fester, R. (eds.), 1989: Global Ecology: Towards a Science of the Biosphere, Academic Press, San Diego, California.

Figure 1. Stromatolites found in the Bathhurst Inlet, Victoria Island, in the Northwest Territories of Canada. (Photograph courtesy of F. H. Campbell, Geological Survey of Canada.)

Figure 2. The details of the microbial community of Lake Ciso drawn here by Christie Lyons, have been reconstructed from ecological and electron microscopical studies (Margulis et al., 1986).

MODELS FOR PRIMITIVE MEMBRANES

Janos H. Fendler
Syracuse University
Syracuse, New York 13244-1200

ABSTRACT

Relatively simple surfactants containing sulfate, phosphate, carboxylate, and quaternary ammonium groups spontaneously self-assemble under plausible prebiotic conditions. Depending on conditions and volume ratios of the headgroups to the tails of these surfactants, simple micelles, oil-drop-like microemulsions, micellar mesophases, or closed vesicles could be formed from them. Completely closed vesicles formed from simple surfactants could be considered as primitive models of cell membranes. They can selectively compartmentalize amino acids and nucleotides and develop into more complex entities capable of recognition and replication.

PREBIOLOGICAL SELF ORGANIZATION OF MATTER
C. Ponnamperuma and F.R. Eirich (Eds.)

ISBN 0-937194-18-2

THE CONSTRUCTION OF MODELS TO PRODUCE DISTRIBUTIONS OF SIMPLE CELL MORPHOLOGIES

Paul K. Strother
Boston University
Boston, Massachusetts 02215

ABSTRACT

The size frequency distributions of populations of rods and spheres from living populations, precambrian fossil populations and computer-generated populations are compared in order to determine the general characteristics of biogenic distributions. Distributions of simple cells which are generated at random are different from those which involve growth and divisional processes. It is possible to construct a cell generator model involving a minimum of assumptions and randomness of division events which should mimic living populations and fossil populations. However, such computer-generated distributions are dissimilar to actual examples. An attempt to model the growth of <u>Eosynechococcus thuleensis</u> from 750 MY cherts in NW Greenland shows that distributions found in the rock record are dependent on sampling factors and possible changes in growth rates during the life cycle of the organism. The recognition and quantification of these factors should enable the construction of a more general growth model which differs from the misleading notion that biogenic populations of simple cells are distributed normally about a mean linear dimension.

PREBIOLOGICAL SELF ORGANIZATION OF MATTER
C. Ponnamperuma and F.R. Eirich (Eds.)

ISBN 0-937194-18-2

MODEL OF THE PROTOMEMBRANE

Aleksander T. Przybylski
Institute for Molecular and Cellular Evolution
University of Miami
521 Anastasia Avenue
Coral Gables, Florida 94063

ABSTRACT

The origin and the mechanism of formation of membrane-like structures are among the crucial questions of the origin of life. A morpho-functional formation of a spherical protomembrane is presented to be an imminent stage after phase separation followed by processes controlling membrane permeability and energy transduction.

The concept presented of the model of the protomembrane is based on the following assumptions:

-- development of the membrane is a result of phase separation and all consecutive physicochemical stages toward morpho-functional formation of the membrane;
-- there are paleobiological data on cellular forms which existed over three billion years ago;
-- there are experimental evidences for chemical formation of membrane lipo-protein constituents and their self-assembly into membranous structures.

The plausible sequences of events during protomembrane formation comprise:

-- preservation of the barrier between the protocell and its environment,
-- establishment of the proton and cation gradients,
-- energy transduction,
-- formation of surface and electrical phenomena on and across the membrane.

Results of simulation of the protomembrane demonstrated the presence of the above features in such membranes and indicate the plausibility of such processes during protomembranogenesis. Since both phospholipid and protein-like materials were synthesized under plausible prebiotic conditions, it is assumed that protomembranes were present (prior to genetically coded and replicated cells) on the prebiotic Earth. Such forms were able to perform the process

PREBIOLOGICAL SELF ORGANIZATION OF MATTER
C. Ponnamperuma and F.R. Eirich (Eds.)

ISBN 0-937194-18-2

of separation and selective transport, combined with the required surface and electrical membrane phenomena.

The concept of the membrane formation has a direct relation to the protein-lipid interrelationship during membrane formation, as well as to the main RNA-protein-first question. Both cellular membrane formation and replication processes during the emergence of the protocell belong to the broader context of the origin of the protocell. In a narrower aspect we may, however, distinguish a pregenetic stage of protocellular formation. The contemporary lipo-protein concept of the membrane structure and the results of the reassembly technique do indicate the possibility of partial morphological and functional capacity of some membrane and cellular elements.

It seems that the appearance of the replicating protocell was preceded by a relatively stable form of a cell-like protostructure.

AUTOCATALYSIS IN THE HYDROLYSIS REACTION OF GUANOSINE 5'-PHOSPHATE-2-METHYLIMIDAZOLIDE

Anastassia Kanavarioti
National Research Council Fellow
Life Sciences Division/Planetary Biology Branch
NASA/Ames Research Center
Moffett Field, California 94035

ABSTRACT

Until recently the role of efficient and selective catalysts in biochemical reactions had been ascribed exclusively to enzymes. However, this notion requires careful reexamination after the demonstration of the catalytic abilities of nucleic acids in a) the non-enzymatic RNA-type template-directed oligonucleotide synthesis (Inoue et al., 1983) and b) the self-splicing reaction of the nuclear pre-rRNA of the ciliated protozoan Tetrahymena thermophilia that occurs in the absence of any proteins (Cech et al., 1981). In both cases the catalytic function was provided by an RNA molecule, and the question arose whether the building blocks of RNA, the nucleotide bases with their various functional groups, can act as acid-base catalysts (Orgel, 1986). The chemistry of related heterocyclic molecules suggests that the nucleotide bases should indeed have some catalytic ability, but until now there has been no direct experimental evidence for this assertion. We now report that in the hydrolysis of guanosine 5'-phosphate-2-methylimidazolide, 2-MeImpG, to form guanosine 5'-phosphate, pG, and 2-methylimidazole, 2-MeIm (Kanavarioti, 1986), both, 2-MeImpG and pG, catalyze the reaction. In other words, the rate constant of 2-MeImpG hydrolysis increases with 2-MeImpG concentration. To the best of our knowledge, this is the first case where a nucleotide derivative, acting as a general acid-base catalyst, catalyzes its own reaction. This autocatalytic ability must be present in the nucleotide moiety, as shown by the fact that pG, but not 2-MeIm, enhances the rate constant of hydrolysis. In the context of the above mentioned recent discovery of RNA molecules acting catalytically, it is not surprising that their building blocks, the mononucleotides, also have catalytic ability. Furthermore, Mg^{2+} and polycytidylic acid (poly(C)) were shown to enhance the rate constant of hydrolysis of 2-MeImpG. The bimolecular rate constants for catalysis of 2-MeImpG hydrolysis by Mg^{2+}, pG and poly(C) are 0.123, 0.35 and approximately 0.5 M^{-1} hr^{-1} at 37°C and pH 7.80 (0.5 M tris.HCl and 1 M NaCl). This compares with a unimolecular rate constant of hydrolysis of 0.045

PREBIOLOGICAL SELF ORGANIZATION OF MATTER
C. Ponnamperuma and F.R. Eirich (Eds.)

ISBN 0-937194-18-2

hr^{-1}. These results have also implications in the context of evaluating the template effect in the template-directed oligomerization reaction of 2-MeImpG on poly(C) acting as the template (Inoue *et al.*, 1982). Since template-directed conditions in the oligomerization of 2-MeImpG require the presence of Mg^{2+} and poly(C), as well as concentrated solutions of 2-MeImpG, it seems that the template effect in the oligomerization reaction is counteracted by the effect of Mg^{2+} and poly(C), enchancing the hydrolysis of the activated monomer (2-MeImpG). This is important because under typical reaction conditions oligomerization is in competition with monomer hydrolysis and hence the efficiency of oligomerization is directly related to the degree of competition between the two reactions.

REFERENCES

Cech, T. R., Zaug, A. J. and Grabowski, P. J., 1981: *Cell, 27*, 487.

Inoue, T. and Orgel, L.E., 1983: *Science, 219*, 859-862.

Inoue, T. and Orgel, L. E., 1982: *J. Mol. Biol., 162*, 201-217.

Kanavarioti, A., 1986: *Origins of Life, 17*, 85-103.

Orgel, L. E., 1986: *J. Theor. Biol., 123*, 127-149.

BIOENERGETICS OF SULFOLOBUS, A THERMOACIDOPHILIC ARCHAEBACTERIUM

Takayoshi Wakagi and Tairo Oshima
Department of Life Science
Tokyo Institute of Technology
Yokohama, Kanagawa 227 Japan

ABSTRACT

Recent progress in the studies on bioenergetic aspects of *Sulfolobus*, a thermoacidophilic archaebacterium, has revealed that (1) glucose metabolism is divergent even in two genuses of thermoacidophiles: the Embden-Meyerhoff pathway is found in *Thermoplasma*, whereas a modified Entner-Doudorff pathway operates in *Sulfolobus*; (2) 2-oxoacid:ferredoxin oxidoreductases act as members of the reductive carboxylic acid cycle, while a multienzyme complex of 2-oxoacid dehydrogenase is absent in all other archaebacteria so far examined; (3) although the respiratory chain systems are different, archaebacteria appear to have a common ATPase, similar to F_0F_1 ATPase on the one hand, and to the vacuole type ATPase on the other; (4) elemental sulfur is oxidized by an enzyme of the oxygenase type. These findings concerning the bioenergetic machinery of archaebacteria suggest a primitive position they could have occupied in early biological evolution.

INTRODUCTION

The archaebacteria are a group of microorganisms which are phylogenically distinct from eukaryotes and eubacteria. They are recognized by a large phylogenic distance from other bacteria, as revealed by comparing S_{ab} values of 16S rRNA (Woose and Fox, 1977; Fox *et al.*, 1980). Archaebacteria are divided into the three major orders: methanogens, extreme halophiles, and thermoacidophiles, including *Sulfolobus* and *Thermoplasma*.

Sulfolobus is a thermoacidophilic archaebacterium which is able to grow in autotrophic and heterotrophic media at high temperatures up to 85°C, and at low pH (about 2) (Brock *et al.*, 1972; Stetter and Zillig, 1985). Under autotrophic growth, the organism assimilates carbon dioxide via a reductive carboxylic acid pathway (Kandler and Stetter, 1981). The bacterium thrives in an acidic environment, and must extrude protons through the cytoplasmic

PREBIOLOGICAL SELF ORGANIZATION OF MATTER
C. Ponnamperuma and F.R. Eirich (Eds.)

ISBN 0-937194-18-2

membrane to keep the inside of the cell neutral. Proton pumping activity of the membrane coupled to respiration, and possible ATP synthesis according to a proton chemiosmotic gradient have been reported (Anemuellar et al., 1985).

In this paper, bioenergetic aspects of Sulfolobus are reviewed and compared to those of other archaebacteria, in terms of sulfur metabolism, glycolysis, tricarboxylic acid cycle, electron transport system and ATPase. The phylogenic status of archaebacteria will be discussed.

SULFUR METABOLISM

Although Sulfolobus is a potent autotroph able to grow on elemental sulfur, little is known about the sulfur metabolism except that sulfur oxidation is performed by an enzyme of the oxygenase type. The sulfur oxygenase that catalyzes the oxidation of elemental sulfur to sulfite has been purified. The enzyme has a molecular weight of 560,000, consisting of 16 identical subunits (35,000 KDa.) (Emmel et al., 1986).

GLUCOSE METABOLISM

The Embden-Meyerhoff pathway of glycolysis, usually found in most eubacteria and eukaryotes, and also in Thermoplasma (Searcy, 1986) is absent in Sulfolobus.

Two main routes have been reported for the glucose break down in the organism (DeRosa et al., 1984; Giardina et al., 1986). One is an ATP-dependent phosphorylation of glucose, followed by isomerization and dephosphorylation (scheme 1). The other is an NAD+-dependent dehydrogenation to give final products of pyruvate and glyceraldehyde (scheme 2).

(1) glucose --> glucose 6-phosphate --> fructose 6-phosphate --> fructose

(2) glucose --> gluconate --> 2-keto-3-dehydrogluconate --> pyruvate + glyceraldehyde

The latter pathway is the same as the pathway of Entner-Doudorff except that the metabolite is not phosphorylated at any step. Glyceraldehyde thus produced must subsequently be phosphorylated and converted to pyruvate. Without these processes, no ATP could be formed during glycolysis. Only pyruvate kinase has been detected so far (Wakagi and Oshima, 1985).

Recently, PPi has been proposed to replace ATP as energy source for various rections (Reeves, 1976; Kulaev _et al._, 1980), including glycolysis. PPi dependent phosphofructokinases (PFKs) have been found in various sources (Wood and Goss, 1985; Bertagnolli _et al._, 1986a, 1986b; Pollack and Williams, 1986). In _Sulfolobus acidocaldarius_, neither ATP or PPi dependent PFK's were detected (Wakagi, unpublished observation). It is nevertheless an interesting question whether archaebacteria show such PPi dependent metabolism at all, since PPi is a probable precursor of ATP in early biological evolution (see Baltscheffsky, this issue). In this context, it is noteworthy that membranes of _Sulfolobus_ catalyze the hydrolysis of ATP, ADP and PPi, as shown later. Finally, the glycolytic pathway of _Sulfolobus_ is simpler and the energy yield of glycolysis is lower than that of the Embden-Meyerhoff pathway.

Another striking feature of glucose metabolism in _Sulfolobus_ is intracellular storage of glycogen whose average chain length is 7. The glucosyl transferase utilizes UDP-glucose as well as ADP-glucose. The enzyme is complexed with glycogen and appears in the form of amorphous bodies occupying a large portion of the cell (Koenig _et al._, 1982).

TRICARBOXYLIC ACID CYCLE

2-OXOACID:FERREDOXIN OXIDOREDUCTASES

The oxidative decarboxylation of 2-oxoacids such as pyruvate and 2-oxoglutarate is a key reaction of the tricarboxylic acid cycle. In aerobic eubacteria and mitochondria, the reaction is catalyzed by a 2-oxoacid dehydrogenase multienzyme complex (large aggregates with mol. wts. of 3 - 7 million) composed of 2-oxoacid decarboxylase, lypoyl-reductase transcarbamylase and dihydrolipoyl dehydrogenase. In contrast, the oxidation of 2-oxoacid is catalyzed by ferredoxin oxidoreductase in all archaebacteria so far tested, as well as in anaerobic, eubacteria and photosynthetic eubacteria (Kerscher and Oesterhelt, 1981 and 1982; Kerscher _et al._, 1982). Multienzyme complexes of 2-oxoacid dehydrogenase are absent from these organisms. It should be noted that a dihydrolipoamide dehydrogenase has been purified in _Halobacterium halobium_ (Danson _et al._, 1984 and 1986).

Pyruvate:ferredoxin oxidoreductase (mol. wt. 256 k) and 2-oxoglutarate:ferredoxin oxidoreductase (mol. wt. 248 k) have been purified from _H. halobium_. These enzymes have also been detected in _Sulfolobus_ and _Thermoplasma_, whose ferredoxins can be classified as an eubacterial type (Kerscher _et al._, 1982).

The 2-oxoacid:ferredoxin oxidoreductases act as accumulators of carbon dioxide in the reductive carboxylic acid cycle in chemolithotrophic eubacteria, and possibly in Sulfolobus (Kandler and Stetter, 1981). These enzymes seem to be the functional precursors of the multienzyme complexes of 2-oxoacid dehydrogenase.

These ferredoxin-linked enzymes may have existed before the divergence of the ur-kingdoms of prokaryotes and eubacteria, and 2-oxoacid dehydrogenase complexes might have emerged at later stages of the evolutionary process, i.e., after the development of oxidative phosphorylation (Kerscher and Oesterheldt, 1981 and 1982).

ISOCITRATE DEHYDROGENASE

Isocitrate dehydrogenase (ICDH) catalyzes oxidative decarboxylation of isocitrate with either NAD+, or NADP+ as a coenzyme. In the reverse reaction (included in reductive carboxylic acid cycle, as referred above), carbon dioxide is assimilated by ICDH. Therefore, the enzyme plays an important role in Sulfolobus. Partially purified ICDH was reported to have both NAD+- and NADP+-dependent activities with almost the same Vmax values, due likely to a single species of the enzyme (Danson and Wood, 1984). However, purified ICDH prefers NADP+ much more (with a Vmax value about ten times larger) than NAD+ (Wakagi and Oshima, manuscript in preparation). The reason for the discrepancy is not clear, as the situation is similar between crude and purified preparations of the enzyme. Existing differences between the strains of Sulfolobus acidocaldarius used might be pointed out.

MALATE DEHYDROGENASE

Malate dehydrogenases (MDHs) from Sulfolobus and Thermoplasma have been purified and show A-side stereospecificity for NADH, as well as for other eubacterial and eukaryotic MDHs (Goerisch et al., 1985).

CITRATE SYNTHASE

Partially purified citrate synthases from Sulfolobus and Thermoplasma have mol. wts. of about 70,000, which is similar to the enzymes from eukaryotes and Gram-positive eubacteria (Grossebueter and Goerisch, 1985; Danson et al., 1985).

OXIDATIVE PHOSPHORYLATION

Sulfolobus is an aerobic organism. Just as in aerobic eubacteria, the ATP synthesis seems to depend on oxidative phosphorylation. NADH is likely to be the respiratory substrate (Anemuellar *et al.*, 1985). Proton transport across the cytoplasmic membrane, coupled to oxidative phosphorylation, seems to be essential not only for the energy requirement, but for the adaptation of acidophiles, such as *Sulfolobus*, to the acidic environment.

RESPIRATORY CHAIN

The composition of the respiratory chain in these organisms is still not clear. Early studies revealed that cytochrome *a* is firmly bound to the cell membrane, but cytochrome *c* is absent in the whole cell. Cytochrome *a* appears to act as a terminal oxidase, because the membranes catalyze the oxidation of cytochrome *c* from various sources, such as horse heart, pig heart, yeast, candida, and a thermophilic bacterium. Cytochrome *c* oxidase activity is inhibited by about 50 percent of 0.1 mM of either azide or cyanide, which can well account for the growth inhibition caused by these poisons (Wakagi and Oshima, 1986 and 1987).

Recently, an NADH dehydrogenase has been purified to homogeneity. The enzyme is a flavoprotein with two identical 50-kDa subunits. Caldariellaquinone, a unique quinone containing the benzothiophene moiety (DeRosa *et al.*, 1977; Collins and Langworthy, 1983), is slowly reduced by the enzyme, suggesting the NADH dehydrogenase to be the first enzyme of the respiratory chain (Wakao *et al.*, 1987).

ATPASE

Most of the membrane bound ATPases are responsible for the ion transport across the membrane. These ATPases are classified into three major classes, designated as "P" (phosphorylated), "V" (vacuolar) and "F" (F_0F_1) according to Pedersen and Carafoli (1987). The phylogenic relationship between these ATPases are currently discussed. The "F" enzyme, namely, F_0F_1 proton translocating ATPase (= ATP synthase) is believed to be an ubiquitous enzyme, since it has been detected in a variety of cells including even strictly anaerobic bacteria that lack a respiratory chain.

ATP synthesis in archaebacteria seems to be coupled to a proton electrochemical potential, as in eubacteria. Dicyclohexylcarbodiimide (DCCD), an inhibitor of F_0F_1 ATPase, inhibited ATP synthesis in methanogens and extreme halophiles, which was thought to indicate the presence of a typical F_0F_1 in these organisms. In contrast, a different type of ATPase, activated by sulfate or sulfite at an optimum pH around 5 and insensitive to DCCD, has been reported for two acidothermophilic archaebacteria, *Thermoplasma* (Searcy and Whatley, 1982) and *Sulfolobus* (Wakagi and Oshima, 1985). Therefore, a diversity of archaebacterial ATPases appears to exist.

Recently, ATPases have bene purified partially or completely from methanogens (Inatomi, 1986), halophiles (Kristjansson *et al.*, 1986; Nanba and Mukohata, 1987) and *Sulfolobus* (Luebben and Schaefer, 1987; Konishi *et al.*, 1987). ATPase purified to homogeneity from *S. acidocaldarius* (Konishi *et al.*, 1987) is a 360 kDa. protein with 3 types of subunits with mol. weights of 69 kDa., 51 kDa. and 28 kDa. The composition of the subunits appears to be 3:3:1, which resembles F_1ATPase. The enzyme activity is enhanced by sulfite anion with an optimum pH of 5, and is inhibited by nitrate and NBD-Cl, an inhibitor of both F_1 and "V" ATPases. DCCD failed to inactivate the enzyme. The activity was not affected by azide, a specific inhibitor of F_1.

Most of these characteristics of *Sulfolobus* ATPase appear to be similar to other archaebacterial ATPases, except for the sensitivity to DCCD. Some of their properties are similar to F_1 ATPase, but others are similar to "V" ATPase. These observations suggest that archaebacterial ATPases are the possible ancestors of "F" and "V" ATPases. To prove the hypothesis, immunological and molecular biological approaches are necessary. Preliminary experiments in our laboratory revealed that antibodies raised against the *Sulfolobus* ATPase cross-reacted with the ATPases from methanogens, extreme halophiles and the yeast vacuole. Also it remains as yet undetermined whether these archaebacterial ATPases are electrogenic and translocate protons.

Membrane fractions of *Sulfolobus* shows an acid ATPase activity. The enzyme was most active at pH 2.5, with ATP, ADP and PPi as substrates. AMP and other phosphoesters so far tested were not hydrolyzed. Therefore, the enzyme seems to attack high energy phosphoesters selectively. The activity was inhibited by sulfate. The acid ATPase (or acid PPase) may be located on the outer surface of the membrane to supply energy for some processes yet unknown, since PPi is able to play an important role in the energy metabolism coupled to the electrochemical potential created by protons (Reeves, 1976; Mansurova *et al.*, 1977; Lahti, 1983).

CONCLUSION

Since the environmental conditions under which the archaebacteria live are very different from one another, these organisms appear to use different modes of energy metabolism. However, the archaebacterial ATPases, as well as the 2-oxoacid:ferredoxin oxidoreductases, are remarkably similar to each other. These enzymes are possible ancestors of their counterparts in other bacteria and eukaryotes. For example, vacuolar ATPase and mitochondrial ATPase may have evolved from a common ancestor, an archaebacterial ATPase.

Archaebacteria are also characterized by unique lipid components (glycerol-isoprenyl ethers) (Langworthy et al., 1985) and a glycoprotein cell wall (Inatomi et al., 1983; Sleytr and Messner, 1983). These features may indicate that some biochemical properties of archaebacteria are relics of the biochemistry of the primitive cells that existed in early phases of the biological evolution, long before the separation of prokaryotes and eukaryotes.

REFERENCES

Anemuellar, S., Leubben, M. and Schaefer, G., 1985: FEBS Lett., 193, 83.

Bertagnolli, B. L., Younathan, E. S., Voll, R. J., Pittman, C. E. and Cook, P. F., 1986: Biochemistry, 25, 4674.

Bertagnolli, B. L., Younathan, E. S., Voll, R. J. and Cook, P. F., 1986: Biochemistry, 25, 4682.

Brock, T. D., Brock, K. M., Belly, R. T. and Weiss, R. L., 1972: Arch. Mikrobiol., 84, 54.

Collins, M. D. and Langworthy, T. A., 1983: System. Appl. Microbiol., 4, 295.

Danson, M. J., Eisenthal, R., Hall, S., Kessell, S. R., and Williams, D. L., 1984: Biochem. J., 218, 811.

Danson, M. J. and Wood, P. A., 1984: FEBS Lett., 172, 289.

Danson, M. J., Black, S. C., Woodland, D. L. and Wood, P. A., 1985: FEBS Lett., 179, 120.

Danson, M. J., McQuattie, A. and Stevenson, K. J., 1986: Biochemistry, 25, 3880.

DeRosa, M., Derosa, S., Gambacorta, A., Minale, L., Thomson, R. H. and Worthington, R. D., 1977: J. Chem. Soc. Perkin. Trans., 1, 653.

DeRosa, M., Gambacorta, A., Nicolaus, B., Giardina, P., Poerio, E. and Buonocore, V., 1984: Biochem. J., 224, 407.

Emmel, T., Sand, W., Koenig, W. and Bock, E., 1986: J. Gen. Microbiol., 132, 3415.

Fox, G., Stackenbrandt, E., Hespell, R. B., Gibson, J., Malinoff, J., Dyer, T. A., Wolfe, R. S., Balsh, W. E., Tanner, R. S., Magrum, L. J., Zablen, L. B., Blakemore, R., Gupts, R., Bohen, L., Lewis, B. J., Stahl, D. A., Luehrsen, K. R., Chen, K. N. and Woese, C. R., 1980: Science, 209, 457.

Giardina, P., DeBiasi, M. G., DeRosa, Gambarcota, A. and Buonocore, V., 1986: Biochem. J., 239, 517.

Goerisch, H., Hartl, T., Grossebueter, W. and Stezowski, J. J., 1985: Biochem. J., 226, 885.

Grossebueter, W. and Goerisch, H., 1985: System. Appl. Microbiol., 6, 119.

Inatomi, K., Ohba, M. and Oshima, T., 1983: Chemistry Lett., 1191.

Inatomi, K., 1986: J. Bacteriol., 167, 837.

Kandler, O. and Stetter, K. O., 1981: Zbl. Bakt. Hyg. I. Abt. Orig., C2, 111.

Kerscher, L. and Oesterheldt, D., 1981: Eur. J. Biochem., 116, 587.

Kerscher, L., Nowitzki, S. and Oesterheldt, D., 1982: Eur. J. Biochem., 128, 223.

Kerscher, L. and Oesterheldt, D., 1982: Trends Biochem Sci., 7, 371.

Koenig, H., Sorko, R., Zillig, W. and Reiter, W. D., 1982: Arch. Microbiol., 132, 297.

Konishi, J., Wakagi, T., Oshima, T. and Yoshida, M., 1987: J. Biochem., 102, 1379.

Kristjansson, H., Sadler, M. H. and Hochstein, L. I., 1986: FEMS Microbiol. Reviews, 39, 151.

Kulaev, I. S., Mansurova, S. E., Burlakova, E. B. and Dukhovich, V. F., 1980: Biosystems, 12, 177.

Lahti, R., 1983: Microbiol. Rev., 47, 69.

Langworthy, T. A., 1982: In I. C. Gunsalus (ed.), The Bacteria, Vol. 8, Academic Press Inc., 459.

Luebben, M. and Schaefer, G., 1987: Eur. J. Biochem., 164, 533.

Mansurova, S. E., Shakhov, Y. A. and Kulaev, I. S., 1977: FEBS Lett., 74, 31.

Nanba, T. and Mukohata, Y., 1987: J. Biochem., 102, in press.

Nicolaus, B., deSimone, A., delPiano, L., Giardina, P., and Lama, L., 1986: Biothech. Lett., 8, 497.

Oshima, T., Ohba, M. and Wakagi, T., 1984: Origins of Life, 14, 665.

Pedersen, P. L. and Carafoli, E., 1987: Trends Biochem. Sci., 12, 146.

Pollack, J. D. and Williams, M. V., 1986: J. Bacteriol., 165, 53.

Reeves, R. E., 1976: Trends Biochem. Sci., 1, 53.

Searcy, D. G., Whatley, F. R., 1982: Zbl. Bakt. Hyg., I. Abt. Orig., C3, 245.

Searcy, D. G., 1986: System. Appl. Microbiol., 7, 198.

Sleytr, U. B. and Messner, P., 1983: Ann. Rev. Microbiol., 37, 311.

Solioz, M., 1984: Trends Biochem. Sci., 7, 309.

Stetter, K. O. and Zillig, W., 1985: In I. C. Gunsalus (ed.), The Bacteria, Vol. 8, Academic Press, Inc., 85.

Thomson, J. W. and Shapiro, B. M., 1981: J. Biol. Chem., 256, 3077.

Wakagi, T. and Oshima. T., 1985: Biochem. Biophys. Acta., 817, 33.

Wakagi, T. and Oshima, T., 1986: System. Appl. Microbiol., 7, 342.

Wakagi, T. and Oshima, T., 1987: Origins of Life, 17, 391.

Wakao, H., Wakagi, T. and Oshima, T., 1987: J. Biochem., 102, 255.

Woese, C. R. and Fox, G. E., 1977: Proc. Natl. Acad. Sci. USA, 74, 5088.

Wood, H. G. and Goss, N. H., 1985: Proc. Natl. Acad. Sci. USA, 82, 312.

ADDENDUM

Danson, M. J., 1988: Adv. Microbial Physiol., 29, 165-231.

Denda, K., Konishi, J., Oshima, T., Date, T. and Yoshida, M., 1988: J. Biol. Chem., 263, 6012-6015.

Moll, R. and Schafer, G., 1988: FEBS Lett., 232, 359-363.

ENERGY TRANSFER IN PRIMITIVE PIGMENTED MEMBRANE SYSTEMS

Bettina Heinz
Institut fur Organische Chemie
der J. W. Goethe Universitaet
Laboratorium Niederrad
6000 Frankfurt AM Main 70 FRG

ABSTRACT

Chromophores such as the N-heterocycles, pteridines and flavins are apparently readily produced under various reaction conditions that could have been likely in a prebiotic world. These chromophores are tested for eventual electron transfer activities when incorporated into a bilayer of vesicles prepared from present-day phospholipids, or in connection with primordial membrane systems based on different amphiphilic molecules (lipid ethers and lipid esters, fatty alcohols, polyglycols, proteinoids and humic acids).

ISBN 0-937194-18-2

APPENDIX 1

AGENDA

Center of Adult Education
University of Maryland
College Park, Maryland 20742

Wednesday, April 8

5:30 p.m.	Registration, Chesapeake Room
6:30 p.m.	Reception, Chesapeake Room

Thursday, April 9

8:00 a.m.	Registration, Volunteer Firefighter's Room

SESSION I

PHYSICS OF PREBIOLOGICAL ORGANIZATION
Chairman: Frederick R. Eirich

9:00 a.m.	WELCOME; Introduction
9:15 a.m.	H. Baltscheffsky "Molecular Co-evolution of Phosphate Compounds and Polypeptides"
9:45 a.m.	J. Oro and A. Lazcano "A Holistic Precellular Organization Model"
10:15 a.m.	W. A. Bonner and E. Rubenstein "Photochemical Origins of Biomolecular Chirality"
10:45 a.m.	COFFEE
11:00 a.m.	J. Rothstein "Entropy and the Evolution of Complexity and Individuality"
11:30 a.m.	R. E. Clapp "Mechanisms of Entropy Disposal: Their Role in Shaping Prebiological Organization"
12:00 p.m.	B. C. Clark "Macrobiont: The Pond as Proto-Organism"

12:30 p.m. LUNCH

SESSION II

ORGANIZATION OF PROTEINS AND NUCLEIC ACIDS I
Chairman: H. Baltscheffsky

Thursday, April 9

1:45 p.m.	Lacey, Jr. "Hydrophobicity as an Organizing Principle in the Origin of the Genetic Code and Protein Synthesis"
2:15 p.m.	H. Yanagawa "Polymerization of Nucleotides in an Aqueous Solution"
2:45 p.m.	D. Usher "Are Ribonucleosides Special?"
3:15 p.m.	TEA
3:30 p.m.	N. Senaratne, M. K. Hobish and C. Ponnamperuma "Direct Interactions Between Amino Acids and Nucleotides as a Possible Explanation for the Origin of the Genetic Code"
4:00 p.m.	F. Eirich "Co-evolution of Proteins and Nucleic Acids"
4:30 p.m.	C. N. Matthews "Simultaneous Origin of Polypeptides and Poly-nucleotides: Hydrogen Cyanide Polymers as Prebiotic Condensing Agents"
5:00 p.m.	END OF SESSION
6:30 p.m.	BANQUET

SESSION III

ORGANIZATION OF PROTEINS AND NUCLEIC ACIDS II
Chairman: S. W. Fox

Friday, April 10

9:15 a.m.	S. W. Fox "The Evolution of Protocells from Endo-genously Ordered Polymerization of Amino Acids"

9:45 a.m. S. Tyagi and C. Ponnamperuma
"A Search for Order Among Prebiotic Peptides"

10:15 a.m. S. L. Goldberg, J. M. Crosby, N. D. Iuseum, U. E. Younes
"Isotacticity: The Foundation of Protein and Organization"

10:45 a.m. COFFEE

11:00 a.m. A. Lazcano, J. Fastag, P. Gariglio, C. Ramirez, J.Oro
"Molecular Evolution of Nucleic Acid Polymers"

11:30 a.m. G. R. Fleischaker
"Three Models of a Minimal Cell"

12:00 p.m. V. Liaca, A. Lazcano, L. M. Rangel, E. Silva, P. Gariglio, J. Oro
"In Search of the Ancestral RNA Polymerase: An Experimental Approach"

12:30 p.m. LUNCH

SESSION IV

PRIMITIVE CELL MODELS
Chairman: J. C. Lacey

Friday, April 10

1:30 p.m. L. Margulis and R. Guerrero
"From Origins of Life to Evolution of Microbial Communities: A Minimalist Approach"

2:00 p.m. J. H. Fendler
"Model of Primitive Membrane and Origin of Life"

2:30 p.m. A. Kanavarioti
"Autocatalysis in the Hydrolysis Reaction of Guanosine 5'-phosphate-2-methylimidazolide"

3:00 p.m. TEA

3:15 p.m. P. K. Strother
"The Construction of Models to Produce a Distribution of Simple Cell Morphologies"

3:45 p.m.	T. Wakagi and T. Oshima "Bioenergetics of Sulfolobus, A Thermoacidophilic Archaebacterium"
4:15 p.m.	B. Heinz "Energy Transfer and Primitive Pigmented Membrane Systems"
4:45 p.m.	GENERAL DISCUSSION
5:15 p.m.	END OF SESSION

APPENDIX 2

LIST OF PARTICIPANTS

Robert K. Arnold
40 South Van Dorn Street
Apartment D-218
Alexandria, VA 22304

Herrick Baltscheffsky
Department of Biochemistry
University of Stockholm
S-10691 Stockholm
Sweden

William A. Bonner
Department of Chemistry
Stanford University
Stanford, CA 94305

Roger E. Clapp
The Mitre Corporation
Bedford, MA 01730

Benton C. Clark
Martin Marietta Aerospace
Planetary Sciences Laboratory
(0560)
P.O. Box 179
Denver, CO 80201

Nathan Dubowsky
Department of Biology
Westchester Community College
75 Grasslands Road
Valhalla, NY 10595

Frederick Eirich
Polytechnic University
of New York
333 Jay Street
Brooklyn, NY 11201

Janos H. Fendler
Department of Chemistry
Syracuse University
Syracuse, NY 13244

Sidney W. Fox
Institute for Molecular and
Cellular Evolution
University of Miami
521 Anastasia
Miami, FL 33177

Stanley I. Goldberg
Department of Chemistry
University of New Orleans
New Orleans, LA 70148

Theodore Gull
Head, Astronomy Division
NASA/Goddard Space Center
Greenbelt, MD 20771

Elliott M. Hartman, Jr.
Westchester Community College
75 Grasslands Road
Valhalla, NY 10595

Bettina Heinz
Institut fur Organische Chemie
der J. W. Goethe Universitaet
Laboratorium Niederrad
6000 Frankfurt Am Main 70
Federal Republic of Germany

Mitchell K. Hobish
MKH & Associates, Inc.
3162 Remington Avenue
Baltimore, MD 21211

Marsha Hollander
Chemistry Department, LCE
University of Maryland
College Park, MD 20742

Yasuhiro Honda
Chemistry Department, LCE
University of Maryland
College Park, MD 20742

Gail R. Fleischaker
Department of Biology
Boston University
Boston, MA 02215

Anastassia Kanavarioti
M/S: 239-4
NASA/Ames Research Center
Moffett Field, CA 94035

James Lacey, Jr.
Department of Biochemistry
CHSB 513
University of Alabama
at Birmingham
University Station
Birmingham, AL 35294

A. Lazcano
Departamento de Microbiologia
Escuela Nacional de Ciencias
Biologicas
IPN, Apdo. Postal 4-870, 06400
Mexico D. F., Mexico

V. Llaca
Departamento de Microbiologia
Escuela Nacionial de Ciencias
Biologicas
IPN, Apdo. Postal 4-870, 06400
Mexico D. F., Mexico

Lynn Margulis
Department of Biology
Boston University
2 Cummington Street
Boston, MA 02215

Clifford Matthews
Department of Chemistry
University of Illinois
Chicago, IL 60680

Rafael Navarro-Gonzalez
Chemistry Department-LCE
University of Maryland
College Park, MD 20742

Ei-Ichiro Ochiai
Department of Chemistry
Juniata College
Huntington, PA 16652

John Oro
Department of Biochemistry
University of Houston
4800 Calhoun Road
Houston, TX 77004

Tairo Oshima
Laboratory of Chemistry for
Natural Products
Tokyo Institute of Technology
Nagatsuta, Yokohama 227
Japan

Cyril Ponnamperuma
Laboratory of Chemical
Evolution
Department of Chemistry
University of Maryland
College Park, MD 20742

Aleksander T. Przybylski
Institute for Molecular and
Cellular Evolution
University of Miami
521 Anastasia Avenue
Coral Gables, FL 33134

Jerome Rothstein
Department of Computer &
Information Science
The Ohio State University
Columbus, OH 43210

John D. Rummel
Acting Program Manager
(Exobiology)
Life Sciences Division
NASA Headquarters (Code EBR)
Washington, DC 20742

Mark P. Staves
Department of Botany
University of Texas at Austin
Austin, TX 78713

Paul K. Strother
Department of Geology
Boston University
Boston, MA 02215

Sanjay Tyagi
Public Health Research
Institute
455 First Avenue
New York, NY 10016

David A. Usher
Department of Chemistry
Cornell University
Ithaca, NY 14853

Robert L. Weintraub
407 Brooks Avenue
Raleigh, NC 27607

Hiroshi Yanagawa
Mitsubishi-Kasei Institute for
Life Sciences
11 Minamiooya
Machida-Shi
Tokyo 194
Japan

Martynas Ycas
Department of Microbiology
State University of New York
766 Irving Avenue
Syracuse, NY 13210

LIST OF AUTHORS